Recent Developments in Fractal Geometry and Dynamical Systems

CONTEMPORARY MATHEMATICS

797

Recent Developments in Fractal Geometry and Dynamical Systems

AMS Special Session
Fractal Geometry and Dynamical Systems
May 14–15, 2022
Virtual

Sangita Jha
Mrinal Kanti Roychowdhury
Saurabh Verma
Editors

AMERICAN MATHEMATICAL SOCIETY

2020 *Mathematics Subject Classification.* Primary 28A33, 28A50, 28A80, 35B41, 37D05, 37F10, 37F12, 37F20.

Library of Congress Cataloging-in-Publication Data

Names: Virtual AMS Special Session on Fractal Geometry and Dynamical Systems (2022 : online) | Jha, Sangita, 1991- editor.

Title: Recent developments in fractal geometry and dynamical systems : Virtual AMS Special Session on Fractal Geometry and Dynamical Systems, May 14–15, 2022 / Sangita Jha, Mrinal Kanti Roychowdhury, Saurabh Verma, editors.

Description: Providence, Rhode Island : American Mathematical Society, [2024] | Series: Contemporary mathematics, 0271-4132 ; volume 797 | Includes bibliographical references.

Identifiers: LCCN 2023053402 | ISBN 9781470472160 (paperback) | ISBN 9781470476106 (ebook)

Subjects: LCSH: Fractals. | Dynamics. | AMS: Measure and integration – Classical measure theory – Spaces of measures, convergence of measures. | Measure and integration – Classical measure theory – Integration and disintegration of measures. | Measure and integration – Classical measure theory – Fractals. | Partial differential equations – Qualitative properties of solutions – Attractors. | Dynamical systems and ergodic theory – Ergodic theory – Relations with number theory and harmonic analysis. | Dynamical systems and ergodic theory – Dynamical systems with hyperbolic behavior – Hyperbolic orbits and sets. | Dynamical systems and ergodic theory – Complex dynamical systems – Polynomials; rational maps; entire and meromorphic functions. | Dynamical systems and ergodic theory – Complex dynamical systems – Combinatorics and topology.

Classification: LCC QA614.86 .V57 2024 | DDC 514/.742–dc23/eng/20231229

LC record available at https://lccn.loc.gov/2023053402

Contemporary Mathematics ISSN: 0271-4132 (print); ISSN: 1098-3627 (online)

DOI: https://doi.org/10.1090/conm/797

∞ The paper used in this book is acid-free and falls within the guidelines
established to ensure permanence and durability.
Visit the AMS home page at https://www.ams.org/

10 9 8 7 6 5 4 3 2 1   24 23 22 21 20 19

We dedicate this volume to our beloved parents

Contents

Preface

This book is a theme issue based on the papers presented virtually at an AMS Special Session on "Fractal Geometry and Dynamical Systems", May 14–15, 2022. The session was attended virtually by 40 researchers. There were total 24 invited talks. The list of participants and conference program can be obtained from *http://www.ams.org/meetings/sectional/2294_special.html*. The organizing committee consisted of Sangita Jha (National Institute of Technology Rourkela, India), M.K. Roychowdhury (University of Texas Rio Grandey Valley, US) and Saurabh Verma (Indian Institute of Information and Technology Allahabad, India). We would like to thank all participants and speakers in the conference, who made possible such a successful and fruitful event.

Fractal geometry has recently expanded into a vast field that encompasses practically all fields of science and engineering. It is the study of the properties of fractal objects, which are commonly referred to as fractals. The notion of fractal naturally arises in the study of non-linear functions. We have been fortunate to have eminent speakers in the field and dynamic enthusiastic researchers contributed to our volume with significant advances including some expository articles. The talks in the conference covered a large range of topics: some of them are concerned with geodesics in generalizations of the Sierpiński carpet by Berkove, Karangozishvili, and Smith and this work is relevant to the isoperimetric problem for fractals. The book includes the non-autonomous dynamics of complex polynomials by Comerford, Radulescu, and Cavanagh. It also includes the theory and applications of polymorphisms by Jorgensen and Tian. In particular, they covered areas of probability theory, analysis of Markov maps, transition operators, joinings of measure spaces, and applications to harmonic analysis of self-similar (fractal) measures. Others deal with the random beta-transformations on fat Sierpiński gasket by Dajani, Li, and Zhang; the connections between the shape and size of a period lattice for an elliptic function and its dynamics by Hawkins and Koss; the correspondence between unicritical polygons and maximally multicritical polygons in a lamination modeling a degree d polynomial with connected Julia set by Burdette, Hale, and Mayer; the theory of IFS measures associated with Bratteli diagrams by Bezuglyi and Jorgensen. Others deal with topological and geometric problems related to dynamical systems: unmating of certain expanding Thurston maps by Wilkerson, ergodicity of falling particle systems by Simanyi, problem of computability in dynamical systems by Burr and Wolf. Applications include pure ergodic theory, combinatorial number theory by Cömez. Fractal interpolation over curves was introduced by Massopust, the construction of fractal functions using weak construction was studied by Verma, Priyadarshi and the dimensional analysis of non-stationary fractal function was discussed by Chandra, Abbas.

Last but not least, we would like to express our gratitude to Christine Thivierge for her assistance with the edition and to the American Mathematical Society for continuing to support our proposal. Without the contributions of the authors and the anonymous referees, who nonetheless did good work, this collection would not have been conceivable. They were given incredibly strict deadlines. Our sincere appreciation to each of you.

Contemporary Mathematics
Volume **797**, 2024
https://doi.org/10.1090/conm/797/15931

Mandelbrot sets for fixed template iterations

Mark Comerford, Anca Rădulescu, and Kieran Cavanagh

Abstract. We study the dynamics of template iterations, consisting of arbitrary compositions of functions chosen from a finite set of polynomials. In particular, we focus on templates using complex unicritical maps in the family $\{z^d + c, c \in \mathbb{C}, d \geq 2\}$. We examine the dependence on parameters of the connectedness locus for a fixed template and show that, for most templates, the connectedness locus moves upper semicontiuously. On the other hand, one does not in general have lower semicontinuous dependence, and we show this by means of a counterexample.

1. Introduction on template iterations

In previous work by the second author, we first introduced questions concerning the asymptotic dynamics of template iterations of two quadratic maps from the family $f_c : \mathbb{C} \to \mathbb{C}, f_c(z) = z^2 + c$ where $c \in \mathbb{C}$ is a constant parameter [**12**, **13**]. This can be viewed as a particular case of the non-autonomous iteration of polynomials, which was first studied in 1991 by Fornaess and Sibony [**10**]. Additional work was done done by Brück et al. [**1**–**3**], by Sumi [**15**–**20**], and by the first author [**5**–**9**].

The framework of template iterated systems can be generalized to apply to any number of generating monic unicritical polynomials of the form $f_{(c,d)} = z^d + c$. More precisely, given any natural number D, any D-tuple $\mathbf{c} = (c_0, \ldots, c_{D-1}) \in \mathbb{C}^D$, any D-tuple of integers $\mathbf{d} = (d_0, \ldots, d_{D-1})$ with $d_i \geq 2$ for each $0 \leq i \leq D-1$, and any sequence $\mathbf{s} = \{s_m\}_{m=1}^{\infty} \in \{0, \ldots, D-1\}^{\infty}$ (which we also call a *template*), we consider the iterated system in which the polynomial $P_m^{\mathbf{s},\mathbf{c},\mathbf{d}}$ iterated at each step m is $f_{(c_i,d_i)} = z^{d_i} + c_i$, if $s_m = i$. We then say a sequence $\{P_m^{\mathbf{s},\mathbf{c},\mathbf{d}}\}_{m=1}^{\infty}$ of monic centered unicritical polynomials is *generated by the D-tuples* $\mathbf{c} = (c_0, \ldots, c_{D-1})$, $\mathbf{d} = (d_0, \ldots, d_{D-1})$ and the template $\mathbf{s}$ if it is constructed in this way. For two fixed iterative times $0 \leq m < n$, we will also need to consider the polynomial composition taken from time m to time n given by

$$Q_{m,n}^{\mathbf{s},\mathbf{c},\mathbf{d}} = f_{(c_{s_n}, d_{s_n})} \circ \cdots \circ f_{(c_{s_{m+1}}, d_{s_{m+1}})} = P_n^{\mathbf{s},\mathbf{c},\mathbf{d}} \circ \cdots \circ P_{m+1}^{\mathbf{s},\mathbf{c},\mathbf{d}}.$$

For convenience we will denote in particular $Q_{0,n}^{\mathbf{s},\mathbf{c},\mathbf{d}}$ by $Q_n^{\mathbf{s},\mathbf{c},\mathbf{d}}$ and we will refer to the D-tuples $\mathbf{c}$ and $\mathbf{d}$ as respectively the *constant* and *degree* vectors associated with our template iteration. If $D = 1$, then clearly we recover the classical situation of

2020 *Mathematics Subject Classification.* Primary 37F10; Secondary 30D05.
Key words and phrases. Dynamical systems, nonautonomous iterations, template iterations.
The second author was supported by the Simons Foundation, Award #523763.

iteration with a single monic centered unicritical polynomial. Note that, in view of [**5**, Theorem 2.1], any nonautonomous sequence of unicritical polynomials will be conjugate in the appropriate sense to a sequence of monic centered unicritical polynomials. However, in general, the members of such a monic centered sequence will not be chosen from among finitely many possibilities as we consider here (even when the members of the original sequence were chosen from among finitely many possible polynomials). Finally, we remark that template iteration using polynomials of varying degrees has already appeared in the literature. See [**9**, Theorem 1.12] where the authors consider all possible template sequences arising from choices among the two polynomials $z^2 - 1$ and z^3.

Let σ denote the shift map on the template space $\{0, \dots, D-1\}^\infty$ and for $m \geq 0$ let $\sigma^{\circ m}$ denote the mth iterate of σ, which has the effect of truncating the first m members of any given sequence (here, we just take $\sigma^{\circ 0}$ to be the identity). For any template iterated system, since the leading terms of the finitely many polynomials $f_{(c_i,d_i)}(z)$, $0 \leq i \leq D-1$ dominate when $|z|$ is large, one can then consider, in a similar fashion as in the traditional case of iterations with a single map, the *iterated basin of infinity at time* m, arising from the truncated sequence $\sigma^{\circ m}(\mathbf{s})$, on which all points escape locally uniformly to infinity under the compositions $Q_{m,n}^{\mathbf{s},\mathbf{c},\mathbf{d}}$ as n tends to infinity. We denote this basin of infinity by $\mathcal{A}_\infty^{\sigma^{\circ m}(\mathbf{s}),\mathbf{c},\mathbf{d}}$ or, more simply, $\mathcal{A}_{\infty,m}^{\mathbf{s},\mathbf{c},\mathbf{d}}$.

The complement of $\mathcal{A}_{\infty,m}^{\mathbf{s},\mathbf{c},\mathbf{d}}$, for each $m \geq 0$, is then the set of points whose orbits do not escape to infinity (which is easily seen to be equivalent to the condition that the orbits be bounded) and is called the *iterated filled Julia set at time* m. We denote this set by $\mathcal{K}^{\sigma^{\circ m}(\mathbf{s}),\mathbf{c},\mathbf{d}}$ or, more simply, $\mathcal{K}_m^{\mathbf{s},\mathbf{c},\mathbf{d}}$:

$$\mathcal{K}^{\sigma^{\circ m}(\mathbf{s}),\mathbf{c},\mathbf{d}} = \mathcal{K}_m^{\mathbf{s},\mathbf{c},\mathbf{d}} = \{z \in \mathbb{C} : Q_{m,n}^{\mathbf{s},\mathbf{c},\mathbf{d}}(z) \text{ is bounded}\}.$$

Finally, for each $m \geq 0$ we have the *iterated Julia set at time* m, which is simply the common boundary of the iterated filled Julia set and basin of infinity and is denoted by $\mathcal{J}^{\sigma^{\circ m}(\mathbf{s}),\mathbf{c},\mathbf{d}}$ or just $\mathcal{J}_m^{\mathbf{s},\mathbf{c},\mathbf{d}}$, i.e.,

$$\mathcal{J}^{\sigma^{\circ m}(\mathbf{s}),\mathbf{c},\mathbf{d}} = \mathcal{J}_m^{\mathbf{s},\mathbf{c},\mathbf{d}} = \partial \mathcal{K}_m^{\mathbf{s},\mathbf{c},\mathbf{d}}.$$

The iterated Julia sets coincide with the iterated Julia sets from standard nonautonomous polynomial iteration and are precisely those points on which the compositions $Q_{m,n}^{\mathbf{s},\mathbf{c},\mathbf{d}}$ fail to give a normal family on any neighbourhood (see, e.g., [**6**] for details).

When $m = 0$, we will refer to the corresponding iterated basin of infinity, filled Julia set, and Julia set as simply the *basin of infinity*, *filled Julia set*, and *Julia set* respectively. These then clearly extend the corresponding definitions from standard polynomial iteration and for simplicity we usually denote them by $\mathcal{A}_\infty^{\mathbf{s},\mathbf{c},\mathbf{d}}$, $\mathcal{K}^{\mathbf{s},\mathbf{c},\mathbf{d}}$, and $\mathcal{J}^{\mathbf{s},\mathbf{c},\mathbf{d}}$, i.e.,

$$\mathcal{A}_\infty^{\mathbf{s},\mathbf{c},\mathbf{d}} = \mathcal{A}_{\infty,0}^{\mathbf{s},\mathbf{c},\mathbf{d}}, \; \mathcal{K}^{\mathbf{s},\mathbf{c},\mathbf{d}} = \mathcal{K}_0^{\mathbf{s},\mathbf{c},\mathbf{d}}, \; \mathcal{J}^{\mathbf{s},\mathbf{c},\mathbf{d}} = \mathcal{J}_0^{\mathbf{s},\mathbf{c},\mathbf{d}}.$$

When the D-tuples $\mathbf{c}$, $\mathbf{d}$ are fixed and there is no danger of confusion, we will omit $\mathbf{c}$, $\mathbf{d}$ from the superscript for simplicity, and use the notation $Q_{m,n}^{\mathbf{s}}$ for the composition maps, $\mathcal{A}_{\infty,m}^{\mathbf{s}}$ for the basin of infinity at time m, and respectively $\mathcal{K}_m^{\mathbf{s}}$ and $\mathcal{J}_m^{\mathbf{s}}$ for the iterated filled Julia sets and the iterated Julia sets at time $m \geq 0$ (with similar notation for the basin of infinity, filled Julia set, and Julia set at time 0).

The term 'iterated' for the iterated basins of infinity, filled Julia sets, and Julia sets is justified in view of the following invariance result, the proof of which is a relatively easy exercise we leave to the reader (see [**6**] for a slightly stronger invariance statement).

LEMMA 1.1 (Complete Invariance). *For any* $\mathbf{s} \in \{0, \ldots, D-1\}^\infty$, $0 \le m < n$, *we have that* $Q^{\mathbf{s}}_{m,n}(\mathcal{A}^{\mathbf{s}}_{\infty,m}) = \mathcal{A}^{\mathbf{s}}_{\infty,n}$, $Q^{\mathbf{s}}_{m,n}(\mathcal{K}^{\mathbf{s}}_m) = \mathcal{K}^{\mathbf{s}}_n$, *and* $Q^{\mathbf{s}}_{m,n}(\mathcal{J}^{\mathbf{s}}_m) = \mathcal{J}^{\mathbf{s}}_n$.

For a fixed template $\mathbf{s} \in \{0, \ldots, D-1\}^\infty$ and an upper bound $d \ge \max_{0\le i\le D-1} d_i$ on degrees (which we introduce as we require compactness), we define the *fixed-template Mandelbrot set* $\mathcal{M}^{\mathbf{s}}$ as a subset of $\mathbb{C}^D \times \{2, \ldots, d\}^D$ defined by:

$$\mathcal{M}^{\mathbf{s}} = \{(\mathbf{c}, \mathbf{d}) \in \mathbb{C}^D \times \{2, \ldots, d\}^D : Q^{\mathbf{s},\mathbf{c},\mathbf{d}}_{m,n}(0) \text{ is bounded for all } 0 \le m < n\}.$$

REMARKS. In the case of template iteration where the polynomials have constant degree, the fixed-template Mandelbrot set $\mathcal{M}^{\mathbf{s}}$ can obviously be identified with a subset of $\mathbb{C}^D$ as is more common in classical polynomial iteration (and as we will do ourselves in Section 3). Since having an unbounded orbit is clearly an open condition, it is easy to see that $\mathcal{M}^{\mathbf{s}}$ is automatically a closed subset of $\mathbb{C}^D \times \{2, \ldots, d\}^D$ (where we endow $\{2, \ldots, d\}^D$ with the discrete topology).

With these extensions, some of the results established in the traditional context of single map iterations are easily generalizable for template iterations. Others are nontrivial to extend, or do not hold true for templates. We already remarked earlier that template iterations is a special case of the nonautonomous iteration of polynomials. We also note that there is a strong connection between template iteration and polynomial semigroups as studied by Sumi and Stankewitz [**14**–**19**], especially for finitely generated polynomial semigroups as considered in [**20**].

Some results for templates follow trivially from the general theory of non-autonomous iteration, but we also have objects and properties which apply only to the context of templates. For example, we showed in prior work that template iterated systems have an escape radius, and that, when $D = 2$ and $d_0 = d_1 = 2$, $M = \max\{2, |c_0|, |c_1|\}$ acts as such an escape radius for any template iteration of the two maps arising from the pair (c_0, c_1) [**12**]. In this paper we want to refine this result and we show that the fixed template Mandelbrot set $\mathcal{M}^{\mathbf{s}}$ is a subset of the set $\overline{\mathbb{D}}(0,2)^D \times \{2, \ldots, d\}^D \subset \mathbb{C}^d \times \{2, \ldots, d\}^D$ for most templates $\mathbf{s}$.

For any $\mathbf{s} \in \{0, \ldots, D-1\}^\infty$, it remains true that the fixed template Mandelbrot set coincides with the connectedness locus for the Julia set corresponding to the same template.

LEMMA 1.2. *The fixed template Mandelbrot set* $\mathcal{M}^{\mathbf{s}}$ *is equal to each of the following sets:*

$$\{(\mathbf{c}, \mathbf{d}) \in \mathbb{C}^D \times \{2, \ldots, d\}^D : \mathcal{K}^{\mathbf{s},\mathbf{c},\mathbf{d}} \textit{ is connected }\},$$
$$\{(\mathbf{c}, \mathbf{d}) \in \mathbb{C}^D \times \{2, \ldots, d\}^D : \mathcal{J}^{\mathbf{s},\mathbf{c},\mathbf{d}} \textit{ is connected }\},$$
$$\{(\mathbf{c}, \mathbf{d}) \in \mathbb{C}^D \times \{2, \ldots, d\}^D : \mathcal{K}^{\mathbf{s},\mathbf{c},\mathbf{d}}_m \textit{ is connected, for all } m \ge 0\},$$
$$\{(\mathbf{c}, \mathbf{d}) \in \mathbb{C}^D \times \{2, \ldots, d\}^D : \mathcal{J}^{\mathbf{s},\mathbf{c},\mathbf{d}}_m \textit{ is connected, for all } m \ge 0\},$$
$$\{(\mathbf{c}, \mathbf{d}) \in \mathbb{C}^D \times \{2, \ldots, d\}^D : 0 \in \mathcal{K}^{\mathbf{s},\mathbf{c},\mathbf{d}}_m \textit{ for all } m \ge 0\},$$
$$\{(\mathbf{c}, \mathbf{d}) \in \mathbb{C}^D \times \{2, \ldots, d\}^D : c_{s_m} \in \mathcal{K}^{\mathbf{s},\mathbf{c},\mathbf{d}}_m \textit{ for all } m \ge 1\}.$$

The proof carries through for nonautonomous (and in particular template) iterations in a similar way to that for single polynomial maps (using Green's functions where a critical point in the basin of infinity gives rise to a level curve which separates the Julia set [**3**]). We will make use of this fact in the following sections.

2. Boundedness of the template Mandelbrot set

It is easy to see from the standard theory of iterations with a single quadratic polynomial that, for some templates $\mathbf{s} = \{s_m\}_{m=1}^{\infty} \in \{0, \ldots, D-1\}^{\infty}$ and degree bounds d, the fixed template Mandelbrot set $\mathcal{M}^{\mathbf{s}}$ is unbounded in $\mathbb{C}^D \times \{2, \ldots, d\}^D$. More precisely, if there exists i, $0 \le i \le D-1$ for which $s_m \ne i$ for all $m \ge 1$, so that $P_m^{\mathbf{s},\mathbf{c},\mathbf{d}} \ne z^{d_i} + c_i$ for all $m \ge 1$, then $\mathcal{M}^{\mathbf{s}}$ will be unbounded in the i-th coordinate as there is no constraint on the corresponding critical value c_i in parameter space. With this in mind, we say that the template $\mathbf{s} = \{s_m\}_{m=1}^{\infty}$ is *full* if this does not occur and that, for each $0 \le i \le d-1$, $s_m = i$ and thus $P_m^{\mathbf{s},\mathbf{c},\mathbf{d}} = z^{d_i} + c_i$ for at least one value of m. We observe that this condition is clearly generic. We will show that, in the case of a full template, $\mathcal{M}^{\mathbf{s}}$ is bounded as one might expect.

We first start with a technical lemma which makes use of a well-known fact from potential theory. In this paper, for $z \in \mathbb{C}$ and $r > 0$, we use the common notation $\mathrm{D}(z, r)$, $\overline{\mathrm{D}}(z, r)$ to denote the open and closed discs respectively with centre z and radius r.

LEMMA 2.1. *For any* $\mathbf{s} \in \{0, \ldots D-1\}^{\infty}$ *and any pair of d-tuples* $(\mathbf{c}, \mathbf{d}) \in \mathcal{M}^{\mathbf{s}}$, *the iterated filled Julia sets* $\mathcal{K}_m^{\mathbf{s},\mathbf{c},\mathbf{d}}$ *satisfy*

$$\mathcal{K}_m^{\mathbf{s},\mathbf{c},\mathbf{d}} \subset \overline{\mathrm{D}}(0, 2), \quad \textit{for all } m \ge 0.$$

PROOF. By Lemma 1.2 since $(\mathbf{c}, \mathbf{d}) \in \mathcal{M}^{\mathbf{s}}$, for each $m \ge 0$ the iterated filled Julia set $\mathcal{K}_m^{\mathbf{s},\mathbf{c},\mathbf{d}}$ is connected. By [**11**, Theorem 5.3.2(a)], we have that $\mathrm{diam}(\mathcal{K}_m^{\mathbf{s},\mathbf{c},\mathbf{d}}) \le 4\,\mathrm{cap}(\mathcal{K}_m^{\mathbf{s},\mathbf{c},\mathbf{d}})$, where $\mathrm{diam}(\mathcal{K}_m^{\mathbf{s},\mathbf{c},\mathbf{d}})$ and $\mathrm{cap}(\mathcal{K}_m^{\mathbf{s},\mathbf{c},\mathbf{d}})$ are respectively the diameter and the logarithmic capacity of the set $\mathcal{K}_m^{\mathbf{s},\mathbf{c},\mathbf{d}}$. Furthermore, since the polynomials $f_{(c_{s_m}, d_{s_m})}$ are monic, $\mathrm{cap}(\mathcal{K}_m^{\mathbf{s},\mathbf{c},\mathbf{d}}) = 1$ (see [**6**, Theorem 1.4(5)] or [**10**, Theorem 2.1]), whence $\mathrm{diam}(\mathcal{K}_m^{\mathbf{s},\mathbf{c},\mathbf{d}}) \le 4$.

Recall that the diameter of a regular n-gon with vertices on the unit circle is 2 for n even and

$$\sqrt{2 - 2\cos\left(\frac{2\pi k}{2k+1}\right)} < 2$$

for $n = 2k+1$ odd. Since $P_{m+1}^{\mathbf{s},\mathbf{c},\mathbf{d}}(z) = f_{(c_{s_{m+1}}, d_{s_{m+1}})}(z) = z^{d_{s_{m+1}}} + c_{s_{m+1}}$, if $d_{s_{m+1}}$ is even, then $\mathcal{K}_m^{\mathbf{s},\mathbf{c},\mathbf{d}}$ is symmetric about 0 under $z \mapsto -z$ and it follows immediately that $\mathcal{K}_m^{\mathbf{s},\mathbf{c},\mathbf{d}} \subset \overline{\mathrm{D}}(0, 2)$. On the other hand, when $d_{s_{m+1}}$ is odd, we have the slightly weaker condition that $\mathcal{K}_m^{\mathbf{s},\mathbf{c},\mathbf{d}} \subset \overline{\mathrm{D}}(0, r)$ where $r = 4/\sqrt{2 - 2\cos(2\pi k/(2k+1))} > 2$, the worst case being when $d_{s_{m+1}} = 3$ which gives $r = 4/\sqrt{3}$. Thus, in any case we can say that for each $m \ge 0$ we must have $\mathcal{K}_m^{\mathbf{s},\mathbf{c},\mathbf{d}} \subset \overline{\mathrm{D}}(0, 4/\sqrt{3})$ (regardless of whether $d_{s_{m+1}}$ is even or odd) and so, by Lemma 1.2, since $(\mathbf{c}, \mathbf{d}) \in \mathcal{M}^{\mathbf{s}}$, we have that

$$|c_{s_m}| \le 4/\sqrt{3}, \quad m \ge 1. \tag{2.1}$$

However, if $m \geq 0$, $|z| > 2$, $d_{s_{m+1}} \geq 3$, and $|c_{m+1}| \leq 4/\sqrt{3}$, we have $|P^{\mathbf{s},\mathbf{c},\mathbf{d}}_{m+1}(z)| = |f_{(c_{s_{m+1}}, d_{s_{m+1}})}(z)| > 8 - 4/\sqrt{3} > 4$ and it follows readily using (2.1) above that z escapes to infinity under iteration using the sequence $\{Q^{\mathbf{s},\mathbf{c},\mathbf{d}}_{m,n}\}_{n=m+1}^{\infty}$. Using this, it follows that $\mathcal{K}^{\mathbf{s},\mathbf{c},\mathbf{d}}_m \subset \overline{\mathrm{D}}(0,2)$ when $d_{s_{m+1}}$ is odd also, which completes the proof. □

LEMMA 2.2. *If* $\mathbf{s} \in \{0, \ldots, D-1\}^{\infty}$ *is a full template, then* $\mathcal{M}^{\mathbf{s}}$ *is a compact subset of* $\overline{\mathrm{D}}(0,2)^D \times \{2, \ldots, d\}^D$.

PROOF. Suppose $(\mathbf{c}, \mathbf{d}) = ((c_0, \ldots, c_{D-1}), (d_0, \ldots, d_{D-1})) \in \mathcal{M}^{\mathbf{s}}$. Then, by Lemma 1.2 it follows that for each $m \geq 1$ the critical value c_{s_m} of $P^{\mathbf{s},\mathbf{c},\mathbf{d}}_m = f_{(c_{s_m}, d_{s_m})}$ belongs to $\mathcal{K}^{\mathbf{s},\mathbf{c},\mathbf{d}}_m$. Since $\mathbf{s}$ is full, it follows that, for each $0 \leq i \leq D-1$, there exists $m_i \geq 1$ such that $s_{m_i} = i$ and $P^{\mathbf{s},\mathbf{c},\mathbf{d}}_{m_i} = z^{d_i} + c_i$, (for some $2 \leq d_i \leq d$) so that $c_i \in \mathcal{K}^{\mathbf{s},\mathbf{c},\mathbf{d}}_{m_i}$. On the other hand, from Lemma 2.1, it follows that $\mathcal{K}^{\mathbf{s},\mathbf{c},\mathbf{d}}_{m_i+1} \subset \overline{\mathrm{D}}(0,2)$. Thus, $(\mathbf{c}, \mathbf{d}) \in \overline{\mathrm{D}}(0,2)^D \times \{2, \ldots, d\}^D$, and, since $\mathcal{M}^{\mathbf{s}}$ is closed (as remarked earlier), while the set $\{2, \ldots, d\}^D$ is finite, $\mathcal{M}^{\mathbf{s}}$ is also compact, which concludes our proof. □

3. Limit behavior for arbitrary templates

On the space of templates $\{0, \ldots, D-1\}^{\infty}$ we consider the ultrametric defined for any two templates $\mathbf{s} = \{s_m\}_{m=1}^{\infty}$, $\tilde{\mathbf{s}} = \{\tilde{s}_m\}_{m=1}^{\infty}$ by

$$\|\mathbf{s} - \tilde{\mathbf{s}}\| = \sum_{k=1}^{\infty} \frac{|s_k - \tilde{s}_k|}{D^k}.$$

We also make use of the product metric on $\mathbb{C}^D \times \{2, \ldots, d\}^D$. To be specific, for two points

$$(\mathbf{c}, \mathbf{d}) = ((c_0, \ldots, c_{D-1}), (d_0, \ldots, d_{D-1})),$$
$$(\tilde{\mathbf{c}}, \tilde{\mathbf{d}}) = ((\tilde{c}_0, \ldots, \tilde{c}_{D-1}), (\tilde{d}_0, \ldots, \tilde{d}_{D-1}))$$

in $\mathbb{C}^D \times \{2, \ldots, d\}^D$, we have

$$d((\mathbf{c}, \mathbf{d}), (\tilde{\mathbf{c}}, \tilde{\mathbf{d}})) = \max\{\max\{|\tilde{c}_i - c_i|, |\tilde{d}_i - d_i|\},\ 0 \leq i \leq D-1\}.$$

We then use the notation

$$\begin{aligned} \mathrm{B}((\mathbf{c}, \mathbf{d}), \delta) &= \mathrm{B}(((c_0, \ldots, c_{D-1}), (d_0, \ldots, d_{D-1})), \delta) \\ &= \mathrm{D}(c_0, \delta) \times \cdots \times \mathrm{D}(c_{D-1}, \delta) \times (d_0 - \delta, d_0 + \delta) \times \cdots \\ &\quad \times (d_{D-1} - \delta, d_{D-1} + \delta) \end{aligned}$$

to denote the open balls with respect to this metric.

The distance between a point $(\mathbf{c}, \mathbf{d}) \in \mathbb{C}^D \times \{2, \ldots, d\}^D$ and a set $B \subset \mathbb{C}^D \times \{2, \ldots, d\}^D$ is defined as:

$$d((\mathbf{c}, \mathbf{d}), B) = \inf\{d((\mathbf{c}, \mathbf{d}), (\tilde{\mathbf{c}}, \tilde{\mathbf{d}})),\ (\tilde{\mathbf{c}}, \tilde{\mathbf{d}}) \in B\}.$$

For two sets $A, B \in \mathbb{C}^D \times \{2, \ldots, d\}^D$, the distance from A to B is defined as:

$$d(A, B) = \sup\{d(\mathbf{c}, B),\ \mathbf{c} \in A\}.$$

Finally, the Hausdorff distance between two sets $A, B \in \mathbb{C}^D \times \{2, \ldots, d\}^D$ is then given by:

$$d_H(A, B) = \max\{d(A, B), d(B, A)\}.$$

THEOREM 3.1. *Let* $\mathbf{s} \in \{0, \ldots, D-1\}^\infty$ *be a full template and consider a sequence of templates* $\{\mathbf{s}^N\}_{N=1}^\infty$ *that converges to* $\mathbf{s}$ *as* $N \to \infty$*. Then*

$$\lim_{N\to\infty} d\left(\mathcal{M}^{\mathbf{s}^N}, \mathcal{M}^{\mathbf{s}}\right) = 0.$$

PROOF. Since $\{\mathbf{s}^N\}_{N=1}^\infty$ converges to $\mathbf{s}$, for any finite $m_0 \geq 1$, there exists N_0 such that the first m_0 components of components of the templates $\mathbf{s}^N$ and $\mathbf{s}$ coincide for all $N \geq N_0$. Since $\mathbf{s}$ is full, from this we see that $\mathbf{s}^N$ must be full for all sufficiently large N. It follows from Lemma 2.2 that $\mathcal{M}^{\mathbf{s}} \subset \overline{\mathrm{D}}(0,2)^D \times \{2, \ldots, d\}^D$ and likewise $\mathcal{M}^{\mathbf{s}^N} \subset \overline{\mathrm{D}}(0,2)^D \times \{2, \ldots, d\}^D$ (for all sufficiently large N).

Suppose that $(\mathbf{c}, \mathbf{d}) = ((c_0, \ldots, c_{D-1}), (d_0, \ldots, d_{D-1})) \in \overline{\mathrm{D}}(0,2)^D \times \{2, \ldots, d\}^D$. From above is not hard to see, in a similar manner to the classical case for a single quadratic polynomial z^2+c, that $R = 2$ acts as an escape radius for the polynomial sequence arising from the D-tuples $(\mathbf{c}, \mathbf{d})$ in conjunction with any template and in particular $\mathbf{s}^N$ and $\mathbf{s}$ (note that the existence of this escape radius for the case $D = d = 2$ is [**12**, Theorem 1.6] while the general case is proved similarly). If in addition we assume that $(\mathbf{c}, \mathbf{d}) \in (\overline{\mathrm{D}}(0,2)^D \times \{2, \ldots, d\}^D) \setminus \mathcal{M}^{\mathbf{s}}$, then at least one of the critical points of the template iteration under $\mathbf{s}$ escapes. More precisely, there exist $m_e \geq 0$, $t_e \geq 1$ such that

$$|Q^{\mathbf{s}}_{m_e, m_e+t_e}(0)| = |f_{(c_{s_{m_e+t_e}}, d_{s_{m_e+t_e}})} \circ \cdots \cdots \circ f_{(c_{s_{m_e+1}}, d_{s_{m_e+1}})}(0)| > 3$$

(where the polynomials $f_{(c_{s_m}, d_{s_m})}$, $m_e + 1 \leq m \leq m_e + t_e$ are constructed using the template $\mathbf{s}$ and the parameter $(\mathbf{c}, \mathbf{d})$).

Again since $\{\mathbf{s}^N\}_{N=1}^\infty$ converges to $\mathbf{s}$, there exists N_0 such that the first $m_e + t_e$ entries of the templates $\mathbf{s}^N$ and $\mathbf{s}$ coincide for all $N \geq N_0$. Hence, in particular,

$$Q^{\mathbf{s}^N}_{m_e, m_e+t_e}(0) = Q^{\mathbf{s}}_{m_e, m_e+t_e}(0)$$

so that

$$|Q^{\mathbf{s}^N}_{m_e, t_e}(0)| > 3, \quad \text{for all } \; N \geq N_0. \tag{3.1}$$

For $(\tilde{\mathbf{c}}, \tilde{\mathbf{d}}) = ((\tilde{c}_0, \ldots, \tilde{c}_{D-1}), (\tilde{d}_0, \ldots, \tilde{d}_{D-1})) \in \mathbb{C}^D \times \{2, \ldots, d\}^D$ and $N \geq N_0$, one can define, similarly to our earlier notation, the compositions

$$\tilde{Q}^{\mathbf{s}^N}_{m_e, m_e+t_e} = \tilde{Q}^{\mathbf{s}}_{m_e, m_e+t_e} := f_{(\tilde{c}_{s_{t_e+m_e}}, \tilde{d}_{s_{t_e+m_e}})} \circ \cdots \cdots \circ f_{(\tilde{c}_{s_{m_e+1}}, \tilde{d}_{s_{m_e+1}})}$$

(where the polynomials $f_{(\tilde{c}_{s_m}, \tilde{d}_{s_m})}$, $m_e + 1 \leq m \leq m_e + t_e$ are constructed using the templates $\mathbf{s}, \mathbf{s}^N$, the parameter $\tilde{\mathbf{c}}$ which is allowed to vary over $\mathbb{C}^D$, and the degree bound d which we regard as fixed).

It is easy to see that these polynomials are jointly continuous in z,and in each $\tilde{c}_i$, $0 \leq i \leq d-1$. In addition, the coefficients for these polynomials (in z) are fixed polynomials in the parameters $\tilde{c}_i$ whose degrees in each variable are at most $d^{t_e - 1}$. Further, for each $0 \leq i \leq d-1$, the absolute values of the coefficients as polynomials in $\tilde{c}_i$ of these coefficient polynomials (with the other variables held fixed) will be locally uniformly bounded. Finally, the degrees of these polynomials (again in z) are at most d^{t_e} and thus uniformly bounded. Hence the partial derivatives in each $\tilde{c}_i$, $0 \leq i \leq d-1$ of these polynomials are locally uniformly bounded as functions of these D variables and there thus exists $\delta > 0$ such that:

$$d((\mathbf{c}, \mathbf{d}), (\tilde{\mathbf{c}}, \tilde{\mathbf{d}})) < \delta \implies |\tilde{Q}^{\mathbf{s}^N}_{m_e, m_e+t_e}(0) - Q^{\mathbf{s}^N}_{m_e, m_e+t_e}(0)| < 1. \tag{3.2}$$

From (3.1) and (3.2), it follows that, for all $(\tilde{\mathbf{c}}, \tilde{\mathbf{d}}) \in \mathrm{B}((\mathbf{c}, \mathbf{d}), \delta)$ and $N \geq N_0$, we have:

$$|\tilde{Q}^{\mathbf{s}^N}_{m_e, m_e+t_e}(0)| \geq |Q^{\mathbf{s}^N}_{m_e, m_e+t_e}(0)| - |\tilde{Q}^{\mathbf{s}^N}_{m_e, m_e+t_e}(0) - Q^{\mathbf{s}^N}_{m_e, m_e+t_e}(0)| > 3 - 1 = 2.$$

Since from above $R = 2$ is an escape radius for any iterated template system with $(\mathbf{c}, \mathbf{d}) \in \overline{\mathrm{D}}(0,2)^D \times \{2, \ldots, d\}^D$, it follows that $(\tilde{\mathbf{c}}, \tilde{\mathbf{d}}) \notin \mathcal{M}^{\mathbf{s}^N}$. Hence we have shown that, given any $(\mathbf{c}, \mathbf{d}) \in (\overline{\mathrm{D}}(0,2)^D \times \{2, \ldots, d\}^D) \setminus \mathcal{M}^{\mathbf{s}}$, we can find $N_0 \in \mathbb{N}$ and $\delta > 0$ such that the ball $\mathrm{B}((\mathbf{c}, \mathbf{d}), \delta)$ is contained in $(\mathbb{C}^D \times \{2, \ldots, d\}^D) \setminus \mathcal{M}^{\mathbf{s}^N}$ for all $N \geq N_0$.

Now let $\varepsilon > 0$ and set

$$N_\varepsilon = \{(\mathbf{c}, \mathbf{d}) \in \overline{\mathrm{D}}(0,2)^D \times \{2, \ldots, d\}^D : d((\mathbf{c}, \mathbf{d}), \mathcal{M}^{\mathbf{s}}) < \varepsilon\}.$$

By Lemma 2.2, we have $\mathcal{M}^{\mathbf{s}} \subset \overline{\mathrm{D}}(0,2)^D \times \{2, \ldots, d\}^D$, so that $\mathcal{M}^{\mathbf{s}} \subset N_\varepsilon$. It follows from above that for any D-tuples $(\mathbf{c}, \mathbf{d}) \in (\overline{\mathrm{D}}(0,2)^D \times \{2, \ldots, d\}^D) \setminus N_\varepsilon$, one can make $\delta > 0$ sufficiently small and $N_0 > 0$ sufficiently large to guarantee that

$$B((\mathbf{c}, \mathbf{d}), \delta) \subset (\mathbb{C}^D \times \{2, \ldots, d\}^D) \setminus \mathcal{M}^{\mathbf{s}^N}, \quad N \geq N_0.$$

Since $(\overline{\mathrm{D}}(0,2)^D \times \{2, \ldots, d\}^D) \setminus N_\varepsilon$ is compact (this is the point in the proof where the existence of the degree bound d is essential), it follows that we can actually choose N_0 large enough such that

$$(\overline{\mathrm{D}}(0,2)^D \times \{2, \ldots, d\}^D) \setminus N_\varepsilon \subset (\mathbb{C}^D \times \{2, \ldots, d\}^D) \setminus \mathcal{M}^{\mathbf{s}^N}, \quad \text{for all } N \geq N_0$$

and, since $\mathcal{M}^{\mathbf{s}^N} \subset \overline{\mathrm{D}}(0,2)^D \times \{2, \ldots, d\}^D$, again from Lemma 2.2, we have in fact that

$$(\mathbb{C}^D \times \{2, \ldots, d\}^D) \setminus N_\varepsilon \subset (\mathbb{C}^D \times \{2, \ldots, d\}^D) \setminus \mathcal{M}^{\mathbf{s}^N}, \quad \text{for all } N \geq N_0.$$

Hence

$$\mathcal{M}^{\mathbf{s}^N} \subset N_\varepsilon, \quad \text{for all } N \geq N_0.$$

In conclusion, since $\mathcal{M}^{\mathbf{s}^N}$ is compact in view of Lemma 2.2 (which, again, requires the existence of the degree bound d), we have

$$d(\mathcal{M}^{\mathbf{s}^N}, \mathcal{M}^{\mathbf{s}}) < \varepsilon, \quad \text{for all } N \geq N_0,$$

which completes the proof. □

Basically, the above theorem states that the template Mandelbrot sets depend upper semicontinuoulsy on the parameter $(\mathbf{c}, \mathbf{d})$ (see Figure 1 for an illustration of how this looks in practice). We remark the similarity between this result and the upper semicontinuous dependence of the filled Julia set in nonautonomous polynomial iteration ([**6**, Theorem 2.4]) as well as the classical iteration with a single polynomial. We will next show that this statement does not remain true in the other direction. In other words, we do not have lower semicontinuous dependence, and thus we do not have continuity with respect to the Hausdorff topology. To be precise, we will show that:

THEOREM 3.2. *Let* $D = d = 2$, *let* $\mathbf{s} = (011111\ldots\ldots)$, *and let* $\{\mathbf{s}^N\}_{N=1}^\infty$ *be any sequence of templates in* $\{0,1\}^\infty$ *such that* $\lim_{N\to\infty} \|\mathbf{s}^N - \mathbf{s}\| = 0$, *and such that* $\mathbf{s}^N \neq \mathbf{s}$ *for infinitely many* N. *Then*

$$d(\mathcal{M}^{\mathbf{s}}, \mathcal{M}^{\mathbf{s}^N}) \not\to 0 \quad \textit{as} \quad N \to \infty.$$

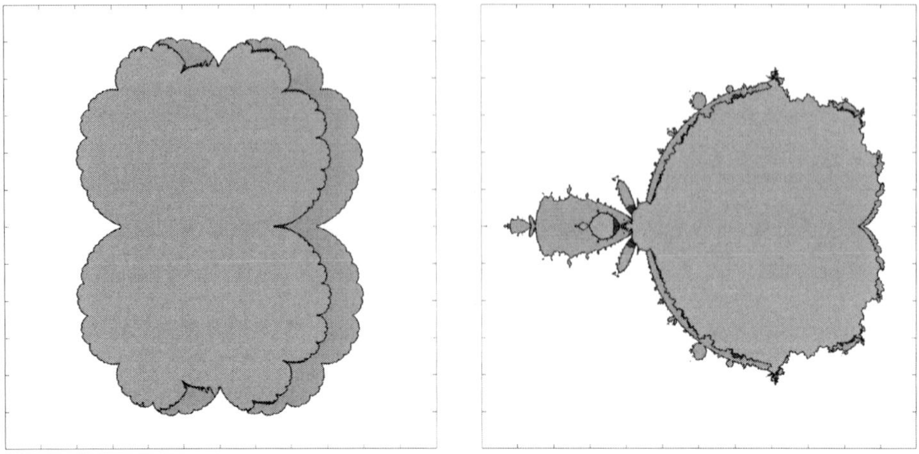

FIGURE 1. Slices of the Mandelbrot set for the template $\mathbf{s} = 0111\ldots$ using two quadratic polynomials. The parameter is $(\mathbf{c}, \mathbf{d}) = ((c_0, c_1), (d_0, d_1)) = ((\frac{1}{2} - \frac{1}{256}, \frac{1}{4} - \frac{1}{256}), (2, 2))$ and slices are taken in the c_0-direction on the left and in the c_1-direction on the right. The Mandelbrot slices for the limit template are shown in green while the approximate slices using periodic approximations to the limit template of periods 20 and 200 are shown in light blue and dark brown, respectively. The pictures were made using a maximum of 400 iterations and a resolution of 1200×1200. The plotting ranges are $[-1.25, 1.25] \times [-1.25, 1.25]$ for the c_0 panel and $[-1.5, 0.5] \times [-1, 1]$ for the c_1 panel.

REMARK. The key to the proof of this result is the asymmetry of the sequence $\mathbf{s}$ above which is a full template and which passes from 0 to 1 but not from 1 to 0. On the other hand, the conditions on the approximating sequences $\mathbf{s}^N$ *force* these sequences to make a transition from 1 to 0 for infinitely many such sequences. Using this, we can make the dynamics sufficiently different (at this point, the reader might like to consult Figure 2) to ensure that all critical points for template iteration using $\mathbf{s}$ with parameters $((c_0, c_1), (2, 2))$ have bounded orbits while at the same time we can ensure that, for a reasonably small neighbourhood of these parameters, there must be a critical point whose orbit escapes if one uses the templates $\mathbf{s}^N$ (again for infinitely many such sequences). Before we can proceed with the proof of this result, we first need a small lemma about stability under perturbation for the basin of infinity. Since we are now in the situation of constant degree 2, for convenience we will suppress the degree vectors in the parameters, referring to (c_0, c_1) instead of $((c_0, c_1), (2, 2))$ as above, using the product metric on $\mathbb{C}^2$ instead of $\mathbb{C}^2 \times \{2\}^2$ for our parameter neighbourhoods, and identifying our parameter spaces with subsets of $\mathbb{C}^2$ instead of $\mathbb{C}^2 \times \{2\}^2$.

LEMMA 3.3. *Let* $0 < \varepsilon_0, \varepsilon_1 \leq \frac{1}{8}$ *and let* $c_0 = \frac{1}{2} - \varepsilon_0$, $c_1 = \frac{1}{4} - \varepsilon_1$. *Then there exist* $\eta > 0$ *and* $\delta > 0$ *depending only on* ε_0 *and* ε_1 *such that, if* $\mathbf{s} \in \{0, 1\}^\infty$ *is any arbitrary template,* $d((c_0, c_1), (\tilde{c}_0, \tilde{c}_1)) < \eta$, *and* $\{\tilde{P}_m^{\mathbf{s}}\}_{m=1}^\infty$ *is the polynomial sequence constructed from the pair* $(\tilde{c}_0, \tilde{c}_1)$ *(and degrees* 2*) according to the template* s, *then*

$$\mathrm{D}(x,\delta) \subset \tilde{A}^{\mathbf{s}}_{\infty}, \ \textit{for all } x \geq \beta + \sqrt{\varepsilon_1},$$

where $\beta := \frac{1}{2} + \sqrt{\epsilon_1}$ *is the repelling fixed point for* f_{c_1} *and* $\tilde{A}^{\mathbf{s}}_{\infty}$ *is the basin of infinity for* $\{\tilde{P}^{\mathbf{s}}_m\}_{m=1}^{\infty}$.

PROOF OF LEMMA 3.3. Let $\mathbf{s} \in \{0,1\}^{\infty}$ be an arbitrary template and let $\mathcal{J}_{c_0}$, $\mathcal{J}_{c_1}$ and $\mathcal{K}_{c_0}$, $\mathcal{K}_{c_1}$ denote the Julia and filled Julia sets respectively for f_{c_0}, f_{c_1} (again the reader is referred to Figure 2 for pictures of these Julia sets and how they are used in our proof). As mentioned above, β is a repelling fixed point for f_{c_1} and one checks that $\mathcal{K}_{c_1} \cap \mathbb{R} = [-\beta, \beta]$. Recall also that, provided $(\tilde{c}_0, \tilde{c}_1) \in \overline{\mathrm{B}}((0,0),2)$, then $R = 2$ is an escape radius for any polynomial sequence $\{\tilde{P}^{\mathbf{s}}_m\}_{m=1}^{\infty}$ as above. Clearly, there exists $N_0 = N_0(\varepsilon_1)$ such that:

$$f_{c_1}^{\circ N_0}(\beta + \sqrt{\varepsilon_1}) > 3.$$

Since $c_0 \geq c_1$ and f_{c_0} and f_{c_1} are both increasing real-valued functions on $(0,\infty)$, this implies that

$$Q^{\mathbf{s}}_{N_0}(x) > 3, \ \text{for all } x \geq \beta + \sqrt{\varepsilon_1} \tag{3.3}$$

where $Q^{\mathbf{s}}_{N_0}$ denotes the finite composition generated by the pair (c_0, c_1) and the first N_0 entries of the template $\mathbf{s}$. We proceed to finish the proof according to two cases depending on the value of the quantity x.

CASE 1. $\beta + \sqrt{\varepsilon_1} \leq x \leq 3$ Consider the collection of compositions $\{\tilde{Q}^{\mathbf{u}}_{N_0}(z)\}$ where $\mathbf{u}$ is any template, and the polynomial sequences $\{\tilde{P}^{\mathbf{u}}_m\}_{m=1}^{\infty}$ (which give rise to the compositions $\{\tilde{Q}^{\mathbf{u}}_{N_0}(z)\}$) are generated according to the template $\mathbf{u}$ and the pair $(\tilde{c}_0, \tilde{c}_1) \in \mathbb{C}^2$ (using degree 2 for each polynomial $\tilde{P}^{\mathbf{u}}_m$). Each member of this collection is jointly differentiable in z, $\tilde{c}_0$, and $\tilde{c}_1$, and the number of possibilities for the first N_0 entries of the template $\mathbf{u}$ is finite (actually 2^{N_0}). In addition, the degrees of these polynomials are all 2^{N_0} and their coefficients (in z) are fixed polynomials in $\tilde{c}_0$, $\tilde{c}_1$ of degree at most 2^{N_0-1} in each variable. Further, for each of these coefficient polynomials, when considered as polynomials in each of $\tilde{c}_0$, $\tilde{c}_1$ separately with the other variable held fixed, then the coefficients of these polynomials in $\tilde{c}_0$, $\tilde{c}_1$ are clearly uniformly bounded (on any bounded subset of $\mathbb{C}^2$). It follows (somewhat) as in the proof of Theorem 3.1 that the partial derivatives of these functions in z, $\tilde{c}_0$, $\tilde{c}_1$ are uniformly bounded on any bounded subset of $\mathbb{C}^3$. From this we can deduce that there exist $\eta, \delta > 0$ (depending only on ε_0 and ε_1) such that, if $(\tilde{c}_0, \tilde{c}_1) \in \mathrm{B}((c_0, c_1), \eta)$, $\beta + \sqrt{\epsilon_1} \leq x \leq 3$, and $|z - x| < \delta$, then $(\tilde{c}_0, \tilde{c}_1) \in \overline{\mathrm{B}}((0,0),2)$ and

$$|\tilde{Q}^{\mathbf{u}}_{N_0}(z) - Q^{\mathbf{u}}_{N_0}(x)| < 1.$$

In conjunction with (3.3), this implies (by the reverse triangle inequality) that

$$\tilde{Q}^{\mathbf{s}}_{N_0}(\mathrm{D}(x,\delta)) \subset \mathbb{C} \setminus \overline{\mathrm{D}}(0,2).$$

Since $(\tilde{c}_0, \tilde{c}_1) \in \overline{\mathrm{B}}((0,0),2)$, $R = 2$ is an escape radius for any template sequence of quadratic polynomials $\{\tilde{P}^{\mathbf{s}}_m\}_{m=1}^{\infty}$ generated using $(\tilde{c}_0, \tilde{c}_1)$ and $\mathbf{s}$ as above. It therefore follows that $\mathrm{D}(x,\delta) \subset \tilde{A}^{\mathbf{s}}_{\infty}$.

CASE 2. $x > 3$. To conclude the proof: if $x > 3$, let η be as before and make δ smaller if needed so that $\delta \leq 1$ (which still leaves the dependence on ε_0 and ε_1 unaltered). Then, since we still have $(\tilde{c}_0, \tilde{c}_1) \in \overline{\mathrm{B}}((0,0),2)$, it follows that $R = 2$ is an escape radius for $\{\tilde{P}^{\mathbf{s}}_m\}_{m=1}^{\infty}$. We then automatically have $\mathrm{D}(x,\delta) \subset \tilde{A}^{\mathbf{s}}_{\infty}$ in this case also, which completes the proof. $\square$

PROOF OF THEOREM 3.2. Clearly the result follows if we show that, for any template sequence $\{\mathbf{s}^N\}_{N=1}^{\infty}$ such that $\mathbf{s}^N \neq \mathbf{s}$ for every $N > 1$, we have

$$\limsup_N d(\mathcal{M}^{\mathbf{s}}, \mathcal{M}^{\mathbf{s}^N}) > 0.$$

We therefore start (without loss of generality) by assuming this is the case and that $\mathbf{s}^N \neq \mathbf{s}$ for every $N \geq 1$.

As before, let $0 < \varepsilon_0, \varepsilon_1 \leq \frac{1}{8}$ and set $c_0 = \frac{1}{2} - \varepsilon_0$, $c_1 = \frac{1}{4} - \varepsilon_1$. Recall that, as a single map under iteration, $f_{c_1}(z) = z^2 + \frac{1}{4} - \varepsilon_1$ has two fixed points: $\alpha := \frac{1}{2} - \sqrt{\varepsilon_1}$ (which is attracting) and $\beta := \frac{1}{2} + \sqrt{\varepsilon_1}$ (which is repelling). As above, the filled Julia set $\mathcal{K}_{c_1}$ of f_{c_1} still satisfies $\mathcal{K}_{c_1} \cap \mathbb{R} = [-\beta, \beta]$. Then

$$f_{c_0}(0) = \frac{1}{2} - \varepsilon_0 \in [-\beta, \beta] \subset \mathcal{K}_{c_1}.$$

Thus, $f_{c_0}(0)$ has bounded orbit under f_{c_1}. Since 0 also has bounded orbit under f_{c_1}, all critical points of the sequence $\{P^{\mathbf{s}}_m\}_{m=1}^{\infty}$ generated by the pair (c_0, c_1) and the template $\mathbf{s}$ (as well as the degree pair $(2,2)$) have bounded orbit and so

$$(c_0, c_1) \in \mathcal{M}^{\mathbf{s}}.$$

Now take ε_0 and ε_1 sufficiently small such that:

$$3\sqrt{\varepsilon_1} - \varepsilon_1 + \varepsilon_0 < \frac{1}{4}. \tag{3.4}$$

Then

$$f_{c_0}(\alpha) = \alpha^2 + \frac{1}{2} - \varepsilon_0 > \beta + \sqrt{\varepsilon_1}. \tag{3.5}$$

Let $\eta, \delta > 0$ (depending on ε_0, ε_1) be as in Lemma 3.3, let $m_0 \geq 0$ be arbitrary, and let $\mathbf{u} = \{u_m\}_{m=1}^{\infty}$ be any template satisfying $u_{m_0+1} = 0$. Hence, using (3.5), Lemma 1.1, and Lemma 3.3, if $d((c_0, c_1), (\tilde{c}_0, \tilde{c}_1)) < \eta$, then

$$\mathrm{D}(f_{c_0}(\alpha), \delta) \subset \tilde{A}^{\mathbf{u}}_{\infty, m_0+1}, \tag{3.6}$$

where the polynomial sequence $\{\tilde{P}^{\mathbf{u}}_m\}_{m=1}^{\infty}$ is generated according to $\mathbf{u}$ and $(\tilde{c}_0, \tilde{c}_1)$ (as well as the degree pair $(2,2)$), while $\tilde{A}^{\mathbf{u}}_{\infty, m_0+1}$ represents the basin of ∞ at time $m_0 + 1$ for this polynomial sequence.

Since $f_{\tilde{c}_0}$ is continuous, it follows that there exists a $\tau_0 > 0$ (also depending on ε_0 and ε_1), such that:

$$f_{c_0}(\mathrm{D}(\alpha, \tau_0)) \subset \mathrm{D}(f_{c_0}(\alpha), \tfrac{\delta}{2}). \tag{3.7}$$

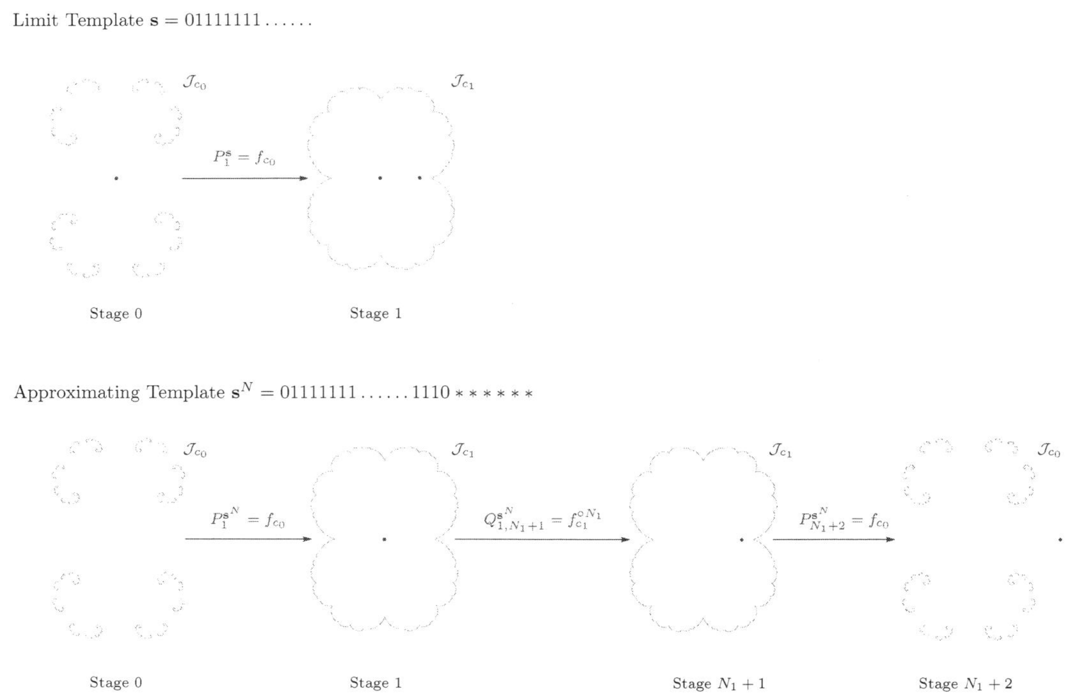

FIGURE 2. Schematic for the proof of Theorem 3.2. The top set of pictures illustrates the behaviour of the sequence$\{P_m^{\mathbf{s}}\}_{m=1}^{\infty}$ of quadratics generated by the pair (c_0, c_1) and the template $\mathbf{s}$. Here the critical point 0 at time 0 is mapped inside $\mathcal{K}_1$ while all subsequent critical points also lie inside this filled Julia set and thus do not escape. In the bottom set of pictures we have the behaviour of the sequence of quadratics $\{\tilde{P}_m^{\mathbf{s}^N}\}_{m=1}^{\infty}$ generated by the pair (c_0, c_1) and the template $\mathbf{s}^N$. In this case the critical point 0 at time 1 remains inside $\mathcal{K}_1$ for N_1 iterations. However, the next polynomial in the template sequence is $P_{N_1+2}^{\mathbf{s}^N} = f_{c_0}$ which maps the iterate of this critical point to the basin of infinity so that it escapes. Much of the proof is devoted to showing this behaviour is stable under perturbation and that there is a lower bound for the size of the perturbation which depends only on the constants ϵ_0, ϵ_1.

Let $\eta' = \min(\eta, \delta/2)$ (which also depends on only ε_0 and ε_1). It follows that, when $|\tilde{c}_0 - c_0| < \eta'$ (so that $|f_{\tilde{c}_0}(z) - f_{c_0}(z)| = |\tilde{c}_0 - c_0| < \eta' \leq \frac{\delta}{2}$ for any $z \in \mathbb{C}$), then, by (3.6) and (3.7)

(3.8)
$$\tilde{Q}^{\mathbf{u}}_{m_0,m_0+1}(\mathrm{D}(\alpha,\tau_0)) = \tilde{P}^{\mathbf{u}}_{m_0+1}(\mathrm{D}(\alpha,\tau_0)) = f_{\tilde{c}_0}(\mathrm{D}(\alpha,\tau_0)) \subset \mathrm{D}(f_{c_0}(\alpha),\delta) \subset \tilde{A}^{\mathbf{u}}_{\infty,m_0+1}.$$

Recall that f_{c_1} has an attracting fixed point at $\alpha = 1/2 - \sqrt{\varepsilon_1}$, with multiplier $2\alpha = 1 - 2\sqrt{\varepsilon_1}$. Since this condition is clearly stable under perturbation, we can make η' smaller if needed (still only depending on ε_0 and ε_1), such that, if $|\tilde{c}_1 - c_1| < \eta'$, then $f_{\tilde{c}_1}$ has a fixed point $\tilde{\alpha}$ such that

$$|\tilde{\alpha} - \alpha| < \tau_0/2 \tag{3.9}$$

while we can at the same time ensure that the multiplier $2\tilde{\alpha}$ of this fixed point has absolute value strictly less than $1 - \sqrt{\varepsilon_1}$. The multiplier of $\tilde{\alpha}$ is thus uniformly bounded below away from 1 in absolute value so that the fixed points $\tilde{\alpha}$ are uniformly attracting and we can clearly choose τ_0 to be smaller if needed (but still depending on ε_0, ε_1) so that the disc $\mathrm{D}(\tilde{\alpha}, \tau_0/2)$ is forward invariant under $f_{\tilde{c}_1}$ so that it lies in the Fatou set for $f_{\tilde{c}_1}$ and hence the immediate basin of $\tilde{\alpha}$ contains a disc about $\tilde{\alpha}$ whose radius is uniformly bounded below away from 0. Further, the immediate basin of each of them must contain the critical point 0 (see, e.g., [**4**, Chapter III Theorem 2.2]). An easy argument using relative compactness of $\mathrm{D}(c_1, \eta')$) shows that there then exists $N_0 > 0$ (depending only on ε_0 and ε_1), such that, in view of (3.9):

$$f_{\tilde{c}_1}^{\circ N}(0) \in \mathrm{D}(\tilde{\alpha}, \tau_0/2) \subset \mathrm{D}(\alpha, \tau_0), \text{ for all } N \geq N_0. \tag{3.10}$$

Now we look at the sequence of templates $\{\mathbf{s}^N\}_{N=1}^{\infty}$. Since, from the assumptions at the start of the theorem, $\|\mathbf{s}^N - \mathbf{s}\| \to 0$ and $\mathbf{s}^N \neq \mathbf{s}$ for all N, it follows that there exists N_1 (which thus tends to infinity as N tends to infinity) such that (for any $(\tilde{c}_0, \tilde{c}_1) \in \mathbb{C}^2$)

- $\tilde{Q}^{\mathbf{s}^N}_{0,1} = \tilde{P}^{\mathbf{s}^N}_1 = f_{\tilde{c}_0}$,
- $\tilde{Q}^{\mathbf{s}^N}_{1,N_1+1} = f_{\tilde{c}_1}^{\circ N_1}$, and
- $\tilde{Q}^{\mathbf{s}^N}_{N_1+1,N_1+2} = \tilde{P}^{\mathbf{s}^N}_{N_1+2} = f_{\tilde{c}_0}$.

In other words, the first member of the sequence $\{\tilde{P}^{\mathbf{s}^N}\}_{m=1}^{\infty}$ is $f_{\tilde{c}_0}$, the next N_1 members are $f_{\tilde{c}_1}$, and the (N_1+2)-nd member is again $f_{\tilde{c}_0}$. Then, by (3.8) and (3.10) (where we let $m_0 = N_1+1$ in (3.8)), it follows that, if we choose N sufficiently large so that we also have $N_1 \geq N_0$, then,

$$\tilde{Q}^{\mathbf{s}^N}_{1,N_1+2}(0) \in \tilde{A}^{\mathbf{s}^N}_{\infty,N_1+2}.$$

It follows that, for N sufficiently large, if $|\tilde{c}_0 - c_0| < \eta'$, then the critical point zero at time one escapes for any template in this sequence, that is:

$$(\tilde{c}_0, \tilde{c}_1) \notin \mathcal{M}^{\mathbf{s}^N}.$$

Hence

$$\mathrm{B}((c_0, c_1), \eta') \subset \mathbb{C}^2 \setminus \mathcal{M}^{\mathbf{s}^N}$$

and thus

$$d((c_0, c_1), \mathcal{M}^{\mathbf{s}^N}) \geq \eta'$$

(see Figure 3) so that

$$\limsup_{N\to\infty} d(\mathcal{M}^{\mathbf{s}}, \mathcal{M}^{\mathbf{s}^N}) > 0$$

as required, from which the result follows. □

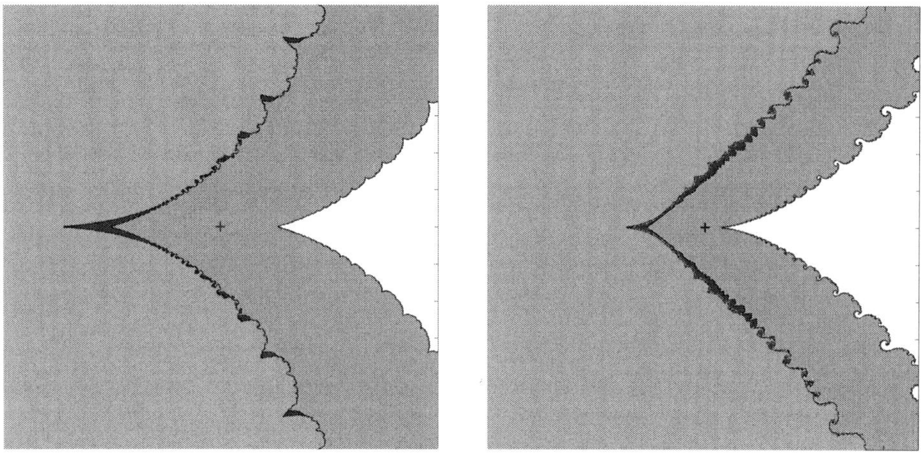

FIGURE 3. Zooms about $(c_0, c_1) = (\frac{1}{2} - \frac{1}{256}, \frac{1}{4} - \frac{1}{256})$ (with degree vector $(2, 2)$) of the pictures in Figure 1 illustrating the lack of lower semicontinuity where $d(\mathcal{M}^{\mathbf{s}}, \mathcal{M}^{\mathbf{s}^N})$ does not tend to zero as N tends to infinity. Here we have $\varepsilon_0 = \varepsilon_1 = \frac{1}{256}$ which satisfies condition (3.4) in the proof of Theorem 3.2 above. Again, slices in the c_0-direction are on the left and in the c_1-direction on the right. The cross marks the point of intersection of these two complex lines at (c_0, c_1). As before, we have a maximum of 400 iterations, and a resolution of 1200×1200. As with Figure 1, the Mandelbrot slices for the limit sequence are in green, with those for the periodic approximations of periods 20 and 200 in light blue and and dark brown respectively. The plotting ranges are $[c_0 - 0.25, c_0 + 0.25] \times [-0.25, 0.25]$ for the c_0 panel and $[c_1 - 0.05, c_1 + 0.05] \times [-0.05, 0.05]$ for the c_1 panel.

References

[1] R. Brück. Geometric properties of julia sets of the composition of polynomials of the form $z^2 + c_n$. *Pacific J. Math*, 198:347–371, 2001.

[2] Rainer Brück, Matthias Büger, and Stefan Reitz, *Random iterations of polynomials of the form $z^2 + c_n$: connectedness of Julia sets*, Ergodic Theory Dynam. Systems **19** (1999), no. 5, 1221–1231, DOI 10.1017/S0143385799141658. MR1721617

[3] Rainer Brück, *Connectedness and stability of Julia sets of the composition of polynomials of the form $z^2 + c_n$*, J. London Math. Soc. (2) **61** (2000), no. 2, 462–470, DOI 10.1112/S0024610799008509. MR1760694

[4] Lennart Carleson and Theodore W. Gamelin, *Complex dynamics*, Universitext: Tracts in Mathematics, Springer-Verlag, New York, 1993, DOI 10.1007/978-1-4612-4364-9. MR1230383

[5] Mark Comerford, *Conjugacy and counterexample in random iteration*, Pacific J. Math. **211** (2003), no. 1, 69–80, DOI 10.2140/pjm.2003.211.69. MR2016591

[6] Mark Comerford, *Hyperbolic nonautonomous Julia sets*, Ergodic Theory Dynam. Systems **26** (2006), no. 2, 353–377, DOI 10.1017/S0143385705000441. MR2218765
[7] Mark Comerford, *Nonautonomous Julia sets with measurable invariant sequences of line fields*, Discrete Contin. Dyn. Syst. **33** (2013), no. 2, 629–642, DOI 10.3934/dcds.2013.33.629. MR2975127
[8] Mark Comerford and Todd Woodard, *Preservation of external rays in nonautonomous iteration*, J. Difference Equ. Appl. **19** (2013), no. 4, 585–604, DOI 10.1080/10236198.2012.662966. MR3040817
[9] Mark Comerford and Todd Woodard, *Orbit portraits in nonautonomous iteration*, Discrete Contin. Dyn. Syst. Ser. S **12** (2019), no. 8, 2253–2277, DOI 10.3934/dcdss.2019144. MR4025983
[10] John Erik Fornæss and Nessim Sibony, *Random iterations of rational functions*, Ergodic Theory Dynam. Systems **11** (1991), no. 4, 687–708, DOI 10.1017/S0143385700006428. MR1145616
[11] Thomas Ransford, *Potential theory in the complex plane*, London Mathematical Society Student Texts, vol. 28, Cambridge University Press, Cambridge, 1995, DOI 10.1017/CBO9780511623776. MR1334766
[12] Anca Rădulescu, Kelsey Butera, and Brandee Williams, *Template iterations of quadratic maps and hybrid Mandelbrot sets*, J. Nonlinear Sci. **31** (2021), no. 1, Paper No. 22, 26, DOI 10.1007/s00332-020-09675-0. MR4214466
[13] Anca Rădulescu and Ariel Pignatelli, *Symbolic template iterations of complex quadratic maps*, Nonlinear Dynam. **84** (2016), no. 4, 2025–2042, DOI 10.1007/s11071-016-2626-3. MR3504280
[14] Rich Stankewitz and Hiroki Sumi, *Dynamical properties and structure of Julia sets of postcritically bounded polynomial semigroups*, Trans. Amer. Math. Soc. **363** (2011), no. 10, 5293–5319, DOI 10.1090/S0002-9947-2011-05199-8. MR2813416
[15] Hiroki Sumi, *Skew product maps related to finitely generated rational semigroups*, Nonlinearity **13** (2000), no. 4, 995–1019, DOI 10.1088/0951-7715/13/4/302. MR1767945
[16] Hiroki Sumi, *Dynamics of sub-hyperbolic and semi-hyperbolic rational semigroups and skew products*, Ergodic Theory Dynam. Systems **21** (2001), no. 2, 563–603, DOI 10.1017/S0143385701001286. MR1827119
[17] Hiroki Sumi, *Semi-hyperbolic fibered rational maps and rational semigroups*, Ergodic Theory Dynam. Systems **26** (2006), no. 3, 893–922, DOI 10.1017/S0143385705000532. MR2237476
[18] Hiroki Sumi, *Dynamics of postcritically bounded polynomial semigroups III: classification of semi-hyperbolic semigroups and random Julia sets which are Jordan curves but not quasicircles*, Ergodic Theory Dynam. Systems **30** (2010), no. 6, 1869–1902, DOI 10.1017/S0143385709000923. MR2736899
[19] Hiroki Sumi, *Dynamics of postcritically bounded polynomial semigroups II: fiberwise dynamics and the Julia sets*, J. Lond. Math. Soc. (2) **88** (2013), no. 1, 294–318, DOI 10.1112/jlms/jdt017. MR3092269
[20] Hiroki Sumi, *The space of 2-generator postcritically bounded polynomial semigroups and random complex dynamics*, Adv. Math. **290** (2016), 809–859, DOI 10.1016/j.aim.2015.12.011. MR3451939

Department of Mathematics, University of Rhode Island, 5 Lippitt Road, Kingston, Rhode Island 02881
Email address: `mcomerford@math.uri.edu`

Department of Mathematics, State University of New York New Paltz, 1 Hawk Road, New Paltz, New York 12561
Email address: `radulesa@newpaltz`

Department of Mathematics, Pennsylvania State University, 104 McAllister Building, University Park, State College, Pennsylvania 16802
Email address: `kvc5984@psu.edu`

Contemporary Mathematics
Volume **797**, 2024
https://doi.org/10.1090/conm/797/15932

Random beta-transformations on fat Sierpinski gaskets

Karma Dajani, Wenxia Li, and Tingyu Zhang

ABSTRACT. We consider the iterated function system (IFS)

$$f_{\vec{q}}(\vec{z}) = \frac{\vec{z}+\vec{q}}{\beta}, \ \vec{q} \in \{(0,0),(1,0),(0,1)\}.$$

As is well known, for $\beta = 2$ the attractor, S_β, is a fractal called the Sierpiński gasket (or sieve) and for $\beta > 2$ it is also a fractal. Our goal is to study random β-transformations on the attractor for this IFS with $1 < \beta \le 3/2$. In this case, S_β is a triangle. We show that all β-expansions of a point $\vec{z}$ in S_β can be generated by a random map K_β defined on $\{0,1\}^{\mathbb{N}} \times \{0,1,2\}^{\mathbb{N}} \times S_\beta$ and K_β has a unique invariant measure of maximal entropy. Furthermore, we show the existence of a K_β-invariant probability measure of the form $m_1 \otimes m_2 \otimes \mu_\beta$, where m_1, m_2 are product measures on $\{0,1\}^{\mathbb{N}}, \{0,1,2\}^{\mathbb{N}}$, respectively, and μ_β is absolutely continuous with respect to the two-dimensional Lebesgue measure λ_2.

1. Introduction

Let $\beta > 1$ and consider the *iterated function system* (IFS):

$$f_{\vec{q}_0}(\vec{z}) = \frac{\vec{z}+\vec{q}_0}{\beta}, \quad f_{\vec{q}_1}(\vec{z}) = \frac{\vec{z}+\vec{q}_1}{\beta}, \quad f_{\vec{q}_2}(\vec{z}) = \frac{\vec{z}+\vec{q}_2}{\beta}, \tag{1.1}$$

where the coordinates of the three points $\vec{q}_0, \vec{q}_1, \vec{q}_2$ are $(0,0),(1,0),(0,1)$, respectively. It is well known that there exists a unique nonempty compact set $S_\beta \subset \mathbb{R}^2$ such that $S_\beta = \cup_{i=0}^2 f_{\vec{q}_i}(S_\beta)$; see [**F**] for further details. The *attractor* for the IFS, S_β, is a Sierpinski gasket. Denote the convex hull of S_β by Δ which is a triangle with vertices at $(0,0), (\frac{1}{\beta-1}, 0)$ and $(0, \frac{1}{\beta-1})$. For every point $\vec{z} \in S_\beta$, there exists a sequence $(a_i)_{i=1}^\infty \in \{\vec{q}_0, \vec{q}_1, \vec{q}_2\}^{\mathbb{N}}$ such that

$$\vec{z} = \lim_{n\to\infty} f_{a_1} \circ \cdots \circ f_{a_n}(\vec{q}_0) = \sum_{i=1}^\infty \frac{a_i}{\beta^i}.$$

We call $(a_i)_{i=1}^\infty$ a *coding* of $\vec{z}$ and $\sum_{i=1}^\infty a_i\beta^{-i}$ a *representation* of $\vec{z}$ in base β, or simple a β-expansion of $\vec{z}$.

2020 *Mathematics Subject Classification.* Primary 37H99, 37A45; Secondary 37A05, 11K55.

Key words and phrases. Fat Sierpinski gasket, random β-transformation, absolutely continuous invariant measures, measures of maximal entropy.

The second and third author were supported by National Natural Science Foundation of China (NSFC) No. 12071148 and Science and Technology Commission of Shanghai Municipality (STCSM) No. 18dz2271000. The third author was also supported by China Scholarship Council No.202006140156 during her visit to Utrecht University.

Let $i \in \{0,1,2\}$. For $\beta > 2$, the images $f_{\vec{q}_i}(\Delta)$ are disjoint. In this case the IFS $\{f_{\vec{q}_i}\}$ satisfies the strong separation condition and each point in S_β has a unique coding. For $\beta = 2$, the sets $f_{\vec{q}_i}(\Delta)$ overlap only at the vertices. Therefore only countably many points in S_β have two codings, and all other points have a unique coding. When $\beta \in (1,2)$, we call S_β a *fat Sierpinski gasket* and we distinguish two cases. For $1 < \beta \leq 3/2$, we have a non-empty double and triple overlaps and $S_\beta = \Delta$, see Figure 1. Furthermore, Lebesgue almost every point in S_β has a continuum of codings (see [**S1**, Theorem 3.5]). For $3/2 < \beta < 2$, there are holes in S_β as well as overlaps, which makes its structure more complex. In [**BMS**], Broomhead et al. described two special types of structures: those in which holes are radially distributed and those that are totally self-similar. Total self-similarity in our case implies $f_{\vec{q}_i}(S_\beta) = f_{\vec{q}_i}(\Delta) \cap S_\beta$. For more results on the Hausdorff dimension of the attractors, see [**KL, HP, SS, JP, H**].

In this article, we focus on the case $1 < \beta \leq 3/2$. In order to capture all possible β-expansions and to describe their statistical properties, we take an ergodic view. We start by defining a map K_β whose iterations generate all possible β-expansions of points in S_β. Dynamical properties of this map give information on the asymptotic properties of these expansions. The definition of our map is motivated from an analogous study of random β-expansions for points on an interval with digits in $\{0,1,\cdots,\lfloor\beta\rfloor\}$, see [**DK2, DV1, DV2**]. Our main aim is to generalise their results, in particular exhibit natural invariant ergodic measures for the random β-transformation K_β.

The rest of the article is organized as follows. In Section 2, we give the definition of the random transformation K_β on $\{0,1\}^{\mathbb{N}} \times \{0,1,2\}^{\mathbb{N}} \times S_\beta$ and prove basic properties. In Section 3, we prove that K_β has a unique invariant measure of maximal entropy. In Section 4, we give a position-dependent random map R on S_β. With two skew product transformations, we establish a connection between R and K_β, and finally prove that K_β has an invariant measure of the form $m_1 \otimes m_2 \otimes \mu_\beta$, where m_1 is the product measure on $\{0,1\}^{\mathbb{N}}$ with weights $\{p, 1-p\}$, m_2 is the product measure on $\{0,1,2\}^{\mathbb{N}}$ with weights $\{s, t, 1-s-t\}$, and μ_β is R-invariant and absolutely continuous with respect to λ_2, the normalized Lebesgue measure on S_β.

2. Random beta-transformations

Given $1 < \beta \leq 3/2$, recall that the fat Sierpinski gasket S_β is the self-similar set in $\mathbb{R}^2$ generated by the IFS (1.1). For every point $\vec{z} \in S_\beta$, there exists a sequence $(a_i)_{i=1}^\infty \in \{\vec{q}_0, \vec{q}_1, \vec{q}_2\}^{\mathbb{N}}$ such that $\vec{z} = \sum_{i=1}^\infty a_i \beta^{-i}$. Notice that S_β and its convex hull Δ are identical, both being an isosceles right triangle. We denote the Borel σ-algebra on S_β by $\mathcal{S}$. We also consider the following ordering of points in the plane. We write $(x_1, y_1) < (x_2, y_2)$ if $x_1 + y_1 < x_2 + y_2$, or $x_1 + y_1 = x_2 + y_2$ and $y_1 < y_2$. Notice that $\vec{q}_0 < \vec{q}_1 < \vec{q}_2$.

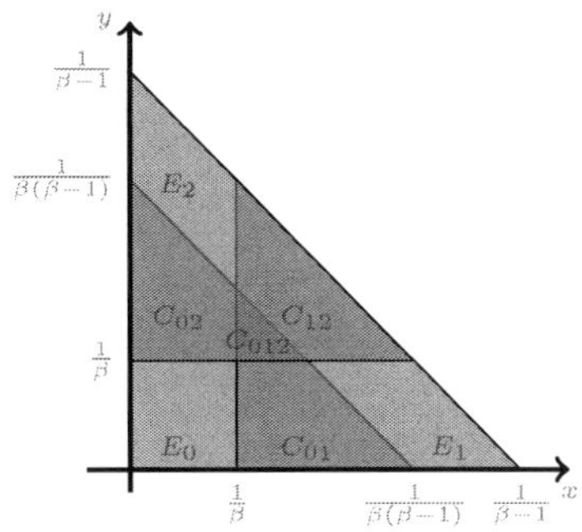

FIGURE 1. S_β for $1 < \beta \le \frac{3}{2}$

Divide S_β into the following sets according to the overlapping structure of $f_{\vec{q}_i}(S_\beta)$ (see Figure 1):

$$\begin{aligned} E_0 &= [0, \frac{1}{\beta}) \times [0, \frac{1}{\beta}), \\ E_1 &= \{(x,y) : 0 \le y < \frac{1}{\beta}, \frac{1}{\beta(\beta-1)} < x + y \le \frac{1}{\beta-1}\}, \\ E_2 &= \{(x,y) : 0 \le x < \frac{1}{\beta}, \frac{1}{\beta(\beta-1)} < x + y \le \frac{1}{\beta-1}\}, \\ C_{01} &= \{(x,y) : x \ge \frac{1}{\beta}, 0 \le y < \frac{1}{\beta}, x + y \le \frac{1}{\beta(\beta-1)}\}, \\ C_{12} &= \{(x,y) : x \ge \frac{1}{\beta}, y \ge \frac{1}{\beta}, \frac{1}{\beta(\beta-1)} < x + y \le \frac{1}{\beta-1}\}, \\ C_{02} &= \{(x,y) : 0 \le x < \frac{1}{\beta}, y \ge \frac{1}{\beta}, x + y \le \frac{1}{\beta(\beta-1)}\}, \\ C_{012} &= \{(x,y) : x \ge \frac{1}{\beta}, y \ge \frac{1}{\beta}, x + y \le \frac{1}{\beta(\beta-1)}\}. \end{aligned} \tag{2.1}$$

Notice that $C_{012} = \{(\frac{2}{3}, \frac{2}{3})\}$ is a single point set if $\beta = 3/2$. These regions specify the digits that our random map assigns to points in S_β. For points in E_i the digit assigned is q_i, while in the double overlapping region C_{ij} we have two choices, q_i or q_j, and in the triple overlapping C_{012} we can choose q_0, q_1 or q_2. The choices will be dictated by either a double-sided coin or a triple-sided coin. To incorporate these choices in our definition of the random map K_β, we introduce two shift spaces representing the required coin tosses.

Let $\Omega = \{0,1\}^{\mathbb{N}}$ with the product σ-algebra $\mathcal{A}$ and $\Upsilon = \{0,1,2\}^{\mathbb{N}}$ with the product σ-algebra $\mathcal{B}$. Define metrics d and ρ on Ω and Υ, respectively, by

$$d((\omega_i),(\omega_i')) = 2^{-\inf\{k:\omega_k \ne \omega_k'\}} \quad \text{and} \quad \rho((v_i),(v_i')) = 2^{-\inf\{k:v_k \ne v_k'\}}.$$

Let $\sigma : \Omega \to \Omega$ and $\sigma' : \Upsilon \to \Upsilon$ be the left shifts.

Throughout the article, the lexicographical ordering on Ω, Υ and $\{\vec{q}_0, \vec{q}_1, \vec{q}_2\}^{\mathbb{N}}$ are all denoted by $\prec$ and $\preceq$. More precisely, for two sequences $(c_i), (d_i)$ we write $(c_i) \prec (d_i)$ if $c_1 < d_1$, or there exists $k \ge 2$ such that $c_i = d_i$ for all $1 \le i < k$ and $c_k < d_k$. Similarly, we write $(c_i) \preceq (d_i)$ if $(c_i) \prec (d_i)$ or $(c_i) = (d_i)$.

Let $C = C_{01} \cup C_{12} \cup C_{02}$ and $E = \cup_{i=0}^{2} E_i$. Define $K_\beta : \Omega \times \Upsilon \times S_\beta \to \Omega \times \Upsilon \times S_\beta$ by

$$K_\beta(\omega, \upsilon, \vec{z}) = \begin{cases} (\omega, \upsilon, \beta\vec{z} - \vec{q}_i), & \text{if } \vec{z} \in E_i, i = 0,1,2, \\ (\sigma\omega, \upsilon, \beta\vec{z} - \vec{q}_i), & \text{if } \omega_1 = 0 \text{ and } \vec{z} \in C_{ij},\ ij \in \{01,12,02\}, \\ (\sigma\omega, \upsilon, \beta\vec{z} - \vec{q}_j), & \text{if } \omega_1 = 1 \text{ and } \vec{z} \in C_{ij},\ ij \in \{01,12,02\}, \\ (\omega, \sigma'\upsilon, \beta\vec{z} - \vec{q}_i), & \text{if } \vec{z} \in C_{012} \text{ and } \upsilon_1 = i \in \{0,1,2\}. \end{cases}$$

The digits are given by

$$d_1 = d_1(\omega, \upsilon, \vec{z}) = \begin{cases} \vec{q}_i, & \text{if } \vec{z} \in E_i, i = 0,1,2, \\ & \text{or } (\omega, \upsilon, \vec{z}) \in \Omega \times \{\upsilon_1 = i\} \times C_{012}, \\ & \text{or } (\omega, \upsilon, \vec{z}) \in \{\omega_1 = 0\} \times \Upsilon \times C_{ij},\ ij \in \{01,12,02\}, \\ \vec{q}_j, & \text{if } (\omega, \upsilon, \vec{z}) \in \{\omega_1 = 1\} \times \Upsilon \times C_{ij},\ ij \in \{01,12,02\}. \end{cases}$$

Then

$$K_\beta(\omega, \upsilon, \vec{z}) = \begin{cases} (\omega, \upsilon, \beta\vec{z} - d_1), & \text{if } \vec{z} \in E, \\ (\sigma\omega, \upsilon, \beta\vec{z} - d_1), & \text{if } \vec{z} \in C, \\ (\omega, \sigma'\upsilon, \beta\vec{z} - d_1), & \text{if } \vec{z} \in C_{012}. \end{cases}$$

Set $d_n = d_n(\omega, \upsilon, \vec{z}) = d_1(K_\beta^{n-1}(\omega, \upsilon, \vec{z}))$, and $\pi_3 : \Omega \times \Upsilon \times S_\beta \to S_\beta$ be the canonical projection onto the third coordinate. Then

$$\pi_3(K_\beta^n(\omega, \upsilon, \vec{z})) = \beta^n \vec{z} - \beta^{n-1} d_1 - \cdots - \beta d_{n-1} - d_n,$$

and rewriting yields

$$\vec{z} = \frac{d_1}{\beta} + \frac{d_2}{\beta^2} + \cdots + \frac{d_n}{\beta^n} + \frac{\pi_3(K_\beta^n(\omega, \upsilon, \vec{z}))}{\beta^n}.$$

Since $\pi_3(K_\beta^n(\omega, \upsilon, \vec{z})) \in S_\beta$ and S_β is a bounded set in $\mathbb{R}^2$, it follows that

$$\|\vec{z} - \sum_{i=1}^{n} \frac{d_i}{\beta^i}\|_1 = \frac{\|\pi_3(K_\beta^n(\omega, \upsilon, \vec{z}))\|_1}{\beta^n} \to 0,$$

where $\|\cdot\|_1$ denotes the L_1 norm, i.e., the sum of the absolute values of the vector elements. This shows that for all $\omega \in \Omega$, $\upsilon \in \Upsilon$ and for all $\vec{z} \in S_\beta$ one has

$$\vec{z} = \sum_{i=1}^{\infty} \frac{d_i}{\beta^i} = \sum_{i=1}^{\infty} \frac{d_i(\omega, \upsilon, \vec{z})}{\beta^i}.$$

For each point $\vec{z} \in S_\beta$, consider the set

$$D_{\vec{z}} = \{(d_1(\omega, \upsilon, \vec{z}), d_2(\omega, \upsilon, \vec{z}), \ldots) : \omega \in \Omega, \upsilon \in \Upsilon\}.$$

Theorem 2.1 shows how the lexicographical ordering on Ω and Υ affect the ordering of the elements in $D_{\vec{z}}$.

THEOREM 2.1. *Suppose $\omega, \omega' \in \Omega, \upsilon, \upsilon' \in \Upsilon$ are such that $\omega \prec \omega'$ and $\upsilon \prec \upsilon'$. Then for $\vec{z} \in S_\beta$,*

$$(d_1(\omega, \upsilon, \vec{z}), d_2(\omega, \upsilon, \vec{z}), \ldots) \preceq (d_1(\omega', \upsilon', \vec{z}), d_2(\omega', \upsilon', \vec{z}), \ldots).$$

PROOF. Let $m := \inf\{i : \omega_i < \omega'_i\}$, $n := \inf\{i : \upsilon_i < \upsilon'_i\}$. Then we have $\omega_m < \omega'_m$ and $\upsilon_n < \upsilon'_n$. Denote by t_i the time of the ith visit to the region $\Omega \times \Upsilon \times C$ of the orbit of $(\omega, \upsilon, \vec{z})$ under K_β. Denote by s_j the time of the jth visit to the region $\Omega \times \Upsilon \times C_{012}$ of the orbit of $(\omega, \upsilon, \vec{z})$ under K_β. One can see that

$K_\beta^{t_i}(\omega, \upsilon, \vec{z})$ hits $\Omega \times \Upsilon \times C$ for the ith time and $K_\beta^{s_j}(\omega, \upsilon, \vec{z})$ hits $\Omega \times \Upsilon \times C_{012}$ for the jth time.

Let $l = \min\{t_m, s_n\}$. Then $\pi_3(K_\beta^i(\omega, \upsilon, \vec{z})) = \pi_3(K_\beta^i(\omega', \upsilon', \vec{z}))$ for $i = 0, \dots, l$. It follows that $d_i(\omega, \upsilon, \vec{z}) = d_i(\omega', \upsilon', \vec{z})$ for $i = 0, \dots, l$.

If $l = \infty$, then $d_i(\omega, \upsilon, \vec{z}) = d_i(\omega', \upsilon', \vec{z})$ for all i. If $l < +\infty$, then $K_\beta^l(\omega, \upsilon, \vec{z}) = K_\beta^l(\omega', \upsilon', \vec{z})$ hits $\Omega \times \Upsilon \times C$ for the mth time or $\Omega \times \Upsilon \times C_{012}$ for the nth time. Since $\omega_m < \omega'_m$ and $\upsilon_n < \upsilon'_n$, then

$$d_{l+1}(\omega, \upsilon, \vec{z}) = d_1(K_\beta^l(\omega, \upsilon, \vec{z})) < d_1(K_\beta^l(\omega', \upsilon', \vec{z})) = d_{l+1}(\omega', \upsilon', \vec{z}). \qquad \square$$

Now we show that any representation of $\vec{z}$ can be generated from the map K_β by choosing appropriate $\omega \in \Omega$ and $\upsilon \in \Upsilon$. We need the following lemma.

LEMMA 2.2. *Let $\beta \in (1, 3/2]$. Let $(x, y) \in S_\beta$ and $(x, y) = \sum_{i=1}^{\infty} a_i \beta^{-i}$ with $a_i \in \{\vec{q}_0, \vec{q}_1, \vec{q}_2\}$ be a representation of (x, y) in base β. One has,*

(i) *If $(x, y) \in E_i$ for some $i \in \{0, 1, 2\}$, then $a_1 = \vec{q}_i$;*
(ii) *If $(x, y) \in C_{ij}$ for some $ij \in \{01, 12, 02\}$, then $a_1 \in \{\vec{q}_i, \vec{q}_j\}$;*
(iii) *If $(x, y) \in C_{012}$, then $a_1 \in \{\vec{q}_0, \vec{q}_1, \vec{q}_2\}$.*

PROOF.

(i) Suppose $a_1 \neq \vec{q}_0$. From $a_1 \in \{\vec{q}_1, \vec{q}_2\}$ and $(x, y) = \sum_{i=1}^{\infty} a_i \beta^{-i}$ we have

$$x \geq \frac{1}{\beta} \text{ or } y \geq \frac{1}{\beta}.$$

Then $(x, y) \notin E_0$.

Suppose $a_1 \neq \vec{q}_1$. If $a_1 = \vec{q}_0$, then

$$x + y = \sum_{i=2}^{\infty} \frac{\| a_i \|_1}{\beta^i} \leq \frac{1}{\beta(\beta - 1)}.$$

If $a_1 = \vec{q}_2$, then $y \geq \frac{1}{\beta}$. In both cases, $(x, y) \notin E_1$.

Suppose $a_1 \neq \vec{q}_2$. By a similar proof we have $(x, y) \notin E_2$.

(ii) If $a_1 = \vec{q}_2$, then $y \geq \frac{1}{\beta}$, which implies $(x, y) \notin C_{01}$. If $a_1 = \vec{q}_0$, then $x + y \leq \frac{1}{\beta(\beta-1)}$, which implies $(x, y) \notin C_{12}$. If $a_1 = \vec{q}_1$, then $x \geq \frac{1}{\beta}$, which implies $(x, y) \notin C_{02}$.

(iii) holds trivially. $\square$

Using the above lemma and a construction similar to the one used in [**DV1**, Theorem 2], we have the following theorem.

THEOREM 2.3. *For $\beta \in (1, 3/2]$, let $\vec{z} \in S_\beta$ and $\vec{z} = \sum_{i=1}^{\infty} a_i \beta^{-i}$ with $a_i \in \{\vec{q}_0, \vec{q}_1, \vec{q}_2\}$ be a representation of $\vec{z}$ in base β. Then there exists an $\omega \in \Omega$ and an $\upsilon \in \Upsilon$ such that $a_i = d_i(\omega, \upsilon, \vec{z})$.*

3. Unique invariant measure of maximal entropy for random beta-transformations

Equip Υ with the uniform product measure $\mathbb{P}$ and recall that σ' is the left shift on Υ. On the set $\Omega \times \Upsilon \times S_\beta$ we consider the product σ-algebra $\mathcal{A} \times \mathcal{B} \times \mathcal{S}$. Define the function $\rho_1 : \Omega \times \Upsilon \times S_\beta \to \{\vec{q}_0, \vec{q}_1, \vec{q}_2\}^{\mathbb{N}}$ by

$$\rho_1(\omega, \upsilon, \vec{z}) = (d_1(\omega, \upsilon, \vec{z}), d_2(\omega, \upsilon, \vec{z}), \dots).$$

Define the function $\rho_2 : \{\vec{q}_0, \vec{q}_1, \vec{q}_2\}^{\mathbb{N}} \to \Upsilon$ by

$$\rho_2(\vec{q}_{b_1}, \vec{q}_{b_2}, \vec{q}_{b_3}, \ldots) = (b_1, b_2, b_3, \ldots).$$

Denote by $\varphi = \rho_2 \circ \rho_1$ which is a function from $\Omega \times \Upsilon \times S_\beta$ to Υ. Then $\varphi \circ K_\beta = \sigma' \circ \varphi$, and φ is surjective from Theorem 2.3.

It is easily seen that φ is measurable. In fact, the inverse image of the cylinder set with the first digit fixed is measurable in $\Omega \times \Upsilon \times S_\beta$:

$$\begin{aligned}
&\varphi^{-1}(\{(b_1, b_2, \ldots) \in \Upsilon : b_1 = 0\}) \\
&\quad = (\Omega \times \Upsilon \times E_0) \cup (\{\omega \in \Omega : \omega_1 = 0\} \times \Upsilon \times (C_{01} \cup C_{02})) \\
&\qquad \cup (\Omega \times \{\upsilon \in \Upsilon : \upsilon_1 = 0\} \times C_{012}), \\
&\varphi^{-1}(\{(b_1, b_2, \ldots) \in \Upsilon : b_1 = 1\}) \\
&\quad = (\Omega \times \Upsilon \times E_1) \cup (\{\omega \in \Omega : \omega_1 = 1\} \times \Upsilon \times C_{01}) \\
&\qquad \cup (\{\omega \in \Omega : \omega_1 = 0\} \times \Upsilon \times C_{12}) \cup (\Omega \times \{\upsilon \in \Upsilon : \upsilon_1 = 1\} \times C_{012}), \\
&\varphi^{-1}(\{(b_1, b_2, \ldots) \in \Upsilon : b_1 = 2\}) \\
&\quad = (\Omega \times \Upsilon \times E_2) \cup (\{\omega \in \Omega : \omega_1 = 1\} \times \Upsilon \times (C_{02} \cup C_{12})) \\
&\qquad \cup (\Omega \times \{\upsilon \in \Upsilon : \upsilon_1 = 2\} \times C_{012}).
\end{aligned}$$

To show that φ is an isomorphism, let

$$Z_1 = \{(\omega, \upsilon, \vec{z}) \in \Omega \times \Upsilon \times S_\beta : K_\beta^n(\omega, \upsilon, \vec{z}) \in \Omega \times \Upsilon \times C \text{ infinitely often}\},$$
$$Z_2 = \{(\omega, \upsilon, \vec{z}) \in \Omega \times \Upsilon \times S_\beta : K_\beta^n(\omega, \upsilon, \vec{z}) \in \Omega \times \Upsilon \times C_{012} \text{ infinitely often}\},$$
$$D_1 = \{(b_1, b_2, \ldots) \in \Upsilon : \sum_{i=1}^{\infty} \frac{\vec{q}_{b_{j+i-1}}}{\beta^i} \in C \text{ for infinitely many } j\},$$
$$D_2 = \{(b_1, b_2, \ldots) \in \Upsilon : \sum_{i=1}^{\infty} \frac{\vec{q}_{b_{j+i-1}}}{\beta^i} \in C_{012} \text{ for infinitely many } j\}.$$

Notice that

$$Z_1 = \cap_{n=1}^{\infty} \cup_{m=n}^{\infty} K_\beta^{-m}(\Omega \times \Upsilon \times C)$$

and

$$Z_2 = \cap_{n=1}^{\infty} \cup_{m=n}^{\infty} K_\beta^{-m}(\Omega \times \Upsilon \times C_{012}),$$

which imply that Z_1 and Z_2 are Borel sets in $\Omega \times \Upsilon \times S_\beta$. Let $Z = Z_1 \cap Z_2$, $D = D_1 \cap D_2$, then we have $K_\beta^{-1}(Z) = Z$, $(\sigma')^{-1}(D) = D$ and $\varphi(Z) = D$. Let $\varphi' = \varphi|_Z$.

LEMMA 3.1. *The map $\varphi' : Z \to D$ is a bimeasurable bijection.*

PROOF. For any sequence $(b_1, b_2, \ldots) \in D$, we can obtain a point

$$\vec{z} = \sum_{i=1}^{\infty} \vec{q}_{b_i} \beta^{-i}.$$

To determine ω and υ, we could define

$$r_1 = \min\{j \geq 1 : \sum_{i=1}^{\infty} \frac{\vec{q}_{b_{j+i-1}}}{\beta^i} \in C\}, \quad r_k = \min\{j > r_{k-1} : \sum_{i=1}^{\infty} \frac{\vec{q}_{b_{j+i-1}}}{\beta^i} \in C\},$$
$$s_1 = \min\{j \geq 1 : \sum_{i=1}^{\infty} \frac{\vec{q}_{b_{j+i-1}}}{\beta^i} \in C_{012}\}, \quad s_k = \min\{j > s_{k-1} : \sum_{i=1}^{\infty} \frac{\vec{q}_{b_{j+i-1}}}{\beta^i} \in C_{012}\}.$$

- If $\sum_{i=1}^{\infty} \vec{q}_{b_{r_k+i-1}} \beta^{-i} \in C_{ij}, ij \in \{01, 12, 02\}$, then $b_{r_k} \in \{i, j\}$ by Lemma 2.2.
  - If $b_{r_k} = i$, we let $\omega_k = 0$.
  - If $b_{r_k} = j$, we let $\omega_k = 1$.
- If $\sum_{i=1}^{\infty} \vec{q}_{b_{s_k+i-1}} \beta^{-i} \in C_{012}$, then $b_{s_k} \in \{0, 1, 2\}$ by Lemma 2.2. If $b_{s_k} = i$, we let $\upsilon_k = i, i \in \{0, 1, 2\}$.

Notice that for any $N > 0$, there exsit $r, s > N$ such that

$$\sum_{i=1}^{\infty} \vec{q}_{b_{r+i-1}} \beta^{-i} \in C \text{ and } \sum_{i=1}^{\infty} \vec{q}_{b_{s+i-1}} \beta^{-i} \in C_{012},$$

which implies that $K_\beta^n(\omega, \upsilon, \vec{z})$ hits both $\Omega \times \Upsilon \times C$ and $\Omega \times \Upsilon \times C_{012}$ infinitely often. Then the infinite sequences $\omega = (\omega_1, \omega_2, \omega_3, \ldots) \in \Omega$ and $\upsilon = (\upsilon_1, \upsilon_2, \upsilon_3, \ldots) \in \Upsilon$ can be uniquely determined. Therefore, we can define the inverse of φ'. Let $(\varphi')^{-1} : D \to Z$ be

$$(\varphi')^{-1}((b_1, b_2, \ldots)) = (\omega, \upsilon, \sum_{i=1}^{\infty} \frac{\vec{q}_{b_i}}{\beta^i}).$$

If $(\omega, \upsilon, \vec{z}) = (\omega', \upsilon', \vec{z'})$ then $\varphi'(\omega, \upsilon, \vec{z}) = \varphi'(\omega', \upsilon', \vec{z'})$. Since $Z = Z_1 \cap Z_2$ is a Borel set in $\Omega \times \Upsilon \times S_\beta$, then we have that $(\varphi')^{-1}$ is measurable (see [**S2**, Theorem 4.5.4]). Hence φ' is a bimeasurable bijection. □

LEMMA 3.2. *If* $1 < \beta < 3/2$, *then* $\mathbb{P}(D) = 1$.

PROOF. Let us first prove $\mathbb{P}(D_2) = 1$. Let $n \geq 1$ and denote a cylinder set in Υ by

$$[\upsilon_1, \upsilon_2, \ldots, \upsilon_n] = \{(b_1, b_2, \ldots) \in \Upsilon : b_i = \upsilon_i, i = 1, \ldots, n\}.$$

Let

$$S_{\beta, \upsilon_1, \upsilon_2, \ldots, \upsilon_n} = \{\vec{z} = \sum_{i=1}^{\infty} \frac{\vec{q}_{b_i}}{\beta^i} : (b_1, b_2, \ldots) \in [\upsilon_1, \upsilon_2, \ldots, \upsilon_n]\}.$$

Notice that $S_{\beta, \upsilon_1, \upsilon_2, \ldots, \upsilon_n}$ is a right triangle with $\sum_{i=1}^{n} \frac{\vec{q}_{b_i}}{\beta^i}$ as its right-angled vertex and a maximum diameter of $\frac{\sqrt{2}}{\beta^n(\beta-1)}$ when $1 < \beta < \frac{3}{2}$. Since $\lim_{n\to\infty} \frac{\sqrt{2}}{\beta^n(\beta-1)} = 0$ and C_{012} has positive Lebesgue measure, then we can find a cylinder set $[c_1, c_2, \ldots, c_N]$ such that $S_{\beta, c_1, c_2, \ldots, c_N} \subset C_{012}$.

Let

$$D' = \{(b_1, b_2, \ldots) \in \Upsilon : b_j b_{j+1} \ldots b_{j+N-1} = c_1 c_2 \ldots c_N \text{ for infinitely many } j\}$$

then $D' \subset D_2$.

Now we show that $\mathbb{P}(D') = 1$. Notice that $D' = \cap_{n=1}^{\infty} \cup_{m=n}^{\infty} \tilde{D}_m$, where $\tilde{D}_m = \{(b_1, b_2, \ldots) \in \Upsilon : b_m b_{m+1} \ldots b_{m+N-1} = c_1 c_2 \ldots c_N\}$. Let $B_n = \cup_{m=n}^{\infty} \tilde{D}_m$. If $(b_1, b_2, \ldots) \in \Upsilon \setminus B_n$, then we have that for any $j \geq n$, $b_j b_{j+1} \ldots b_{j+N-1} \neq c_1 c_2 \ldots c_N$. Clearly,

$$\Upsilon \setminus B_n \subseteq B' := \{(b_1, b_2, \ldots) : b_{n+kN} \ldots b_{n+(k+1)N-1} \neq c_1 c_2 \ldots c_N, k = 0, 1, 2, \ldots\}.$$

Since $\mathbb{P}(\Upsilon \setminus B_n) \leq \mathbb{P}(B') = \lim_{k\to\infty}(1 - 1/3^N)^k = 0$, then $\mathbb{P}(B_n) = 1$. It follows from $D' = \cap_{n-1}^{\infty} B_n$ and $B_1 \supseteq B_2 \supseteq \cdots$ that $\mathbb{P}(D') = \lim_{n\to\infty} B_n = 1$. Then we get $\mathbb{P}(D_2) = 1$.

To prove that $\mathbb{P}(D_1) = 1$, we can use a similar approach. Here we construct a specific cylinder. Define

$$\vec{z}_l = (x_l, y_l) = \frac{\vec{q}_1}{\beta} + \frac{\vec{q}_0}{\beta^2} + \cdots + \frac{\vec{q}_0}{\beta^l} + \frac{\vec{q}_{b_1}}{\beta^{l+1}} + \frac{\vec{q}_{b_2}}{\beta^{l+2}} + \cdots .$$

Then we have $x_l \geq \frac{1}{\beta}$, and

$$0 \leq y_l \leq \sum_{i=1}^{\infty} \frac{1}{\beta^{l+i}} = \frac{1}{\beta^l(\beta-1)},$$
$$\frac{1}{\beta} \leq x_l + y_l \leq \frac{1}{\beta} + \sum_{i=1}^{\infty} \frac{1}{\beta^{l+i}} = \frac{1}{\beta} + \frac{1}{\beta^l(\beta-1)}.$$

Since $\lim_{l\to\infty} \frac{1}{\beta^l(\beta-1)} = 0$, then there exists $L > 0$ such that for any $l \geq L$,

$$0 \leq y_l < \frac{1}{\beta}, x_l + y_l \leq \frac{1}{\beta(\beta-1)}.$$

It follows that $(x_l, y_l) \in C_{01}$ for any $l \geq L$. Let

$$D'' = \{(b_1, b_2, \ldots) \in \Upsilon : b_j b_{j+1} \ldots b_{j+L-1} = 1\ \underbrace{00\ldots0}_{L-1 \text{ times}} \text{ for infinitely many } j\}, \tag{3.1}$$

then $D'' \subset D_1$. Since $\mathbb{P}(D'') = 1$, we have $\mathbb{P}(D_1) = 1$. Therefore, $\mathbb{P}(D) = 1$. □

Theorem 3.3 can be obtained from Lemmas 3.1 and 3.2.

THEOREM 3.3. *Let $\beta \in (1, 3/2]$ and set $\nu_\beta(A) = \mathbb{P}(\varphi(Z \cap A))$. The dynamical systems $(\Omega \times \Upsilon \times S_\beta, \mathcal{A} \times \mathcal{B} \times \mathcal{S}, \nu_\beta, K_\beta)$ and $(\Upsilon, \mathcal{B}, \mathbb{P}, \sigma')$ are isomorphic.*

REMARK 3.4.

(i) Notice that Lemma 3.2 and Theorem 3.3 remain true if one replaces $\mathbb{P}$ by any other non-uniform product measure on Υ giving a positive weight to each symbol.
(ii) Since $\mathbb{P}$ is the unique measure of maximal entropy on Υ, the above theorem implies that any other K_β-invariant measure with support Z has entropy strictly less than $\log 3$. We now investigate the entropy of K_β-invariant measure μ for which $\mu(Z^c) > 0$.

Divide Z^c into three Borel sets as follows:

$$\begin{aligned} Z^c &= (Z_1 \cap Z_2)^c \\ &= (Z_1^c \setminus Z_2^c) \cup (Z_2^c \setminus Z_1^c) \cup (Z_1^c \cap Z_2^c) \\ &= (Z_2 \setminus Z_1) \cup (Z_1 \setminus Z_2) \cup (Z_1^c \cap Z_2^c) \\ &:= Z_3 \cup Z_4 \cup Z_5, \end{aligned}$$

where

$$\begin{aligned}
Z_3 =& Z_2 \setminus Z_1 \\
=& \{(\omega, \upsilon, \vec{z}) \in \Omega \times \Upsilon \times S_\beta : K_\beta^n(\omega, \upsilon, \vec{z}) \in \Omega \times \Upsilon \times C \text{ for finitely many } n \\
& \text{and } K_\beta^n(\omega, \upsilon, \vec{z}) \in \Omega \times \Upsilon \times C_{012} \text{ infinitely often}\}, \\
Z_4 =& Z_1 \setminus Z_2 \\
=& \{(\omega, \upsilon, \vec{z}) \in \Omega \times \Upsilon \times S_\beta : K_\beta^n(\omega, \upsilon, \vec{z}) \in \Omega \times \Upsilon \times C_{012} \text{ for finitely many } n, \\
& \text{and } K_\beta^n(\omega, \upsilon, \vec{z}) \in \Omega \times \Upsilon \times C \text{ infinitely often}\}, \\
Z_5 =& Z_1^c \cap Z_2^c \\
=& \{(\omega, \upsilon, \vec{z}) \in \Omega \times \Upsilon \times S_\beta : K_\beta^n(\omega, \upsilon, \vec{z}) \in \Omega \times \Upsilon \times C \text{ for finitely many } n, \\
& \text{and } K_\beta^n(\omega, \upsilon, \vec{z}) \in \Omega \times \Upsilon \times C_{012} \text{ for finitely many } n\}.
\end{aligned}$$

We first prove Lemma 3.5.

LEMMA 3.5. *Let* $\beta \in (1, 3/2)$. *Let* μ_3 *be a* K_β*-invariant measure for which* $\mu_3(Z_3) = 1$. *Then* $h_{\mu_3}(K_\beta) < \log 3$. *Similarly, let* μ_4 *and* μ_5 *be* K_β*-invariant measures for which* $\mu_4(Z_4) = \mu_5(Z_5) = 1$. *Then* $h_{\mu_4}(K_\beta), h_{\mu_5}(K_\beta) < \log 3$ *also holds.*

PROOF. Let

$$H_3 = \{\vec{z} = \sum_{i=1}^{\infty} \frac{\vec{q}_{b_i}}{\beta^i} \in S_\beta : \sum_{i=1}^{\infty} \frac{\vec{q}_{b_{j+i-1}}}{\beta^i} \text{ belongs to } C_{012} \text{ for infinitely many } j, \\ \text{and never belongs to } C\}.$$

Then $\Omega \times \Upsilon \times H_3 \subseteq K_\beta^{-1}(\Omega \times \Upsilon \times H_3)$ and $\cup_{i=0}^{\infty} K_\beta^{-i}(\Omega \times \Upsilon \times H_3) = Z_3$. It follows that $\mu_3(Z_3) = \lim_{i\to\infty} \mu_3(K_\beta^{-i}(\Omega \times \Upsilon \times H_3)) = 1$. Since μ_3 is K_β-invariant, then

$$\mu_3(\Omega \times \Upsilon \times H_3) = \mu_3(K_\beta^{-1}(\Omega \times \Upsilon \times H_3)) = \mu_3(K_\beta^{-2}(\Omega \times \Upsilon \times H_3)) = \cdots = 1.$$

Thus it is enough to study the entropy with respect to μ_3 of the map K_β restricted to $\Omega \times \Upsilon \times H_3$. Let π_1, π_2, π_3 be the canonical projection onto the three coordinates respectively. Notice that the action of the transformation K_β on the first coordinate is an identity, which implies that K_β is essentially a product transformation $I_\Omega \times K'_\beta$, where $K'_\beta = (\pi_2 \circ K_\beta) \times (\pi_3 \circ K_\beta)$ on $\Upsilon \times H_3$ and I_Ω is the identity on Ω. Since $(\upsilon, \vec{z}) \in \Upsilon \times S_\beta$ and $\omega \in \Omega$ are independent, and $h_\mu(I_\Omega) = 0$ for any measure μ on $(\Omega, \mathcal{A})$, we see that $h_{\mu_3}(K_\beta) = h_{\mu'_3}(K'_\beta)$, where $\mu'_3(B \times H) = \mu_3(\Omega \times B \times H)$ for $B \in \mathcal{B}, H \in (H_3 \cap \mathcal{S})$.

Let

$$D_3 = \{(b_1, b_2, \dots) \in \Upsilon : \sum_{i=1}^{\infty} \frac{\vec{q}_{b_{j+i-1}}}{\beta^i} \text{ belongs to } C_{012} \text{ for infinitely many } j, \\ \text{and never belongs to } C\}.$$

Define a map ϕ from $(\Upsilon \times H_3, \mathcal{B} \times (H_3 \cap \mathcal{S}), \mu'_3, K'_\beta)$ to $(D_3, D_3 \cap \mathcal{B}, \mu'_3 \circ \phi^{-1}, \sigma')$ as

$$\phi(\upsilon, \vec{z}) = \rho_2(\rho_1(0^\infty, \upsilon, \vec{z})).$$

Since $\vec{z} \in H_3$, then ϕ is well defined and bijective. ϕ is measurable and the inverse is also measurable (see [**S2**, Theorem 4.5.4]). Finally, ϕ preserves the measure and

$\phi \circ K'_\beta = \sigma' \circ \phi$. Then ϕ is an isomorphism and it follows that

$$h_{\mu_3}(K_\beta) = h_{\mu'_3}(K'_\beta) = h_{\mu'_3 \circ \phi^{-1}}(\sigma') \leq h_{\mathbb{P}}(\sigma') = \log 3.$$

Since $\mathbb{P}$ is the unique measure of maximal entropy on D_3, to show $h_{\mu_3}(K_\beta) < \log 3$, it is enough to prove that $\mu'_3 \circ \phi^{-1} \neq \mathbb{P}$. This is done by contradiction. If $\mu'_3 \circ \phi^{-1} = \mathbb{P}$, then

$$\mathbb{P}(D_3) = \mu'_3(\Upsilon \times H_3) = \mu_3(\Omega \times \Upsilon \times H_3) = 1.$$

Since $D_3 \subset (D'')^c$, where D'' is defined as in (3.1), then $\mathbb{P}((D'')^c) = 1$, which is a contradiction to $\mathbb{P}(D'') = 1$. Therefore, $h_{\mu_3}(K_\beta) < \log 3$. Let

$$\begin{aligned} H_4 = \{\vec{z} = \sum_{i=1}^{\infty} \frac{\vec{q}_{b_i}}{\beta^i} \in S_\beta : \sum_{i=1}^{\infty} \frac{\vec{q}_{b_{j+i-1}}}{\beta^i} &\text{ belongs to } C \text{ for infinitely many } j, \\ &\text{and never belongs to } C_{012}\}, \\ H_5 = \{\vec{z} = \sum_{i=1}^{\infty} \frac{\vec{q}_{b_i}}{\beta^i} \in S_\beta : \sum_{i=1}^{\infty} \frac{\vec{q}_{b_{j+i-1}}}{\beta^i} &\text{ never belongs to } C \\ &\text{and never belongs to } C_{012}\}, \\ = \{\vec{z} \in S_\beta : \vec{z} \text{ has a unique } \beta\text{-expansion}\}. \end{aligned}$$

Then it follows that

$$\mu_4(\Omega \times \Upsilon \times H_4) = \mu_5(\Omega \times \Upsilon \times H_5) = 1.$$

We can also obtain that $h_{\mu_4}(K_\beta), h_{\mu_5}(K_\beta) < \log 3$ using the similar method. $\square$

From Lemma 3.5 we can obtain the upper bound of the entropy of K_β-invariant measure for which Z^c has positive measure.

LEMMA 3.6. *Let $\beta \in (1, 3/2]$. Let μ be a K_β-invariant measure for which $\mu(Z^c) > 0$. Then $h_\mu(K_\beta) < \log 3$.*

PROOF. Notice that Z, Z_3, Z_4 and Z_5 are pairwise disjoint and the union is $\Omega \times \Upsilon \times S_\beta$. Since Z, Z_3, Z_4 and Z_5 are K_β-invariant, then there exist K_β-invariant probability measures μ_{12}, μ_3, μ_4 and μ_5 concentrated on Z, Z_3, Z_4 and Z_5, respectively, such that

$$\mu = (1 - \alpha_3 - \alpha_4 - \alpha_5)\mu_{12} + \alpha_3\mu_3 + \alpha_4\mu_4 + \alpha_5\mu_5,$$

where $0 \leq \alpha_3, \alpha_4, \alpha_5 \leq 1$ and $0 < \alpha_3 + \alpha_4 + \alpha_5 \leq 1$. Then

$$h_\mu(K_\beta) = (1 - \alpha_3 - \alpha_4 - \alpha_5)h_{\mu_{12}}(K_\beta) + \alpha_3 h_{\mu_3}(K_\beta) + \alpha_4 h_{\mu_4}(K_\beta) + \alpha_5 h_{\mu_5}(K_\beta).$$

Since $h_{\mu_{12}}(K_\beta) \leq \log 3$ by Remark 3.4 and $h_{\mu_3}(K_\beta), h_{\mu_4}(K_\beta), h_{\mu_5}(K_\beta) < \log 3$ by Lemma 3.5, then the result follows. $\square$

Now we obtain the main result in this section.

THEOREM 3.7. *Let $\beta \in (1, 3/2)$. The measure $\nu_\beta(A) = \mathbb{P}(\varphi(Z \cap A))$ is the unique K_β-invariant measure of maximal entropy.*

REMARK 3.8. The measure ν_β is not self-similar, but the projection in the third coordinate is a self-similar measure defined on S_β. To be more precise, define

$h : \Upsilon \to S_\beta$ by $h(b_1, b_2, \dots) = \sum_{i=1}^{\infty} \frac{\vec{q}_{b_i}}{\beta^i}$ and consider the commuting diagram

$$\begin{array}{ccc} \Omega \times \Upsilon \times S_\beta & \xrightarrow{\pi_2} & S_\beta \\ & \searrow \varphi & \uparrow h \\ & & \Upsilon, \end{array}$$

Then, $\nu_\beta \circ \pi_2^{-1}$ satisfies $\nu_\beta \circ \pi_2^{-1} = \frac{1}{3}\sum_{i=0}^{2} \nu_\beta \circ f_i^{-1}$.

4. An absolutely continuous invariant measure for random beta-transformations

We start by recalling that for $\beta = 3/2$, the region $C_{012} = \{(\frac{2}{3}, \frac{2}{3})\}$ is a point. As a result the analysis for this case is slightly different from the one conducted for a general $\beta \in (1, 3/2)$. In this section, we concentrate on the case $\beta \in (1, 3/2)$ and in Remark 4.9, we give a brief description of the case $\beta = 3/2$.

Endow $\Omega = \{0,1\}^{\mathbb{N}}$ with the product measure m_1 giving the symbol 0 probability p and the symbol 1 probability $1-p$, and $\Upsilon = \{0,1,2\}^{\mathbb{N}}$ with the product measure m_2 giving the symbol 0 probability s, the symbol 1 probability t and the symbol 2 probability $1-s-t$. Consider the measure space $(S_\beta, \mathcal{S}, \lambda_2)$, where λ_2 is the normalized Lebesgue measure. In this section we will prove that K_β has an invariant measure of the form $m_1 \otimes m_2 \otimes \mu_\beta$, where μ_β is absolutely continuous with respect to λ_2. We will show the result by several steps.

STEP 1. A position-dependent random transformation R.

Bahsoun and Góra [**BG**] gave a sufficient condition for the existence of an absolutely continuous invariant measure for a random map with position-dependent probabilities on a bounded domain of $\mathbb{R}^N$. We take some of their results a little further.

For $k = 1, \dots, K$, let $\tau_k : S_\beta \to S_\beta$ be piecewise one-to-one and C^2, non-singular transformations on a common partition $\mathcal{P}$ of $S_\beta : \mathcal{P} = \{S_1, \dots, S_q\}$ and $\tau_{k,i} = \tau_k|_{S_i}, i = 1, \dots, q$. Let $p_k : S_\beta \to [0,1]$ be piecewise C^1 functions such that $\sum_{k=1}^{K} p_k = 1$. Denote by $R = \{\tau_1, \dots, \tau_K; p_1(\vec{z}), \dots, p_K(\vec{z})\}$ the position-dependent random map, i.e., $R(\vec{z}) = \tau_k(\vec{z})$ with probability $p_k(\vec{z})$. Define the transition function for R as follows:

$$\mathbf{P}(\vec{z}, A) = \sum_{k=1}^{K} p_k(\vec{z}) \mathbb{1}_A(\tau_k(\vec{z})),$$

where A is any measurable set and $\mathbb{1}_A$ denotes the indicator function of the set A.

The iteration of R is denoted by $R^n := \{\tau_{k_1 k_2 \cdots k_n}; p_{k_1 k_2 \cdots k_n}\}, k_1 k_2 \cdots k_n \in \{1, 2, \dots, K\}^n$, where $\tau_{k_1 k_2 \cdots k_n}(\vec{z}) = \tau_{k_n} \circ \tau_{k_{n-1}} \circ \cdots \circ \tau_{k_1}(\vec{z})$ and

$$p_{k_1 k_2 \cdots k_n}(\vec{z}) = p_{k_n}(\tau_{k_{n-1}} \circ \cdots \circ \tau_{k_1}(\vec{z})) \cdot p_{k_{n-1}}(\tau_{k_{n-2}} \circ \cdots \circ \tau_{k_1}(\vec{z})) \cdots p_{k_1}(\vec{z}).$$

The transition function $\mathbf{P}$ induces an operator $\mathbf{P}_*$ on the set of probability measures on $(S_\beta, \mathcal{S})$ defined by

$$\mathbf{P}_*\mu(A) = \int \mathbf{P}(\vec{z}, A) d\mu(\vec{z}) = \sum_{k=1}^{K} \int_{\tau_k^{-1}(A)} p_k(\vec{z}) d\mu(\vec{z}) = \sum_{k=1}^{K}\sum_{i=1}^{q} \int_{\tau_{k,i}^{-1}(A)} p_k(\vec{z}) d\mu(\vec{z}).$$

We say that the measure μ is R-invariant iff $\mathbf{P}_*\mu = \mu$.

If μ has density f with respect to λ_2, then $\mathbf{P}_*\mu$ also has a density which we denote by $P_R f$, i.e.,

$$\int_A P_R f(\vec{z})d\lambda_2(\vec{z}) = \sum_{k=1}^{K}\sum_{i=1}^{q}\int_{\tau_{k,i}^{-1}(A)} p_k(\vec{z})f(\vec{z})d\lambda_2(\vec{z}).$$

We call P_R the Perron–Frobenius operator of the random map R and it has very useful properties[**BG**]:

(i) P_R is linear;
(ii) P_R is nonnegative;
(iii) $P_R f = f \iff \mu = f \cdot \lambda_2$ is R-invariant;
(iv) $\|P_R f\|_1 \le \|f\|_1$, where $\|\cdot\|_1$ denotes the L^1 norm;
(v) $P_{R\circ T} = P_R \circ P_T$. In particular, $P_R^N = P_{R^N}$.

Let each S_i be a bounded closed domain having a piecewise C^2 boundary of finite 1-dimensional measure. Assume that the faces of ∂S_i meet at angles bounded uniformly away from 0 and the probabilities $p_k(\vec{z})$ are piecewise C^1 functions on the partition $\mathcal{P}$. We assume:

CONDITION (A).

$$\max_{1\le i\le q}\sum_{k=1}^{K} p_k(\vec{z})\|D\tau_{k,i}^{-1}(\tau_{k,i}(\vec{z}))\| < c < 1,$$

where $D\tau_{k,i}^{-1}(\vec{z})$ is the derivative matrix of $\tau_{k,i}^{-1}$ at $\vec{z}$.

Using the multidimensional notion of variation [**G**]:

$$V(f) = \int_{\mathbb{R}^N}\|Df\|d\lambda_N = \sup\{\int_{\mathbb{R}^N} f\mathrm{div}(g)d\lambda_N : g = (g_1,\dots,g_N)\in C_0^1(\mathbb{R}^N,\mathbb{R}^N) \text{ and } |g(x)|\le 1 \text{ for } x\in\mathbb{R}^N\}$$

where $f \in L_1(\mathbb{R}^N)$ has bounded support, Df denotes the gradient of f in the distributional sense, $\mathrm{div}(g) = \nabla\cdot g = \frac{\partial g_1}{\partial x_1}+\frac{\partial g_2}{\partial x_2}+\cdots+\frac{\partial g_N}{\partial x_N}$ is the divergence operator, and $C_0^1(\mathbb{R}^N,\mathbb{R}^N)$ is the space of continuously differentiable functions from $\mathbb{R}^N$ into $\mathbb{R}^N$ having compact support. Consider the Banach space [**G**, Remark 1.12],

$$BV(S_\beta) = \{f \in L_1(S_\beta) : V(f) < +\infty\},$$

with the norm $\|f\|_{BV} = \|f\|_{L_1} + V(f)$.

Fix $1 \le i \le q$. Let F denote the set of singular points of ∂S_i. At any $x \in F$ we construct the largest cone having a vertex at x and which lies completely in S_i. Let $\theta(x)$ denote the angle subtended at the vertex of this cone. Then define

$$\gamma(S_i) = \min_{x\in F}\theta(x).$$

Since the faces of ∂S_i meet at angles bounded uniformly away from 0, $\gamma(S_i) > 0$. Let $\alpha(S_i) = \pi/2 + \gamma(S_i)$ and $a(S_i) = |\cos(\alpha(S_i))|$.

Now we start at points $y \in F$, where the minimal angle $\gamma(S_i)$ is attained, defining L_y to be central rays of the largest regular cones contained in S_i. Then we extend this field of segments to a C^1 field of segments $L_y, y \in \partial S_i$, every L_y being a central ray of a regular cone contained in S_i, with angle subtended at the vertex

y greater than or equal to $\beta(S_i)$. We make L_y short enough to avoid overlapping. Let $\delta(y)$ be the length of $L_y, y \in \partial S_i$. By the compactness of ∂S_i we have

$$\delta(S_i) := \inf_{y \in \partial(S_i)} \delta(y) > 0.$$

Let $\vec{z}$ be a point in ∂S_i and $J_{k,i}$ the Jacobian of $\tau_{k|S_i}$ at $\vec{z}$.

We recall the following two theorems.

THEOREM 4.1 ([**BG**, Theorem 6.3]). *If R is a random map which satisfies Condition (A), then*

$$V(P_R f) \leq c(1+1/a)V(f) + (M + \frac{c}{a\delta})\|f\|_1 \quad \textit{for all } f \in BV(S_\beta),$$

where $a = \min\{a(S_i) : i = 1, \dots, q\} > 0, \delta = \min\{\delta(S_i) : i = 1 \dots, q\} > 0, M_{k,i} = \sup_{\vec{z} \in S_i}(Dp_k(\vec{z}) - \frac{DJ_{k,i}}{J_{k,i}} p_k(\vec{z}))$ *and* $M = \sum_{k=1}^{K} \max_{1 \leq i \leq q} M_{k,i}$.

THEOREM 4.2 ([**BG**, Theorem 6.4]). *If R is a random map which satisfies Condition (A) and $c(1+1/a) < 1$, then the random map R preserves a measure which is absolutely continuous with respect to Lebesgue measure. Furthermore, the associated random Perron Frobenius operator P_R is quasi compact.*

Now, let R be a random map which is given by $\{\tau_1, \dots, \tau_6; p_1(\vec{z}), \dots, p_6(\vec{z})\}$ where

$$\tau_1(\vec{z}) = \begin{cases} \beta\vec{z} - \vec{q}_i, & \text{if } \vec{z} \in E_i, i \in \{0,1,2\} \\ \beta\vec{z} - \vec{q}_i, & \text{if } \vec{z} \in C_{ij}, ij \in \{01,02,12\} \\ \beta\vec{z}, & \text{if } \vec{z} \in C_{012}, \end{cases}$$

$$\tau_2(\vec{z}) = \begin{cases} \beta\vec{z} - \vec{q}_i, & \text{if } \vec{z} \in E_i, i \in \{0,1,2\} \\ \beta\vec{z} - \vec{q}_i, & \text{if } \vec{z} \in C_{ij}, ij \in \{01,02,12\} \\ \beta\vec{z} - \vec{q}_1, & \text{if } \vec{z} \in C_{012}, \end{cases}$$

$$\tau_3(\vec{z}) = \begin{cases} \beta\vec{z} - \vec{q}_i, & \text{if } \vec{z} \in E_i, i \in \{0,1,2\} \\ \beta\vec{z} - \vec{q}_i, & \text{if } \vec{z} \in C_{ij}, ij \in \{01,02,12\} \\ \beta\vec{z} - \vec{q}_2, & \text{if } \vec{z} \in C_{012}, \end{cases}$$

$$\tau_4(\vec{z}) = \begin{cases} \beta\vec{z} - \vec{q}_i, & \text{if } \vec{z} \in E_i, i \in \{0,1,2\} \\ \beta\vec{z} - \vec{q}_j, & \text{if } \vec{z} \in C_{ij}, ij \in \{01,02,12\} \\ \beta\vec{z}, & \text{if } \vec{z} \in C_{012}, \end{cases}$$

$$\tau_5(\vec{z}) = \begin{cases} \beta\vec{z} - \vec{q}_i, & \text{if } \vec{z} \in E_i, i \in \{0,1,2\} \\ \beta\vec{z} - \vec{q}_j, & \text{if } \vec{z} \in C_{ij}, ij \in \{01,02,12\} \\ \beta\vec{z} - \vec{q}_1, & \text{if } \vec{z} \in C_{012}, \end{cases}$$

$$\tau_6(\vec{z}) = \begin{cases} \beta\vec{z} - \vec{q}_i, & \text{if } \vec{z} \in E_i, i \in \{0,1,2\} \\ \beta\vec{z} - \vec{q}_j, & \text{if } \vec{z} \in C_{ij}, ij \in \{01,02,12\} \\ \beta\vec{z} - \vec{q}_2, & \text{if } \vec{z} \in C_{012}, \end{cases}$$

The probabilities are defined as follows.

$$\begin{aligned} &p_1(\vec{z}) = p \cdot s, && p_4(\vec{z}) = (1-p) \cdot s, \\ &p_2(\vec{z}) = p \cdot t, && p_5(\vec{z}) = (1-p) \cdot t, \\ &p_3(\vec{z}) = p \cdot (1-s-t), && p_6(\vec{z}) = (1-p) \cdot (1-s-t). \end{aligned} \tag{4.1}$$

We have Lemma 4.3.

LEMMA 4.3. *For any $\vec{z} \in S_\beta$ and $n \in \mathbb{N}$, $\sum_{k_1k_2\cdots k_n\in\{1,\ldots,6\}^n} p_{k_1k_2\ldots k_n}(\vec{z}) = 1$.*

PROOF. We prove this lemma by induction. For $n = 1$, $p_1(\vec{z}) + \cdots + p_6(\vec{z}) = 1$. Assume it is true for $n = m$, i.e. for any $\vec{z} \in S_\beta$,

$$\sum_{k_1k_2\cdots k_m\in\{1,\ldots,6\}^n} p_{k_1k_2\ldots k_m}(\vec{z}) = 1.$$

For $n = m + 1$,

$$\begin{aligned}
&\sum_{k_1k_2\cdots k_{m+1}\in\{1,\ldots,6\}^{m+1}} p_{k_1k_2\ldots k_{m+1}}(\vec{z}) \\
=&\sum_{k_1k_2\cdots k_{m+1}\in\{1,\ldots,6\}^{m+1}} p_{k_{m+1}}(\tau_{k_m}\circ\cdots\circ\tau_{k_1}(\vec{z})) \\
&\cdot p_{k_m}(\tau_{k_{m-1}}\circ\cdots\circ\tau_{k_1}(\vec{z}))\cdots p_{k_1}(\vec{z}) \\
=&p_1(\vec{z})\sum_{k_2\cdots k_{m+1}\in\{1,\ldots,6\}^m} p_{k_2\ldots k_{m+1}}(\tau_1(\vec{z})) \\
&+ p_2(\vec{z})\sum_{k_2\cdots k_{m+1}\in\{1,\ldots,6\}^m} p_{k_2\ldots k_{m+1}}(\tau_2(\vec{z})) \\
&+\cdots+ p_6(\vec{z})\sum_{k_2\cdots k_{m+1}\in\{1,\ldots,6\}^m} p_{k_2\ldots k_{m+1}}(\tau_6(\vec{z})) \\
=&p_1(\vec{z})+\cdots+p_6(\vec{z}) \\
=&1.
\end{aligned}$$

□

To prove the existence of an absolutely continuous invariant measure(acim), we would like to use Theorem 4.2. This cannot be done directly since R does not satisfy the hypothesis of the theorem, however a higher iterate of R does. For the convenience of the reader we supply a complete proof.

THEOREM 4.4. *Let $R = \{\tau_1, \ldots, \tau_6; p_1(\vec{z}), \ldots, p_6(\vec{z})\}$, then R admits an acim.*

PROOF. Denote the partition (2.1) by $\mathcal{P}$ with

$$S_1 = \overline{E}_0, S_2 = \overline{E}_1, S_3 = \overline{E}_2, S_4 = \overline{C}_{01}, S_5 = \overline{C}_{12}, S_6 = \overline{C}_{02}, S_7 = \overline{C}_{012}.$$

Consider the iteration of the random map, R^n, the corresponding partition is $\vee_{i=0}^{n-1} R^{-i}\mathcal{P}$, where

$$R^{-i}\mathcal{P} = \vee_{k_1k_2\cdots k_i\in\{1,\ldots,6\}^i} \tau^{-1}_{k_1k_2\ldots k_i}\mathcal{P}.$$

For a set $P_i \in \vee_{i=0}^{n-1} R^{-i}\mathcal{P}$ and a sequence $k_1\ldots k_n \in \{1,\ldots,6\}^n$, let $\tau_{k_1\ldots k_n,i} = \tau_{k_1\ldots k_n}|_{P_i}$ and $M_{k_1\ldots k_n,i} = \sup_{\vec{z}\in P_i}(Dp_{k_1\ldots k_n}(\vec{z}) - \frac{DJ_{k_1\ldots k_n,i}}{J_{k_1\ldots k_n,i}} p_{k_1\ldots k_n,i}(\vec{z}))$, where $J_{k_1\ldots k_n,i}$ is the Jacobian of $\tau_{k_1\ldots k_n,i}$. Let

$$M_n = \sum_{k_1\ldots k_n\in\{0,1,2\}^n} \max_{P_i\in\vee_{i=0}^{n-1}R^{-i}\mathcal{P}} M_{k_1\ldots k_n,i} \quad \text{and} \quad \delta_n = \min_{P_i\in\vee_{i=0}^{n-1}R^{-i}\mathcal{P}} \delta(P_i).$$

For any set $P_i \in \vee_{i=0}^{n-1} R^{-i}\mathcal{P}$, the derivative matrix of $\tau^{-1}_{k_1k_2\ldots k_n}$ is equal to

$$\begin{bmatrix} \frac{1}{\beta^n} & 0 \\ 0 & \frac{1}{\beta^n} \end{bmatrix}.$$

Using Lemma 4.3 we have

$$\max_{P_i\in\vee_{i=0}^{n-1}R^{-i}\mathcal{P}}\sum_{k_1k_2\cdots k_n\in\{1,\dots,6\}^n} p_{k_1k_2\dots k_n}(\vec{z})\|D(\tau_{k_1k_2\dots k_n}|_{P_i})^{-1}\| = \frac{\sqrt{2}}{\beta^n} < \frac{2\sqrt{2}}{\beta^n} := c_n.$$

For the partition $\vee_{i=0}^{n-1}R^{-i}\mathcal{P}$, we have $a_n = \sqrt{2}/2$ (Here a_n refers to a in Theorem 4.1). Let

$$r_n = c_n(1+\frac{1}{a_n}) = \frac{2\sqrt{2}+4}{\beta^n}, \quad R_n = M_n + \frac{c_n}{a_n\delta_n}.$$

We can find $l > \log(2\sqrt{2}+4)/\log\beta$ such that $r_l < 1$. Fix this l and let $C_1 = \max\{r_1, r_2, \dots, r_{l-1}\}, C_2 = \max\{R_1, R_2, \dots, R_{l-1}\}$. For any integer n, we have $n = jl + i$, where $0 \le i \le l-1$. Notice that $P_{R^n} = (P_{R^l})^j P_{R^i}$. Apply Theorem 4.1 on R^l, then we get

$$\begin{aligned}
V(P_{R^n}f) &= VP_{R^l}^j(P_{R^i}f)\\
&\le r_l\cdot VP_{R^l}^{j-1}(P_{R^i}f) + R_l\|f\|_1\\
&\le r_l\cdot(r_l\cdot VP_{R^l}^{j-2}(P_{R^i}f) + R_l\|f\|_1) + R_l\|f\|_1\\
&\dots\\
&\le r_l^jV(P_{R^i}f) + (r_l^{j-1}+r_l^{j-2}+\dots+r_l+1)R_l\|f\|_1\\
&\le r_l^j(C_1V(f)+C_2\|f\|_1) + (r_l^{j-1}+r_l^{j-2}+\dots+r_l+1)R_l\|f\|_1\\
&= C_1r_l^jV(f) + (C_2r_l^j + r_l^{j-1}+r_l^{j-2}+\dots+r_l+1)R_l\|f\|_1\\
&\le C_1r_l^jV(f) + (C_2+\frac{1}{1-r_l})R_l\|f\|_1.
\end{aligned}$$

By definition of the norm $\|\cdot\|_{BV}$,

$$\begin{aligned}
\|P_{R^n}f\|_{BV} &= \|P_{R^n}f\|_1 + V(P_{R^n}f)\\
&\le \|f\|_1 + C_1r_l^jV(f) + (C_2+\frac{1}{1-r_l})R_l\|f\|_1.
\end{aligned}$$

Then the result follows by the technique in [**GB**, Theorem 1]. We write some details for completeness. From the above inequality it follows that the set $\{P_R^n\mathbf{1}\}_{n\ge l}$ is uniformly bounded, where $\mathbf{1}$ is the constant function equal to 1 on S_β. Hence P_R has a nontrivial fixed point $\mathbf{1}^*$ which is the density of an acim by the Kakutani–Yoshida Theorem (see [**K, Y**]). □

STEP 2. For the skew product transformation R' on $S_\beta\times[0,1)$.

Let $(I, \mathcal{B}(I), \lambda_1)$ be the unit interval $I = [0,1)$, with $\mathcal{B}(I)$ the Borel σ-algebra on I and λ_1 being Lebesgue measure on $(I,\mathcal{B}(I))$. Let $Y = S_\beta\times I$ and the set J_k be given by $J_k = \{(\vec{z},w)\colon \sum_{i<k}p_i(\vec{z}) \le w < \sum_{i\le k}p_i(\vec{z})\}$. Define maps $\varphi_k\colon J_k\to I$ by

$$\varphi_k(\vec{z},w) = \frac{1}{p_k(\vec{z})}w - \frac{\sum_{r=1}^{k-1}p_r(\vec{z})}{p_k(\vec{z})}.$$

Define the skew product transformation $R' : S_\beta\times I\to S_\beta\times I$ by

$$R'(\vec{z},w) = (\tau_k(\vec{z}), \varphi_k(\vec{z},w))$$

for $(\vec{z},w)\in J_k$.

Since $p_k(\vec{z})$ is defined as in (4.1), then we have

$$\varphi_1(\vec{z},w)=\frac{w}{ps},\qquad \varphi_4(\vec{z},w)=\frac{w-p}{(1-p)s},$$
$$\varphi_2(\vec{z},w)=\frac{w-ps}{pt},\qquad \varphi_5(\vec{z},w)=\frac{w-p-(1-p)s}{(1-p)t},$$
$$\varphi_3(\vec{z},w)=\frac{w-ps-pt}{p(1-s-t)},\quad \varphi_6(\vec{z},w)=\frac{w-p-(1-p)s-(1-p)t}{(1-p)(1-s-t)}.$$

We denote $p_k(\vec{z})$ and $\varphi_k(\vec{z},w)$ by p_k and $\varphi_k(w)$, respectively, since each $p_k(\vec{z})$ is a constant. Therefore,

$$R'(\vec{z},w)=\begin{cases}(\tau_1(\vec{z}),\varphi_1(w)), & \text{if } w\in[0,ps),\\ (\tau_2(\vec{z}),\varphi_2(w)), & \text{if } w\in[ps,ps+pt),\\ (\tau_3(\vec{z}),\varphi_3(w)), & \text{if } w\in[ps+pt,p),\\ (\tau_4(\vec{z}),\varphi_4(w)), & \text{if } w\in[p,p+(1-p)s),\\ (\tau_5(\vec{z}),\varphi_5(w)), & \text{if } w\in[p+(1-p)s,p+(1-p)s+(1-p)t),\\ (\tau_6(\vec{z}),\varphi_6(w)), & \text{if } w\in[p+(1-p)s+(1-p)t,1).\end{cases}$$

Denote by μ_β an acim for the position-dependent random transformation $R=\{\tau_1,\dots,\tau_6;p_1,\dots,p_6\}$, which means μ_β is R-invariant and absolutely continuous with respect to Lebesgue measure λ_2 in $\mathbb{R}^2$. We start by recalling [**BBQ**, Lemma 3.2].

LEMMA 4.5. *μ_β is invariant for the random map R if and only if $\mu_\beta\otimes\lambda_1$ is invariant for the skew product R'.*

STEP 3. For the skew product transformation R_β on $\Omega\times\Upsilon\times S_\beta$.

Define the *skew product transformation* R_β on $\Omega\times\Upsilon\times S_\beta$ as follows:

$$R_\beta(\omega,\upsilon,\vec{z})=\begin{cases}(\sigma\omega,\sigma'\upsilon,\beta\vec{z}-\vec{q_i}), & \text{if } \vec{z}\in E_i, i\in\{0,1,2\}\\ (\sigma\omega,\sigma'\upsilon,\beta\vec{z}-\vec{q_i}), & \text{if } \vec{z}\in C_{ij}, ij\in\{01,02,12\} \text{ and } \omega_1=0\\ (\sigma\omega,\sigma'\upsilon,\beta\vec{z}-\vec{q_j}), & \text{if } \vec{z}\in C_{ij}, ij\in\{01,02,12\} \text{ and } \omega_1=1\\ (\sigma\omega,\sigma'\upsilon,\beta\vec{z}-\vec{q_i}), & \text{if } \vec{z}\in C_{012}, \upsilon_1=i, i\in\{0,1,2\}.\end{cases}$$

LEMMA 4.6. *$(S_\beta\times I,\mathcal{S}\times\mathcal{B}(I),\mu_\beta\otimes\lambda_1,R')$ and $(\Omega\times\Upsilon\times S_\beta,\mathcal{A}\times\mathcal{B}\times\mathcal{S},m_1\otimes m_2\otimes\mu_\beta,R_\beta)$ are isomorphic.*

PROOF. Let $\pi_2: S_\beta\times I\to I$ be the canonical projection onto the second coordinate. Consider the map $\varphi=\pi_2\circ R'$ on $(I,\mathcal{B}(I),\lambda_1)$. One can see that $\varphi(w)=\varphi_k(w)$ for $w\in I_k$, where $I_1=[0,p_1)$ and $I_k=\left[\sum_{i=1}^{k-1}p_i,\sum_{i=1}^{k}p_i\right)$ for $2\le k\le 6$. Define

$$l(w)=\frac{1}{p_k}\quad\text{and}\quad h(w)=\frac{\sum_{i=1}^{k-1}p_i}{p_k}$$

for $w\in I_k$. It follows that $\varphi(w)=l(w)\cdot w-h(w)$. Let

$$l_n=l_n(w):=l(\varphi^{n-1}(w))\quad\text{and}\quad h_n=h_n(w):=h(\varphi^{n-1}(w)).$$

For $w\in[0,1)$, we can write the generalized Lüroth series (GLS) of w, which is

$$w=\frac{h_1}{l_1}+\frac{h_2}{l_1l_2}+\cdots+\frac{h_n}{l_1\cdots l_n}+\cdots.$$

Consider the system $\{\{0,1,2,3,4,5\}^{\mathbb{N}}, \mathcal{C}, m, \sigma''\}$, where $\mathcal{C}$ is the product σ-algebra, σ'' is the left shift and m is the product measure with weights $\{p_1, \ldots, p_6\}$ as in (4.1). Let $\phi_1 : I \to \{0,1,2,3,4,5\}^{\mathbb{N}}$ be given by

$$\phi_1 : w = \sum_{n=1}^{\infty} \frac{h_i}{l_1 l_2 \cdots l_i} \mapsto (\gamma_1, \gamma_2, \ldots,),$$

where $\gamma_n = \gamma_n(w), n \geq 1$ is defined as follows:

$$\gamma_n := \gamma_n(w) = k - 1 \iff \varphi^{n-1}(w) \in I_k,$$

for $k \in \{1,2,3,4,5,6\}$. It is known that φ preserves the Lebesgue measure λ_1 and ϕ_1 is an isomorphism between the two dynamical systems $(I, \mathcal{B}(I), \lambda_1, \varphi\}$ and $\{\{0,1,2,3,4,5\}^{\mathbb{N}}, \mathcal{C}, m, \sigma''\}$. See [**BBDK**] for more details.

Next we give a map ϕ_2 from $\{\{0,1,2,3,4,5\}^{\mathbb{N}}, \mathcal{C}, m, \sigma''\}$ to $\{\Omega \times \Upsilon, \mathcal{A} \times \mathcal{B}, m_1 \otimes m_2, \sigma \times \sigma'\}$. Let $h_1 : \{0,1,2,3,4,5\} \to \{0,1\}$ and $h_2 : \{0,1,2,3,4,5\} \to \{0,1,2\}$ be given by

$$h_1(x) = \begin{cases} 0, & \text{if } x = 0,1,2, \\ 1, & \text{if } x = 3,4,5, \end{cases} \quad h_2(x) = \begin{cases} 0, & \text{if } x = 0,3, \\ 1, & \text{if } x = 1,4, \\ 2, & \text{if } x = 2,5. \end{cases}$$

Define $\phi_2 : \{0,1,2,3,4,5\}^{\mathbb{N}} \to \Omega \times \Upsilon$ by $\phi_2(\gamma) = (\omega, \upsilon)$, where

$$\begin{aligned} \omega &= (h_1(\gamma_1), h_1(\gamma_2), h_1(\gamma_3), \ldots) := \tilde{h}_1(\gamma), \\ \upsilon &= (h_2(\gamma_1), h_2(\gamma_2), h_2(\gamma_3), \ldots) := \tilde{h}_2(\gamma). \end{aligned}$$

One can see that ϕ_2 maps a cylinder of rank n in $\{0,1,2,3,4,5\}^{\mathbb{N}}$ to the product of two cylinders of the same rank n in $\Omega \times \Upsilon$. It follows that ϕ_2 is a bimeasurable bijection. From the definition of the product measure, we can get the measure preservingness on cylinders. Finally, it is easy to see that $\phi_2 \circ \sigma'' = (\sigma \times \sigma') \circ \phi_2$. Therefore, ϕ_2 is an isomorphism.

Now let $\phi : S_\beta \times I \to \Omega \times \Upsilon \times S_\beta$ be given by

$$\phi(\vec{z}, w) = (\tilde{h}_1(\phi_1(w)), \tilde{h}_2(\phi_1(w)), \vec{z}).$$

In fact, $\phi = \iota \circ (I_{S_\beta} \times (\phi_2 \circ \phi_1))$, where I_{S_β} is the identity map on S_β and $\iota(\vec{z}, \omega, \upsilon) = (\omega, \upsilon, \vec{z})$ is a transformation that only changes the order of coordinates. Since $\phi_2 \circ \phi_1$ preserves the dynamics of $\pi_2 \circ R$ and $\sigma \times \sigma'$, i.e.,

$$(\phi_2 \circ \phi_1) \circ (\pi_2 \circ R) = (\sigma \times \sigma') \circ (\phi_2 \circ \phi_1),$$

we have that $\phi \circ R' = R_\beta \circ \phi$. Therefore, the result follows. □

STEP 4. For the random transformation K_β on $\Omega \times \Upsilon \times S_\beta$.

Define a skew product transformation R_β as follows:

$$R_\beta(\omega, \upsilon, \vec{z}) = \begin{cases} (\sigma\omega, \sigma'\upsilon, \tau_1(\vec{z})), & \text{if } \omega_1 = 0, \upsilon_1 = 0, \\ (\sigma\omega, \sigma'\upsilon, \tau_2(\vec{z})), & \text{if } \omega_1 = 0, \upsilon_1 = 1, \\ (\sigma\omega, \sigma'\upsilon, \tau_3(\vec{z})), & \text{if } \omega_1 = 0, \upsilon_1 = 2, \\ (\sigma\omega, \sigma'\upsilon, \tau_4(\vec{z})), & \text{if } \omega_1 = 1, \upsilon_1 = 0, \\ (\sigma\omega, \sigma'\upsilon, \tau_5(\vec{z})), & \text{if } \omega_1 = 1, \upsilon_1 = 1, \\ (\sigma\omega, \sigma'\upsilon, \tau_6(\vec{z})), & \text{if } \omega_1 = 1, \upsilon_1 = 2, \end{cases}$$

Let μ be an arbitrary probability measure on S_β. We will show that any product measure of the form $m_1 \otimes m_2 \otimes \mu$ is K_β-invariant if and only if it is R_β-invariant.

LEMMA 4.7. *$m_1 \otimes m_2 \otimes \mu \circ K_\beta^{-1} = m_1 \otimes m_2 \otimes \mu \circ R_\beta^{-1} = m_1 \otimes m_2 \otimes \nu$, where*

$$\begin{aligned}\nu =& ps \cdot \mu \circ \tau_1^{-1} + pt \cdot \mu \circ \tau_2^{-1} + p(1-s-t) \cdot \mu \circ \tau_3^{-1} \\ &+ (1-p) \cdot s \cdot \mu \circ \tau_4^{-1} + (1-p) \cdot t \cdot \mu \circ \tau_5^{-1} + (1-p) \cdot (1-s-t) \cdot \mu \circ \tau_6^{-1}.\end{aligned}$$

PROOF. Denote by C_1 and C_2 arbitrary cylinders in Ω and Υ, respectively. Let S be a closed set in S_β. It suffices to verify that the measures coincide on sets of the form $C_1 \times C_2 \times S$, because the collection of these sets forms a generating π-system. Let $[i, C_1] = \{\omega_1 = i\} \cap \sigma^{-1}(C_1)$ for $i = 0, 1$ and $[i, C_2] = \{\upsilon_1 = i\} \cap (\sigma')^{-1}(C_2)$ for $i = 0, 1, 2$. Notice that

$$\begin{aligned}\tau_1^{-1}(S) \cap E = \tau_2^{-1}(S) \cap E = \cdots = \tau_6^{-1}(S) \cap E, \\ \tau_1^{-1}(S) \cap C = \tau_2^{-1}(S) \cap C = \tau_3^{-1}(S) \cap C, \\ \tau_4^{-1}(S) \cap C = \tau_5^{-1}(S) \cap C = \tau_6^{-1}(S) \cap C, \\ \tau_1^{-1}(S) \cap C_{012} = \tau_4^{-1}(S) \cap C_{012}, \\ \tau_2^{-1}(S) \cap C_{012} = \tau_5^{-1}(S) \cap C_{012}, \\ \tau_3^{-1}(S) \cap C_{012} = \tau_6^{-1}(S) \cap C_{012}.\end{aligned}$$

We can divide $K_\beta^{-1}(C_1 \times C_2 \times S)$ into the union of some disjoint sets as follows:

$$\begin{aligned}&K_\beta^{-1}(C_1 \times C_2 \times S) \\ &\quad = C_1 \times C_2 \times (\tau_1^{-1}(S) \cap E) \cup [0, C_1] \times C_2 \times (\tau_1^{-1}(S) \cap C) \\ &\quad\quad \cup [1, C_1] \times C_2 \times (\tau_4^{-1}(S) \cap C) \cup C_1 \times [0, C_2] \times (\tau_1^{-1}(S) \cap C_{012}) \\ &\quad\quad \cup C_1 \times [1, C_2] \times (\tau_2^{-1}(S) \cap C_{012}) \cup C_1 \times [2, C_2] \times (\tau_3^{-1}(S) \cap C_{012})\end{aligned}$$

Hence,

$$\begin{aligned}&m_1 \otimes m_2 \otimes \mu \circ K_\beta^{-1}(C_1 \times C_2 \times S) \\ &\quad = m_1(C_1)m_2(C_2)\mu(\tau_1^{-1}(S) \cap E) \\ &\quad\quad + p \cdot m_1(C_1)m_2(C_2)\mu(\tau_1^{-1}(S) \cap C) \\ &\quad\quad + (1-p) \cdot m_1(C_1)m_2(C_2)\mu(\tau_4^{-1}(S) \cap C) \\ &\quad\quad + s \cdot m_1(C_1)m_2(C_2)\mu(\tau_1^{-1}(S) \cap C_{012}) \\ &\quad\quad + t \cdot m_1(C_1)m_2(C_2)\mu(\tau_2^{-1}(S) \cap C_{012}) \\ &\quad\quad + (1-s-t) \cdot m_1(C_1)m_2(C_2)\mu(\tau_3^{-1}(S) \cap C_{012}) \\ &\quad = ps \cdot m_1(C_1)m_2(C_2)\mu(\tau_1^{-1}(S)) \\ &\quad\quad + pt \cdot m_1(C_1)m_2(C_2)\mu(\tau_2^{-1}(S)) \\ &\quad\quad + p(1-s-t) \cdot m_1(C_1)m_2(C_2)\mu(\tau_3^{-1}(S)) \\ &\quad\quad + (1-p)s \cdot m_1(C_1)m_2(C_2)\mu(\tau_4^{-1}(S)) \\ &\quad\quad + (1-p)t \cdot m_1(C_1)m_2(C_2)\mu(\tau_5^{-1}(S)) \\ &\quad\quad + (1-p)(1-s-t) \cdot m_1(C_1)m_2(C_2)\mu(\tau_6^{-1}(S)) \\ &\quad = m_1 \otimes m_2 \otimes \nu(C_1 \times C_2 \times S).\end{aligned}$$

On the other hand,

$$\begin{aligned}R_\beta^{-1}&(C_1 \times C_2 \times S)\\ &= [0, C_1] \times [0, C_2] \times \tau_1^{-1}(S) \cup [0, C_1] \times [1, C_2] \times \tau_2^{-1}(S)\\ &\quad \cup [0, C_1] \times [2, C_2] \times \tau_3^{-1}(S) \cup [1, C_1] \times [0, C_2] \times \tau_4^{-1}(S)\\ &\quad \cup [1, C_1] \times [1, C_2] \times \tau_5^{-1}(S) \cup [1, C_1] \times [2, C_2] \times \tau_6^{-1}(S).\end{aligned}$$

Therefore, we complete the proof. □

Now we give the main result in this section.

THEOREM 4.8. *Let $\beta \in (1, 3/2)$. Then K_β has an invariant measure of the form $m_1 \otimes m_2 \otimes \mu_\beta$, where μ_β is absolutely continuous with respect to λ_2.*

PROOF. By Theorem 4.4, Lemma 4.5, Lemma 4.6, and Lemma 4.7, we complete the proof. □

REMARK 4.9. When $\beta = 3/2$, $C_{012} = \{(\frac{2}{3}, \frac{2}{3})\}$ is a point. We modify the definition of R, R', R_β, and give relevant conclusions.

(i) Let $R = \{\tau_1, \tau_2; p_1(\vec{z}), p_2(\vec{z})\}$ be a position-dependent random transformation on S_β, where

$$\tau_1(\vec{z}) = \begin{cases} \beta\vec{z} - \vec{q}_i, & \text{if } \vec{z} \in E_i, i \in \{0,1,2\},\\ \beta\vec{z} - \vec{q}_i, & \text{if } \vec{z} \in C_{ij}, ij \in \{01, 02, 12\},\\ \beta\vec{z}, & \text{if } \vec{z} = (\frac{2}{3}, \frac{2}{3}),\end{cases}$$

$$\tau_2(\vec{z}) = \begin{cases} \beta\vec{z} - \vec{q}_i, & \text{if } \vec{z} \in E_i, i \in \{0,1,2\},\\ \beta\vec{z} - \vec{q}_j, & \text{if } \vec{z} \in C_{ij}, ij \in \{01, 02, 12\},\\ \beta\vec{z}, & \text{if } \vec{z} = (\frac{2}{3}, \frac{2}{3}),\end{cases}$$

and $p_1(\vec{z}) = p_2(\vec{z}) = 1/2$ for $\vec{z} \in S_\beta$. Similar to Theorem 4.4, it is not difficult to prove that R has an acim μ_β.

(ii) By [**BBQ**, Lemma 3.2], $\mu_\beta \otimes \lambda_1$ is invariant for the skew product R', where

$$R'(\vec{z}, w) = \begin{cases} (\tau_1(\vec{z}), \frac{w}{p}), & \text{if } w \in [0, p),\\ (\tau_2(\vec{z}), \frac{w-p}{1-p}), & \text{if } w \in [p, 1).\end{cases}$$

(iii) Define the *skew product transformation* R_β on $\Omega \times S_\beta$ as follows:

$$R_\beta(\omega, \vec{z}) = \begin{cases} (\sigma\omega, \beta\vec{z} - \vec{q}_i), & \text{if } \vec{z} \in E_i, i \in \{0,1,2\},\\ (\sigma\omega, \beta\vec{z} - \vec{q}_i), & \text{if } \vec{z} \in C_{ij}, ij \in \{01, 02, 12\}, \omega_1 = 0,\\ (\sigma\omega, \beta\vec{z} - \vec{q}_j), & \text{if } \vec{z} \in C_{ij}, ij \in \{01, 02, 12\}, \omega_1 = 1,\\ (\sigma\omega, \beta\vec{z}), & \text{if } \vec{z} = (\frac{2}{3}, \frac{2}{3}).\end{cases}$$

Then we have that the dynamical systems $(S_\beta \times I, \mathcal{S} \times \mathcal{B}(I), \mu_\beta \otimes \lambda_1, R')$ and $(\Omega \times S_\beta, \mathcal{A} \times \mathcal{S}, m_1 \otimes \mu_\beta, R_\beta)$ are isomorphic. The proof is similar and easier than that of Lemma 4.6.

(iv) Let μ be an arbitrary probability measure on S_β and let $\tilde{K}_\beta : \Omega \times S_\beta \to \Omega \times S_\beta$ be given by

$$\tilde{K}_\beta(\omega, \vec{z}) = \begin{cases} (\omega, \beta\vec{z} - \vec{q}_i), & \text{if } \vec{z} \in E_i, i \in \{0,1,2\} \\ (\sigma\omega, \beta\vec{z} - \vec{q}_i), & \text{if } \vec{z} \in C_{ij}, ij \in \{01, 02, 12\}, \omega_1 = 0 \\ (\sigma\omega, \beta\vec{z} - \vec{q}_j), & \text{if } \vec{z} \in C_{ij}, ij \in \{01, 02, 12\}, \omega_1 = 1 \\ (\omega, \beta\vec{z}), & \text{if } \vec{z} = (\frac{2}{3}, \frac{2}{3}). \end{cases}$$

It is easy to check that

$$m_1 \otimes \mu \circ \tilde{K}_\beta^{-1} = m_1 \otimes \mu \circ R_\beta^{-1} = m_1 \otimes \nu,$$

where

$$\nu = p \cdot \mu \circ \tau_1^{-1} + (1-p) \cdot \mu \circ \tau_2^{-1}$$

by using the same method of calculation in Lemma 4.7. Therefore, it follows from (i)–(iv) that $\tilde{K}_\beta$ has an invariant measure of the form $m_1 \otimes \mu_\beta$.

Acknowledgments

The authors would like to thank Simon Baker for his suggestions on improving the range of β in Lemma 3.2.

References

[BBDK] Jose Barrionuevo, Robert M. Burton, Karma Dajani, and Cor Kraaikamp, *Ergodic properties of generalized Lüroth series*, Acta Arith. **74** (1996), no. 4, 311–327, DOI 10.4064/aa-74-4-311-327. MR1378226

[BBQ] Wael Bahsoun, Christopher Bose, and Anthony Quas, *Deterministic representation for position dependent random maps*, Discrete Contin. Dyn. Syst. **22** (2008), no. 3, 529–540, DOI 10.3934/dcds.2008.22.529. MR2429852

[BG] Wael Bahsoun and Paweł Góra, *Position dependent random maps in one and higher dimensions*, Studia Math. **166** (2005), no. 3, 271–286, DOI 10.4064/sm166-3-5. MR2110096

[BMS] Dave Broomhead, James Montaldi, and Nikita Sidorov, *Golden gaskets: variations on the Sierpiński sieve*, Nonlinearity **17** (2004), no. 4, 1455–1480, DOI 10.1088/0951-7715/17/4/017. MR2069714

[DK1] Karma Dajani and Cor Kraaikamp, *From greedy to lazy expansions and their driving dynamics*, Expo. Math. **20** (2002), no. 4, 315–327, DOI 10.1016/S0723-0869(02)80010-X. MR1940010

[DK2] Karma Dajani and Cor Kraaikamp, *Random β-expansions*, Ergodic Theory Dynam. Systems **23** (2003), no. 2, 461–479, DOI 10.1017/S0143385702001141. MR1972232

[DV1] Karma Dajani and Martijn de Vries, *Measures of maximal entropy for random β-expansions*, J. Eur. Math. Soc. (JEMS) **7** (2005), no. 1, 51–68, DOI 10.4171/JEMS/21. MR2120990

[DV2] Karma Dajani and Martijn de Vries, *Invariant densities for random β-expansions*, J. Eur. Math. Soc. (JEMS) **9** (2007), no. 1, 157–176, DOI 10.4171/JEMS/76. MR2283107

[F] Kenneth Falconer, *Fractal geometry*, 3rd ed., John Wiley & Sons, Ltd., Chichester, 2014. Mathematical foundations and applications. MR3236784

[G] Enrico Giusti, *Minimal surfaces and functions of bounded variation*, Monographs in Mathematics, vol. 80, Birkhäuser Verlag, Basel, 1984, DOI 10.1007/978-1-4684-9486-0. MR775682

[GB] P. Góra and A. Boyarsky, *Absolutely continuous invariant measures for piecewise expanding C^2 transformation in $\mathbf{R}^N$*, Israel J. Math. **67** (1989), no. 3, 272–286, DOI 10.1007/BF02764946. MR1029902

[HP] Boris Hasselblatt and Donald Plante Jr., *On the interior of "fat" Sierpiński triangles*, Exp. Math. **23** (2014), no. 3, 285–309, DOI 10.1080/10586458.2014.900731. MR3255940

[H] Michael Hochman, *On self-similar sets with overlaps and inverse theorems for entropy*, Ann. of Math. (2) **180** (2014), no. 2, 773–822, DOI 10.4007/annals.2014.180.2.7. MR3224722

[JP] Thomas Jordan and Mark Pollicott, *Properties of measures supported on fat Sierpinski carpets*, Ergodic Theory Dynam. Systems **26** (2006), no. 3, 739–754, DOI 10.1017/S0143385705000696. MR2237467

[K] Shizuo Kakutani, *Ergodic theorems and the Markoff process with a stable distribution*, Proc. Imp. Acad. Tokyo **16** (1940), 49–54. MR2049

[KL] Derong Kong and Wenxia Li, *Critical base for the unique codings of fat Sierpinski gasket*, Nonlinearity **33** (2020), no. 9, 4484–4511, DOI 10.1088/1361-6544/ab8baf. MR4127786

[S1] Nikita Sidorov, *Combinatorics of linear iterated function systems with overlaps*, Nonlinearity **20** (2007), no. 5, 1299–1312, DOI 10.1088/0951-7715/20/5/013. MR2312394

[SS] Károly Simon and Boris Solomyak, *On the dimension of self-similar sets*, Fractals **10** (2002), no. 1, 59–65, DOI 10.1142/S0218348X02000963. MR1894903

[S2] S. M. Srivastava, *A course on Borel sets*, Graduate Texts in Mathematics, vol. 180, Springer-Verlag, New York, 1998, DOI 10.1007/978-3-642-85473-6. MR1619545

[Y] Kôsaku Yosida, *The Markoff process with a stable distribution*, Proc. Imp. Acad. Tokyo **16** (1940), 43–48. MR2048

Department of Mathematics, Utrecht University, Fac Wiskunde en informatica and MRI, Budapestlaan 6, P.O. Box 80.000, 3508 TA Utrecht, the Netherlands
Email address: `k.dajani1@uu.nl`

Department of Mathematics, Shanghai Key Laboratory of PMMP, East China Normal University, Shanghai 200241, People's Republic of China
Email address: `wxli@math.ecnu.edu.cn`

Department of Mathematics, Shanghai Key Laboratory of PMMP, East China Normal University, Shanghai 200241, People's Republic of China
Current address: College of Humanities and Law, Shanghai Business School, Shanghai 200235, People's Republic of China
Email address: `tingyuzhangecnu@163.com`

Contemporary Mathematics
Volume **797**, 2024
https://doi.org/10.1090/conm/797/15933

Polymorphisms, their associated operator theory, self-similar fractals, and harmonic analysis

Palle E. T. Jorgensen and James F. Tian

Dedicated to the memory of Robert S. Strichartz.

Abstract. In this paper we develop rigorously the theory and applications of Polymorphisms. The list of topics covered includes: areas of probability theory, analysis of Markov maps, transition operators, joinings of measure spaces, and applications to harmonic and stochastic analysis of self-similar (fractal) measures.

1. Introduction

The notion, and the study of, polymorphisms was pioneered by Anatoly M. Vershik, see [**Ver05**] and the papers cited below. In rough outline, the ideas involved are as follows: While functions serve as transformations between sets, *multi-valued maps*, and other multi-valued generalizations, are needed in the study of diverse measure-theoretic version of dynamics. Polymorphisms were introduced by Vershik, and studied systematically. They were motivated by diverse applications (see, e.g., [**HJW19a, JST20, JT20, JT21, BJ21, JT22a, BJ22**]). This study includes *Markov maps*, transition operators, or two-dimensional distributions of Markov processes, or joinings of measure spaces. Parallel notions in other areas of mathematics are the following: *correspondences* in algebra; *bifibrations* in differential geometry, or *Markov maps* in probability theory. In physics, the notion of polymorphism corresponds to the "*coarse-graining*" *approach* to dynamics, where maps send a point to measures.

Our paper aims to present a unified approach to *polymorphisms*, stressing the underlying operator theory, harmonic analysis, probability theory, self-similar fractals, and stochastic processes; see, e.g., [**CHlQ$^+$21, RS16, JP99, JP98, JP95**].

Organization of the paper. In Sections 2 and 3 we present our analytic tools for the study of polymorphisms, with an aim to applications to probability theory. Section 4, *The nonsymmetric case*, is motivated as follows: In machine learning, e.g., optimization involving statistical regression analysis, it is important

2020 *Mathematics Subject Classification.* Primary 46N30, 47N30, 28A50, 28A80; Secondary 28D05, 37A30, 60G15, 60G60, 60H05, 46N50.

Key words and phrases. Polymorphisms, fractals, IFS-measures, Sierpinski constructions, probability theory, regression, Hilbert space, harmonic analysis, Gaussian fields, transition operators, Wasserstein-metric, covariance, Krein–Feller operators, symbolic dynamics, Perron–Frobenius.

to allow for a modification of the notion of polymorphism such as to allow for different measure spaces in the two factors, see Definition 2.1. Section 5 deals with polymorphisms arising in questions from probability theory and statistical inference. Section 6, *generalized Brownian motion*, addresses the following: we consider polymorphisms that arise from selfsimilar IFS-measures ν, we show that the induced generalized Brownian motion $W^{(\nu)}$ satisfies an analogous self-similarity property. It is induced from the given IFS property for ν itself.

2. Polymorphisms in the measure category

Vershik introduced the notion of polymorphisms as a generalization of a variety of constructions in probability and dynamical systems; see [**SV08, SV09, CV13**].

The purpose of the present section is to develop connections to operator theory, and to harmonic analysis, see, e.g., [**Jor18**].

DEFINITION 2.1. Let $(X, \mathscr{B})$ be a measurable space, i.e., a set X with a specified σ-algebra $\mathscr{B}$. A given positive measure ν on $X \times X$ relative to the product σ-algebra $\mathscr{B}_2$ is said to be a polymorphism for μ_1 and μ_2 when μ_i are the respective marginal measures

$$\mu_i := \nu \circ \pi_i^{-1} \tag{2.1}$$

relative to the projections

$$\pi_i\left((x_1, x_2)\right) = x_i, \quad i = 1, 2, \tag{2.2}$$

see Figure 2.1.

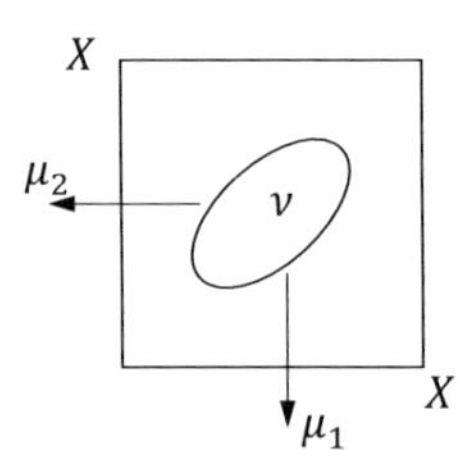

FIGURE 2.1. Measures on product spaces and marginal measures.

Let us recall that a doubly stochastic measure (DSM) is a Borel probability measure defined on the unit square $[0, 1]^2$, with the property that both of its marginals are equal to the Lebesgue measure on the interval $[0, 1]$. DSMs form a weak-* compact and convex set, which encompasses all joint distributions on the unit square. It is noteworthy that DSMs serve as a precursor to the notion of polymorphisms (cf. [**Isb55, Rot62, Ryf68, BS70, Day73, Los82**]; see Remark 2.7).

LEMMA 2.2. *Let $\mu_1 \leftarrow \nu \rightarrow \mu_2$ be a polymorphism (see Figure 2.2). Then the operators*

$$V_i : L^2(X, \mu_i) \rightarrow L^2(X \times X, \nu), \quad V_i f := f \circ \pi_i, \tag{2.3}$$

for $i = 1, 2$, are isometries.

PROOF. The isometric property of V_i in (2.3) follows from the choices (2.1)–(2.2): For all $f \in L^2(X, \mu_i)$, one has

$$\|V_i f\|^2_{L^2(X \times X, \nu)} = \int_{X \times X} |f \circ \pi_i|^2 \, d\nu = \int_X |f|^2 \, d\nu \circ \pi_i^{-1} = \int_X |f|^2 \, d\mu_i = \|f\|^2_{L^2(X, \mu_i)} .$$

□

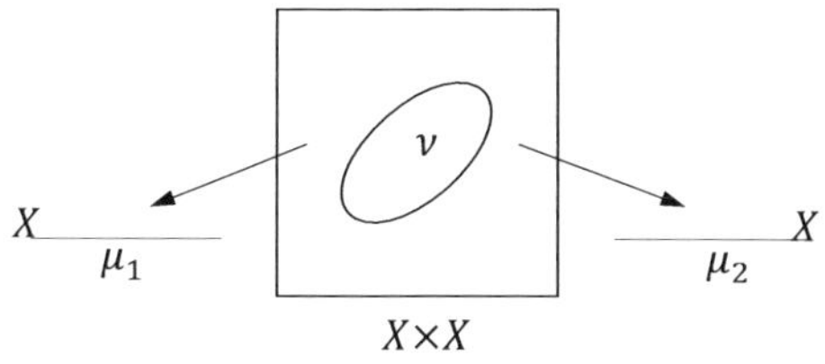

FIGURE 2.2. A polymorphism

LEMMA 2.3. *Assume $(X, \mathscr{B})$ is a standard measurable space. Let $\mu_1 \leftarrow \nu \rightarrow \mu_2$ be a polymorphism, then the marginal measures μ_i yield conditional measures $\nu(\cdot \mid x_i)$ defined as follows:*

$$\nu(dx_1 dx_2) = \mu_1(dx_1)\, \nu(dx_2 \mid x_1) = \mu_2(dx_2)\, \nu(dx_1 \mid x_2), \tag{2.4}$$

or equivalently, with ν specified on product sets $A \times B$, $A, B \in \mathscr{B}$:

$$\nu(A \times B) = \int_A \mu_1(dx_1)\, \nu(B \mid x_1) = \int_B \mu_2(dx_2)\, \nu(A \mid x_2). \tag{2.5}$$

Formulas (2.4) and (2.5) are closely related to the Bayes rules from probability, and the slice decompositions of measures on product spaces, arising in ergodic theory. The Bayes rules (of probability and statistic inference) describe the probability of events, based on prior knowledge of, or making precise, conditions that govern, or might be related to, the events in question, hence conditional distributions.

Our present discussion will stay away from certain technical issues. Nonetheless, for experts, we wish to state that we follow Vershik in assuming that our spaces are standard measure spaces. And we recall that papers of Rokhlin on disintegrations do address this issue in depth, i.e., the context of "separable" measure spaces. In fact, we do not have a short proof covering regular conditional measures for general spaces. Our formulation Lemma 2.3 and sequel is assuming appropriate properties for the underlying measure spaces, i.e., we are in the framework of standard measure spaces. However, regarding this point we have elected a presentation which avoids completeness, offering instead what may be study material for freshmen. Readers interested in more detail are referred to [**BJ21**, **BJ22**, **Ver05**], and to the papers cited there.

LEMMA 2.4. *Let $\mu_1 \leftarrow \nu \rightarrow \mu_2$ be a polymorphism, and let V_i, $i = 1, 2$, be the isometries introduced in (2.3). We then have the following for the respective adjoint operators:*

$$(V_1^* \psi)(x) = \int_X \psi(x, y)\, \nu(dy \mid x), \text{ and} \tag{2.6}$$

$$(V_2^* \psi)(y) = \int_X \psi(x, y)\, \nu(dx \mid y) \tag{2.7}$$

valid for all $\psi \in L^2(X \times X, \nu)$, *where* $\nu(\cdot \mid x)$ *and* $\nu(\cdot \mid y)$ *are the conditional measures introduced in* (2.4).

PROOF. We shall include the details here only for (2.6). The other case is similar.

We use the basic property of the adjoint operator V_1^* as follows. Let $f \in L^2(X, \mu_1)$, and $\psi \in L^2(X \times X, \nu)$, then

$$\begin{aligned}\langle V_1 f, \psi \rangle_{L^2(\nu)} &= \iint_{X\times X} f(x)\,\psi(x,y)\,\nu(dxdy) \\ &\underset{\text{by }(2.4)}{=} \iint_{X\times X} f(x)\,\psi(x,y)\,\mu_1(dx)\,\nu(dy \mid x) \\ &\underset{\text{Fubini}}{=} \int_X f(x)\left(\int_X \psi(x,y)\,\nu(dy\mid x)\right)\mu_1(dx) \\ &= \langle f, V_1^*\psi\rangle_{L^2(\mu_1)}\end{aligned}$$

and conclusion (2.6) follows.

Here we have assumed that the respective Hilbert spaces consist of real valued functions, but the extension to the complex case is immediate, and left to the reader. □

COROLLARY 2.5. *Let* $\mu_1 \leftarrow \nu \rightarrow \mu_2$ *be a polymorphism, and introduce the corresponding conditional measures* $\nu(\cdot \mid x)$ *and* $\nu(\cdot \mid y)$ *as in* (2.4)–(2.5), *see Figure* 2.4. *Then, for all* $f \in L^2(\mu_2)$ *we have (see* (2.9)*)*

$$(V_1^* V_2 f)(x) = \int_X f(y)\,\nu(dy \mid x) \tag{2.8}$$

(2.9)

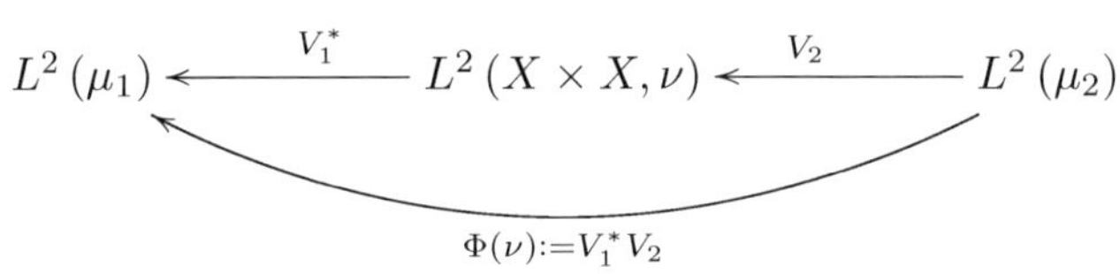

FIGURE 2.3. Polymorphisms as contractive operators

Moreover,

$$\Phi(\nu) := V_1^* V_2 \tag{2.10}$$

is contractive from $L^2(\mu_2)$ *into* $L^2(\mu_1)$.

PROOF. Let the measures be as specified above. Then the conclusions (2.9)–(2.10) follow immediately from (2.6) with the substitution $\psi(x,y) = f(y)$, defined for $(x,y) \in X \times X$, i.e., $(Vf)(x,y) = f \circ \pi_2(x,y) = f(y)$.

The contractivity of the operator in (2.10) follows from the observation that V_1^* is a coisometry. □

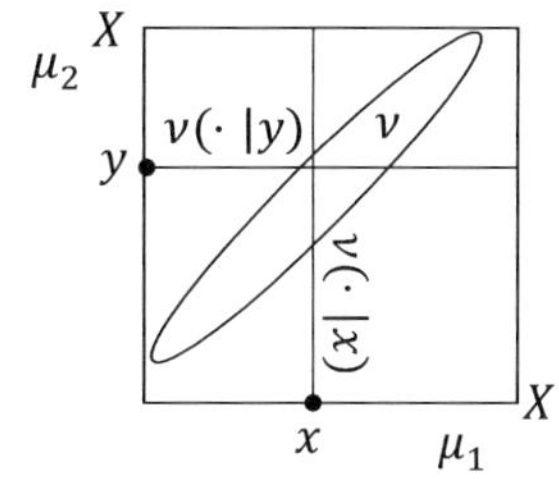

FIGURE 2.4. Polymorphisms and conditional measures

REMARK. Note that, if the measure ν in the diagram (2.9) is assumed to be a probability measure, then the corresponding operator in (2.8) is a Markov transition operator, corresponding to initial distribution μ_1.

COROLLARY 2.6. *Consider three measures μ_i, $i = 1, 2, 3$, and two polymorphisms*

$$\mu_1 \leftarrow \nu \rightarrow \mu_2 \quad \textit{and} \quad \mu_2 \leftarrow \nu' \rightarrow \mu_3 \tag{2.11}$$

with operators (see (2.10)*)*

$$L^2(\mu_1) \xleftarrow{\Phi(\nu)} L^2(\mu_2) \quad \textit{and} \quad L^2(\mu_2) \xleftarrow{\Phi(\nu')} L^2(\mu_3).$$

Then the composition

$$\Phi(\nu\nu') = \Phi(\nu)\,\Phi(\nu'), \tag{2.12}$$

(see Figure 2.5*) defines a semigroup operation for polymorphisms as follows:*

$$(\nu\nu')(A \mid x) = \int_X \nu(dy \mid x)\,\nu'(A \mid y) \tag{2.13}$$

valid for all $A \in \mathscr{B}$, and $x \in X$.

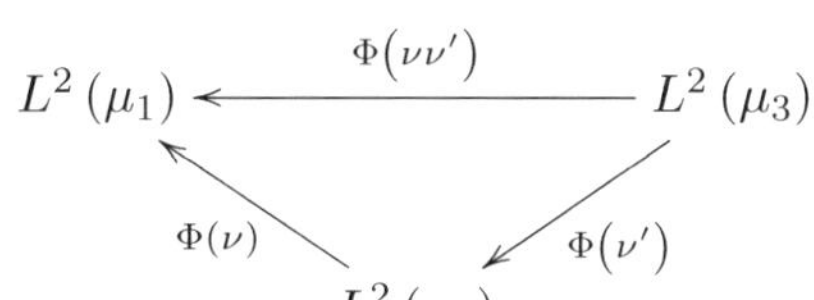

FIGURE 2.5. The operators of the product formula for polymorphisms.

PROOF. The RHS in (2.12) arises by combining the two operators $V_1^* V_2$ for the respective polymorphisms ν and ν', i.e., for $f \in L^2(\mu_3)$, we have:

$$\begin{aligned} &V_1^*(\nu)\,V_2(\nu)\,V_1^*(\nu')\,V_2(\nu')(f)(x) \\ &= \iint_{X\times X} \nu(dy \mid x)\,\nu'(dz \mid y)\,f(z) \\ &\underset{\text{Fubini}}{=} \int_X \left(\int_X \nu(dy \mid x)\,\nu'(dz \mid y)\right) f(z), \end{aligned}$$

and conclusion (2.13) follows. $\square$

EXAMPLE. Consider a fixed measure space $(X, \mathscr{B}, m)$ where m is a positive measure defined on a σ-algebra $\mathscr{B}$. Then for $A, B \in \mathscr{B}$, set

$$\mu(A \times B) := m(A \cap B)$$

and extend μ to a measure on the corresponding product measure space $(X \times X, \mathscr{B}_2)$. Then the operator $\Phi(\mu) = V_1^* V_2 : L^2(X, m) \to L^2(X, m)$ is the identity operator, i.e., $(\Phi(\mu)(f))(x) = f(x)$, for all $f \in L^2(X, \mathscr{B}, m)$, m-a.e. x; write $\Phi(\mu) = I_{L^2(m)}$.

PROOF. Immediate from the reasoning in Lemma 2.3 and Corollary 2.5. □

This family of "diagonal" measures is important as they define covariance kernels for generalized Brownian motion. They reflect the property of "independent increments" in a general context. See the discusses in Section 6, especially (6.2) and (6.3).

REMARK 2.7. Given a system on $X \times X$ as specified in Definition 2.1 and Lemma 2.3, it is important to note that for a fixed pair $\{\mu_i\}_{i=1,2}$ on X, there are many solutions ν to equation (2.1). This fact is important in metric geometry and optimal transport (see, e.g., [**PB22**, **GP22**]).

Fix μ_i, $i = 1, 2$, positive measures on $(X, \mathscr{B}_X)$ and set

$$\text{Poly}(\mu_1, \mu_2) = \left\{\nu \in \mathfrak{M}_+(X \times X) : \nu \circ \pi_i^{-1} = \mu_i\right\} \tag{2.14}$$

see (2.1). Then the set in (2.14) is nonempty and convex.

In optimal transport, one considers a positive cost function (assumed measurable), $C : X \times X \to \mathbb{R}_+$, and the Wasserstein-metric d_C on $\mathfrak{M}_+(X)$, where

$$d_C(\mu_1, \mu_2) := \inf\left\{\int_{X \times X} C(x_1, x_2)\, \nu(dx_1 dx_2) : \nu \in \text{Poly}(\mu_1, \mu_2)\right\}.$$

CONCLUSION. An application of the operators in Corollaries 2.5 and 2.6 leads to the following: Combining the present results above and ideas in recent papers on graphs (see [**AJL17**, **BJ22**, **Jor18**]), one notes that, for every countable set S of measure spaces, and a choice of systems of measures on edges constructed out of pairs from S, and forming product spaces, there is an associated graph G, and a graph representation with linking of the edges in G via Corollary 2.6. Hence the edge operators link pairs of vertices as illustrated in Figure 2.3. Every such choice of vertex measure spaces, and associated edge-operators as per in Figure 2.3 constitutes a graph representation, see Figure 2.6.

For example, let $\rho \in Poly(\mu_1, \mu_2)$, $\nu \in Poly(\mu_2, \mu_3)$, $\mu \in Poly(\mu_3, \mu_4)$ and $\theta \in Poly(\mu_1, \mu_4)$. Then a graphical representation is as follows:

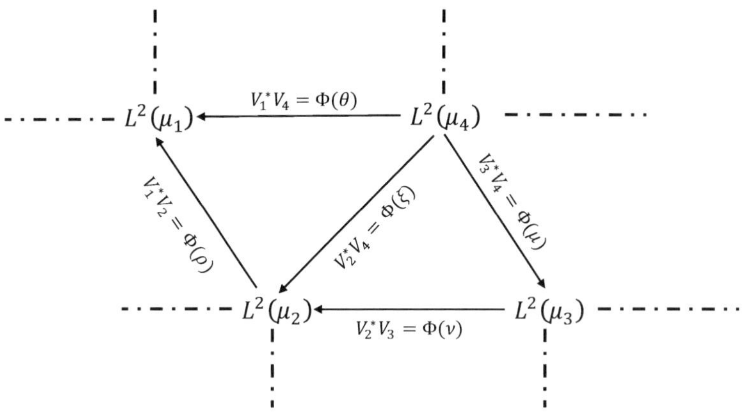

FIGURE 2.6. A graph representations of measure spaces $L^2(\mu_i)$.

This gives the correspondence: vertices $V \longrightarrow$ measures $\mathfrak{M}$, edges $E \longrightarrow$ polymorphisms $\mathscr{P}$. For $v_{i,j} \in \text{Poly}(m_i, m_j)$, we have

$$m_i \xleftarrow[V_i^* V_j = \Phi(v_{i,j})]{v_{i,j}} m_j$$

2.1. Operators. In the present subsection, we study the following relationship between pairs of measures μ on $(X, \mathscr{B}_X)$ and ν on $(Y, \mathscr{B}_Y)$, where two measure spaces are fixed, a measurable function $H : Y \to X$ is specified, and it is assumed that

$$\nu \circ H^{-1} = \mu. \tag{2.15}$$

In this setting, the following holds:

(1) $L^2(\mu) \ni f \xmapsto{V_H} f \circ H \in L^2(\nu)$ is isometric, and for all $\psi \in L^2(\nu)$,
(2)

$$(V_H^* \psi)(x) = \frac{d\left((\psi d\nu) \circ H^{-1}\right)}{d\mu}(x) \tag{2.16}$$

holds for μ-a.e. x.

We note that in (2.16), $\frac{d}{d\mu}$ is a Krein–Feller derivative, i.e., an L^2-extension of the Radon–Nikodym derivative; see, e.g. [**JT22b**].

PROOF. Let $f \in L^2(X, \mu)$ and $\psi \in L^2(Y, \nu)$, then

$$\begin{aligned} \langle V_H f, \psi \rangle_\nu &= \int_Y (f \circ H) \psi d\nu \\ &= \int_X f d(\psi d\nu) \circ H^{-1} \\ &= \int_X f \left(\frac{d(\psi d\nu) \circ H^{-1}}{d\mu} \right) d\mu. \end{aligned}$$

To justify the last step in the computation we must establish the following estimate, valid for all finite systems: $J_i \in \mathscr{B}_X$, $J_i \cap J_j = \emptyset$ if $i \neq j$,

$$\begin{aligned}
\sum_i \frac{\left|\int_{H^{-1}(J_i)} \psi d\nu\right|^2}{\mu(J_i)} &\underset{\text{by Schwarz}}{\leq} \sum \frac{\nu\left(H^{-1}(J_i)\right) \int_{H^{-1}(J_i)} |\psi|^2 d\nu}{\mu(J_i)} \\
&\underset{\text{by (2.15)}}{=} \sum_i \int_{H^{-1}(J_i)} |\psi|^2 d\nu \\
&\leq \int_Y |\psi|^2 d\nu < \infty.
\end{aligned}$$

□

3. Iterated function system measures as polymorphisms

In this section we turn to an analysis of the IFS (iterated function system) measures (see, e.g., [**Hut81, Hut95, DJ07, BHS08, HJW19b, JS18**]). We shall address the case where given a system of contractive mappings, affine or conformal, there are two associated fixed-point problems, one for compact sets, and the other for probability measures. For a fixed IFS, the set in question arises as the support of an associated IFS-measure ν. We outline probabilistic features of these constructions and their fractal properties. We specifically show that these planar Sierpinski measures ν are *slice-singular* (Definition 3.1).

We shall utilize the material developed above to analyze the Sierpinski gaskets and their associated IFS measures, denoted by ν, i.e., a 2D IFS measure on $I \times I$, with $I = [0, 1]$. Recall the three maps τ_i,

$$\begin{cases} \tau_1(x,y) = \left(\frac{x}{2}, \frac{y}{2}\right) \\ \tau_2(x,y) = \left(\frac{x+1}{2}, \frac{y}{2}\right) \\ \tau_3(x,y) = \left(\frac{x}{2}, \frac{y+1}{2}\right) \end{cases} \tag{3.1}$$

and the IFS condition for ν,

$$\nu = \frac{1}{3}\left(\nu \circ \tau_1^{-1} + \nu \circ \tau_2^{-1} + \nu \circ \tau_3^{-1}\right). \tag{3.2}$$

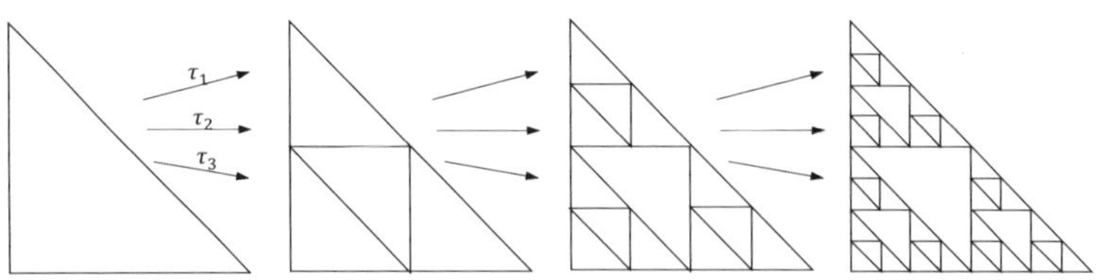

FIGURE 3.1. SG = the tree-graph of the Sierpinski gasket. See Figures 3.2, 3.4, 3.5. 3.6, and 3.7, and the related discussion there.

Let N be a fixed integer, and let $(X, \mathscr{B})$ be a measurable space with endomorphisms $\tau_i : X \to X$, $1 \leq i \leq N$. Set

$$\Omega = \prod_{\mathbb{N}} \{1, 2, \cdots, N\} \tag{3.3}$$

to be the infinite product space with its usual cylinder σ-algebra $\mathscr{C}$. For $p = (p_i)_{i=1}^N$, $p_i > 0$, $\sum_1^N p_i = 1$, set

$$\mathbb{P} = \mathbb{P}_p = \bigtimes_{\mathbb{N}} (p_i)$$

to be the Kakutani infinite-product measure; see **[Kak43]**.

Assume that the following intersection is a singleton for all, $\omega \mapsto x(\omega)$

$$\omega = (i_1, i_2, i_3, \cdots) \in \Omega :$$

$$\{x(\omega)\} = \bigcap_{n=1}^{\infty} \bigcap_{\omega|_n = (i_1, i_2, \cdots, i_n)} \left(\tau_{i_n} \circ \tau_{i_{n-1}} \circ \cdots \circ \tau_{i_1}\right)(X), \tag{3.4}$$

and then set

$$A(\omega) := x(\omega) \,(= \text{the intersection point in } (3.4)),$$

so $\Omega \xrightarrow{A} X$, then we proved in **[JST20]** that the distribution

$$\nu_A := \mathbb{P} \circ A^{-1} \tag{3.5}$$

is an IFS measure. More precisely, the measure in (3.5) satisfies

$$\nu_A = \sum_{i=1}^{N} p_i \nu_A \circ \tau_i^{-1}. \tag{3.6}$$

As noted, the IFS measure in (3.6) includes the Sierpinski-measures discussed above.

While the Sierpinski constructions in Figure 3.2 are better known as self-affine planar sets, it is in fact the corresponding *measures* which are important for algorithms and for frame-harmonic analysis. As it turns out, the particular affine maps (see (3.9), (3.10), and Figure 3.4) going into the Sierpinski constructions are in fact special cases of a more general family of iterated function systems (IFS).

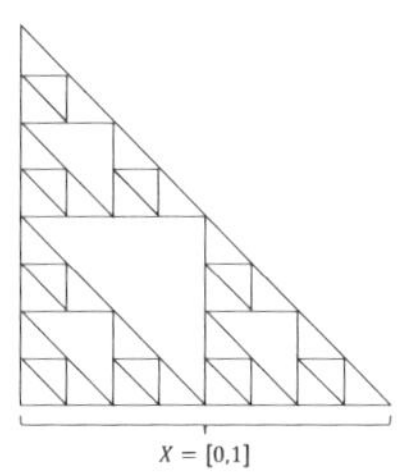

FIGURE 3.2. The Sierpinski gasket as a polymorphism.

For a fixed IFS, the set in question arises as the support of an associated IFS-measure ν. In particular, we show that these planar Sierpinski measures ν are *slice-singular.*

DEFINITION 3.1. A Borel measure ν on $I^2 := [0,1] \times [0,1]$ is called *slice singular* iff (Def.)

(1) $\xi = \nu \circ \pi_1^{-1}$ is singular, where π_1 is the projection onto the first coordinate; and
(2) for ξ-a.e. x, the measure $\sigma^x(\cdot)$ is singular, where σ^x is the conditional measure given by $d\nu = \int \sigma^x(dy)\, d\xi(x)$.

"Singular" is defined relative to Lebesgue measure.

THEOREM 3.2. *If ν is slice singular on I^2, then $\{e_n\}_{n\in\mathbb{N}_0^2}$ has dense span in $L^2(\nu)$, where $e_n(x)=e^{i2\pi(n_1x_1+n_2x_2)}$, for all $n=(n_1,n_2)\in\mathbb{N}_0^2$, and $x=(x_1,x_2)\in I^2$.*

PROOF. We begin with a general guide to the arguments making up the proof. We are faced with the problem of deriving conclusions in 2D, so the Hilbert space $L^2(I^2,\nu)$, from a system of 1D slices, i.e., measures derived from ν, but each with support in a copy of the 1D interval I. So the given slice-singular polymorphism measure ν is supported on I^2. We make use of the F.&M. Riesz theorem, not for ν in 2D, but for these slices, i.e., the slice-intervals. Since the F.&M. Riesz theorem applies in 1D, we may conclude that, for every singular measure γ, the 1D-Fourier frequencies e_m, $m\in\mathbb{N}_0$ span a dense subspace in $L^2(I,\gamma)$.

We shall show that, if $\langle F,e_n\rangle_{L^2(\nu)}=0$, $\forall n\in\mathbb{N}_0^2$, then $F=0$ ν-a.e. But

$$\begin{aligned}\langle F,e_n\rangle_{L^2(\nu)} &= \int_0^1 e_{n_1}(x)\left(\int_0^1 e_{n_2}(y)\overline{F(x,y)}\sigma^x(dy)\right)d\xi(x)\\ &= 0,\ \forall n=(n_1,n_2)\in\mathbb{N}_0^2\\ &\Downarrow\quad(\text{since }\xi\text{ is singular})\end{aligned}\tag{3.7}$$

$$\begin{aligned}&\int_0^1 e_{n_2}(y)\overline{F(x,y)}\sigma^x(dy)=0,\ a.a.\ x,\ \forall n_2\in\mathbb{N}_0\\ &\Downarrow\quad(\text{since }\sigma^x(\cdot)\text{ is singular a.a. }x)\\ &F(x,y)=0,\ a.a.\ (x,y)\text{ w.r.t. }\nu.\end{aligned}\tag{3.8}$$

This gives the desired conclusion that $\{e_n\}_{n\in\mathbb{N}_0^2}$ is total in $L^2(\nu)$. □

Given a probability measure ν on I^d where $I=[0,1]$, a key property that ν may, or may not, have is that the Fourier frequencies $\{e_n\}_{n\in\mathbb{N}^d}$ are *total* in $L^2(\nu)$, i.e., that the closed span of $\{e_n\}_{n\in\mathbb{N}^d}$ is $L^2(\nu)$.

The result in $d=1$, that, if ν on I is singular, then the set $\{e_n\}_{n\in\mathbb{N}_0}$ is total in $L^2(\nu)$, fails for $d=2$. There are examples when ν on I^2 is positive, singular w.r.t. the 2D Lebesgue measure, but $\{e_n\}_{n\in\mathbb{N}_0^2}$ is *not* total in $L^2(\nu)$.

EXAMPLE 3.3. Take $\nu=\lambda_1\times\mu$ (see Figure 3.3), where λ_1 is Lebesgue measure and μ is a singular measure in I, then $\{e_n\}_{n\in\mathbb{N}_0^2}$ is *not* total in $L^2(\nu)$.

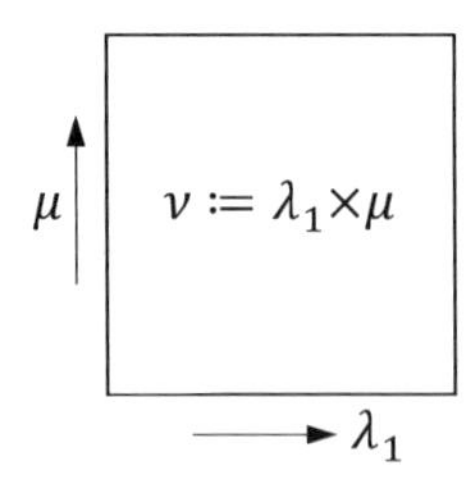

FIGURE 3.3. $\lambda_1=$ Lebesgue, $\mu\perp\lambda_1$

We shall show that the Sierpinski gasket SG (realized as the support of an affine IFS-measures ν) is slice singular, see Figure 3.7. So, for the Sierpinski polymorphism ν, we will conclude below that the 2D Fourier exponentials $\{e_n\}_{n\in\mathbb{N}_0^2}$ are

FIGURE 3.4. Construction of the Sierpinski gasket.

indeed total in $L^2(\mathrm{SG}, \nu)$. The proof makes use of Theorem 3.2, as well as the results below.

Let the alphabets for the Sierpinski gasket SG be

$$B = \{b_0, b_1, b_2\} := \left\{ \begin{bmatrix} 0 \\ 0 \end{bmatrix}, \begin{bmatrix} 1 \\ 0 \end{bmatrix}, \begin{bmatrix} 0 \\ 1 \end{bmatrix} \right\}. \tag{3.9}$$

Set

$$M = \begin{bmatrix} 2 & 0 \\ 0 & 2 \end{bmatrix}, \text{ and } \quad \tau_j(x) = M^{-1}(x + b_j). \tag{3.10}$$

The Sierpinski gasket (Figure 3.4) is the IFS attractor W satisfying

$$W = \bigcup_{j=0}^{2} \tau_j(W).$$

We have the random variable $B^{\mathbb{N}} \xrightarrow{X} W$, given by

$$\omega = (b_{i_1}, b_{i_2}, b_{i_3}, \cdots) \longmapsto x = \sum_{k=1}^{\infty} M^{-k} b_{i_k}. \tag{3.11}$$

As a Cantor set, W (the Sierpinski gasket) is the boundary of the tree symbol representation; see Figure 3.5.

Recall that every $\omega \in B^{\mathbb{N}}$ is an infinite word $\omega = (b_{i_1}, b_{i_2}, b_{i_3}, \cdots)$, with $i_k \in \{0, 1, 2\}$. Setting $\omega\big|_n = (b_{i_1}, \cdots, b_{i_n})$, a finite truncated word, and $\tau_{\omega|_n} = \tau_{i_n} \circ \cdots \circ \tau_{i_1}$; then $\bigcap_n \tau_{\omega|_n}(W) = \{x\}$, i.e., the intersection is a singleton. We set $X(\omega) = x$.

Let p be the probability distribution on B, where

$$p = \left(\frac{1}{3}, \frac{1}{3}, \frac{1}{3} \right). \tag{3.12}$$

Let $\mathbb{P} = \times_1^{\infty} p$, and $\nu = \mathbb{P} \circ X^{-1}$ be the corresponding IFS measure, i.e., ν is the unique Borel probability measure on W, s.t.

$$\nu = \frac{1}{3} \sum_{j=0}^{2} \nu \circ \tau_j^{-1}. \tag{3.13}$$

REMARK 3.4.

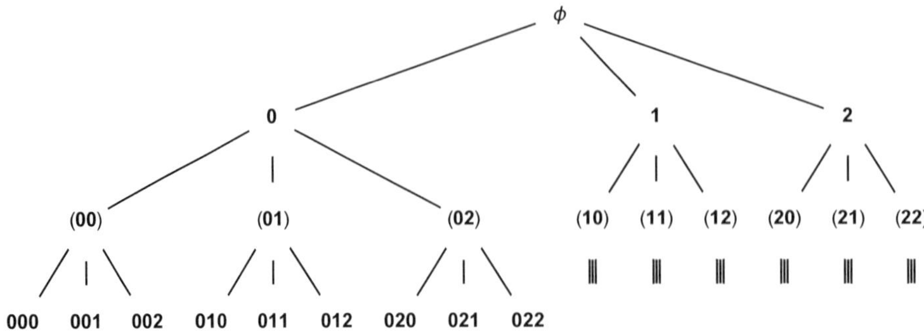

FIGURE 3.5. Symbol representations of infinite words.

(1) The Hausdorff dimension of W is $\ln 3/\ln 2$, where $3 = \#\{B\}$ and $2 =$ scaling number.
(2) Let O_j be the triangles removed from the jth iteration (Figure 3.4), and let $O = \bigcup_{j=1}^{\infty} O_j$. Then,

$$\lambda_2(O) - \lambda_2\left(\bigcup_{j=1}^{\infty} O_j\right) = \frac{1}{2}\left[\frac{1}{4} + \frac{3}{4^2} + \frac{3^2}{4^3} + \cdots\right] = \frac{1}{2};$$

and so $\lambda_2(W) = 0$, where λ_2 denotes the 2D Lebesgue measure. Note that $\nu(W) = 1$.

LEMMA 3.5. *Let W be the Sierpinski gasket, and ν be the corresponding IFS measure. Let $\widehat{\nu}$ be the Fourier transform of v, i.e., $\widehat{\nu}(\lambda) := \int_W e^{i2\pi\lambda\cdot x} d\nu(x)$. Then*

$$\widehat{\nu}(\lambda) = \frac{1}{3}\left[1 + e^{i\pi\lambda_1} + e^{i\pi\lambda_2}\right]\widehat{\nu}(\lambda/2), \tag{3.14}$$

where $\lambda = (\lambda_1, \lambda_2) \in \mathbb{R}^2$.

PROOF. Immediate from (3.13). More specifically, we have

$$\begin{aligned}
\widehat{\nu}(\lambda) &= \frac{1}{3}\sum_{j=0}^{2}\int_W e^{i2\pi\lambda\cdot\tau_j(x)} d\nu(x) \\
&= \frac{1}{3}\Big(\int_W e^{i2\pi\lambda\cdot x/2} d\nu(x) + \int_W e^{i2\pi\lambda/2\cdot(x+(1,0))} d\nu(x) + \int_W e^{i2\pi\lambda/2\cdot(x+(0,1))} d\nu(x)\Big) \\
&= \frac{1}{3}\left(1 + e^{i\pi\lambda_1} + e^{i\pi\lambda_2}\right)\widehat{\nu}(\lambda/2),
\end{aligned}$$

which is the assertion (3.14). □

By general theory, the IFS measure ν as in (3.13) has a disintegration

$$d\nu = \int_0^1 \sigma^x(dy)\, d\xi(x), \tag{3.15}$$

where

$$\xi = \nu \circ \pi_1^{-1} \tag{3.16}$$

with *supp* $(\xi) \subset [0,1]$. Note, if $S \subset [0,1]$ is a measurable subset, then

$$\xi(S) = \nu(\{(x,y) \mid x \in S\}). \tag{3.17}$$

LEMMA 3.6. *Let W be the Sierpinski gasket. Then points in W are represented as random power series*

$$\begin{bmatrix} x \\ y \end{bmatrix} \in W \Longleftrightarrow \begin{cases} x = \sum_{k=1}^{\infty} \varepsilon_k 2^{-k} \\ y = \sum_{k=1}^{\infty} \eta_k 2^{-k} \end{cases} \tag{3.18}$$

where $(\varepsilon_k), (\eta_k)$ are defined on $\Omega = \{0,1\}^{\mathbb{N}}$, i.e., the binary probability space.

Moreover, ε_k is i.i.d. on $\{0,1\}$, $k \in \mathbb{N}$, with distribution $(2/3, 1/3)$. That is, $Prob\,(\varepsilon_k = 0) = 2/3$, and $Prob\,(\varepsilon_k = 1) = 1/3$. The same conclusion holds for η_k as well.

PROOF. This follows from (3.11) and (3.12).

In detail, let $X : B^{\mathbb{N}} \to W$ be the random variable from (3.11), $X(\omega) = \sum_{k=1}^{\infty} M^{-k} b_{i_k}$, for all $\omega \in B^{\mathbb{N}}$; then

$$\begin{aligned} W \ni \begin{bmatrix} x \\ y \end{bmatrix} &= X(\omega) \\ &= \sum_{k=1}^{\infty} \begin{bmatrix} 2^{-k} & 0 \\ 0 & 2^{-k} \end{bmatrix} b_{i_k} \\ &= \sum_{k=1}^{\infty} \begin{bmatrix} 2^{-k} & 0 \\ 0 & 2^{-k} \end{bmatrix} \left\{ \begin{bmatrix} 0 \\ 0 \end{bmatrix}, \begin{bmatrix} 1 \\ 0 \end{bmatrix}, \begin{bmatrix} 0 \\ 1 \end{bmatrix} \right\} \\ &= \begin{bmatrix} \sum_{k=1}^{\infty} 2^{-k} \varepsilon_k(x) \\ \sum_{k=1}^{\infty} 2^{-k} \eta_k(x) \end{bmatrix}, \end{aligned}$$

where

$$\begin{aligned} Pr\,(\varepsilon_k = 0) &= Pr\,(\eta_k = 0) = 2/3, \\ Pr\,(\varepsilon_k = 1) &= Pr\,(\eta_k = 1) = 1/3. \end{aligned}$$

□

PROPOSITION 3.7 (See also Figures 3.6 and 3.7, and Remark 3.9). *We have the following conditional probabilities:*

$$\begin{aligned} Pr\,(\eta_k = 0 \mid \varepsilon_k = 0) &= 1/2, \\ Pr\,(\eta_k = 1 \mid \varepsilon_k = 0) &= 1/2, \\ Pr\,(\eta_k = 0 \mid \varepsilon_k = 1) &= 1. \end{aligned}$$

PROOF. One checks that

$$\begin{aligned} Pr\,(\eta_k = 0) &= Pr\,(\eta_k = 0 \mid \varepsilon_k = 0)\, Pr\,(\varepsilon_k = 0) \\ &\quad + Pr\,(\eta_k = 0 \mid \varepsilon_k = 1)\, Pr\,(\varepsilon_k = 1) = \frac{1}{2} \cdot \frac{2}{3} + 1 \cdot \frac{1}{3} = \frac{2}{3}, \end{aligned} \tag{3.19}$$

$$\begin{aligned} Pr\,(\eta_k = 1) &= Pr\,(\eta_k = 1 \mid \varepsilon_k = 0)\, Pr\,(\varepsilon_k = 0) \\ &\quad + Pr\,(\eta_k = 1 \mid \varepsilon_k = 1)\, Pr\,(\varepsilon_k = 1) = \frac{1}{2} \cdot \frac{2}{3} + 0 \cdot \frac{1}{3} = \frac{1}{3}. \end{aligned} \tag{3.20}$$

See the diagram in Figure 3.6. □

LEMMA 3.8. *Let ν be the IFS measure of the Sierpinski gasket as above, and $\xi = \nu \circ \pi_1{}^{1}$ be as in (3.15)–(3.17), so that ν has the disintegration in (3.15).*

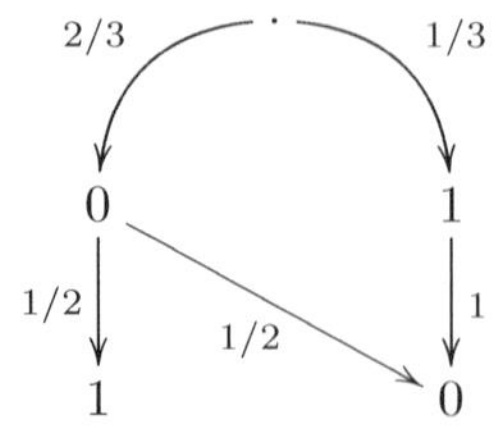

FIGURE 3.6. Transition probabilities

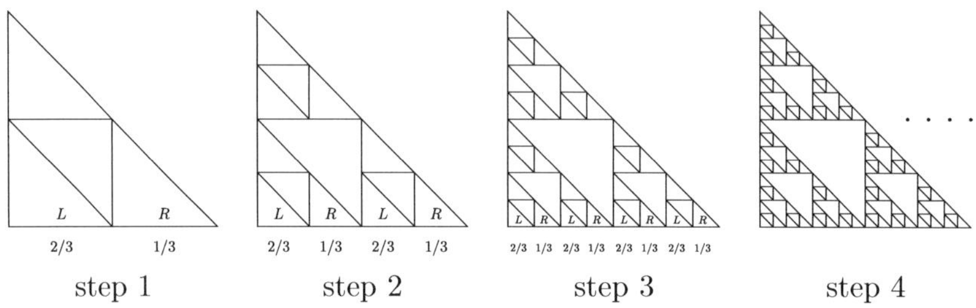

$$\xi = \nu \circ \pi_1^{-1} = \times_1^\infty \{2/3, 1/3\}$$
$$Pr(\varepsilon_k \in L) = 2/3, \quad Pr(\varepsilon_k \in R) = 1/3$$

FIGURE 3.7. The measure ξ, or $d\xi(x)$ as an infinite product measure.

(1) *Then the measure* ξ *is singular and nonatomic. More precisely,* ξ *is the product measure* $\times_1^\infty p$ *defined on* $\{0,1\}^{\mathbb{N}}$*, where* $p := \{2/3, 1/3\}$ *on* $\{0,1\}$*.*
(2) *For* ξ*-a.e.* x*, the measure* $\sigma^x(dy)$ *(in the* y*-variable) is singular. Hence* ν *is slice singular (see Definition 3.1), and* $\{e_n\}_{n\in\mathbb{N}_0^2}$ *is total in* $L^2(\nu)$*.*

PROOF. For all points $(x,y) \in W$, let $x = \sum_{k=1}^\infty \varepsilon_k 2^{-k}$, $y = \sum_{k=1}^\infty \eta_k 2^{-k}$ be as in (3.18). Then (3.19) & (3.20) hold for all $x \in I$.

Therefore, we get the product measure $\xi = \times_1^\infty \{2/3, 1/3\}$ on the space $\Omega = \times_1^\infty \{0,1\}$; see Figure 3.7. By contrast, $\lambda = \times_1^\infty \{1/2, 1/2\}$ is Lebesgue measure; hence ξ and λ are mutually singular by Kakutani's theorem.

Note that, for a.e. x, the measure $\sigma^x(dy)$ is the middle interval gap supported on $A(x) = \{y \mid (x,y) \in W\}$, and we conclude that $\sigma^x(dy)$ is singular w.r.t. Lebesgue measure for a.e. x. By Theorem 3.2, it follows that $\{e_n\}_{n\in\mathbb{N}_0^2}$ is total in $L^2(\nu)$. □

REMARK 3.9. There is a Markov chain associated with the transition probabilities (see Figure 3.6). Note that

$$\begin{bmatrix} 2/3 & 1/3 \end{bmatrix} \begin{bmatrix} 1/2 & 1/2 \\ 1 & 0 \end{bmatrix} = \begin{bmatrix} 2/3 & 1/3 \end{bmatrix},$$

so the conditional expectation can be expressed as a Perron–Frobenius problem with the row vector $\begin{bmatrix} 2/3 & 1/3 \end{bmatrix}$ as a left Perron–Frobenius vector.

As another example, consider the fractal Eiffel Tower W_{Ei} (see Figure 3.9). In this case, we have

$$M = \begin{bmatrix} 2 & 0 & 0 \\ 0 & 2 & 0 \\ 0 & 0 & 2 \end{bmatrix}, \quad B = \left\{ \begin{bmatrix} 0 \\ 0 \\ 0 \end{bmatrix}, \begin{bmatrix} 1 \\ 0 \\ 0 \end{bmatrix}, \begin{bmatrix} 0 \\ 1 \\ 0 \end{bmatrix}, \begin{bmatrix} 0 \\ 0 \\ 1 \end{bmatrix} \right\},$$

and $p = (1/4, 1/4, 1/4, 1/4)$. It follows that each coordinate of points in W_{Ei} has representation $\sum_{k=1}^{\infty} \varepsilon_k 2^{-k}$, where $\{\varepsilon_k\}$ is i.i.d. with $Pr\,(\varepsilon_k = 0) = 3/4$, and $Pr\,(\varepsilon_k = 1) = 1/4$. The transition probabilities are given by Figure 3.8. One checks

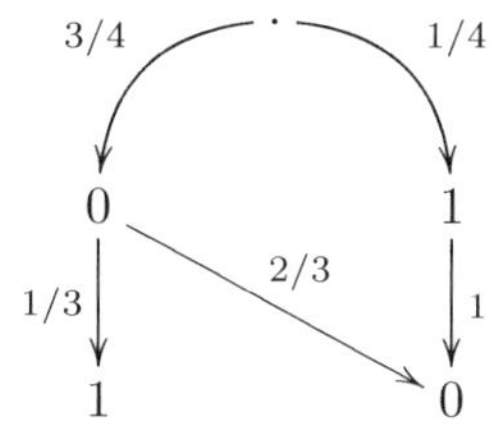

FIGURE 3.8. Transition probabilities

that

$$\begin{bmatrix} 3/4 & 1/4 \end{bmatrix} \begin{bmatrix} 2/3 & 1/3 \\ 1 & 0 \end{bmatrix} = \begin{bmatrix} 3/4 & 1/4 \end{bmatrix}.$$

CONJECTURE 3.10 (The role of Perron–Frobenius). *Given an affine contractive IFS measure ν supported in $[0,1]^d$, let $T = (T_{ij})$ be the corresponding Markov transition matrix. Then the following are equivalent:*

(1) *The Fourier frequencies $\{e_n\}_{n \in \mathbb{N}_0^d}$ are total in $L^2(\nu)$.*
(2) *The Perron–Frobenius vector v ($vT = v$, or $\sum_j v_j T_{ji} = v_i$) is non-constant, i.e., not proportional to $(1, 1, \cdots, 1)$.*

REMARK 3.11 (The Sierpinski carpet). In the above, we carried out all the detailed computation justifying our conclusions for the case of the Sierpinski gasket, Figure 3.10(A). Recall that Figure 3.10(B) represents the Sierpinski carpet, a close

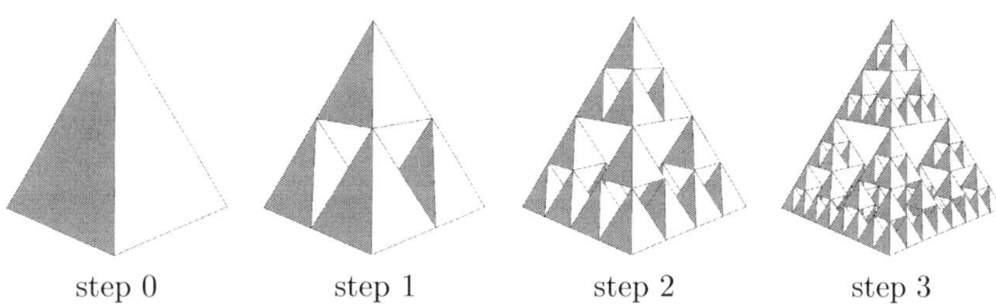

$$\xi = \nu \circ \pi_1^{-1} = \times_1^{\infty} \{3/4, 1/4\}$$
$$Pr\,(\varepsilon_k = 0) = 3/4, \quad Pr\,(\varepsilon_k = 1) = 1/4$$

FIGURE 3.9. Construction of the fractal Eiffel Tower.

cousin; and the reader will be able to fill in details from inside the section, spelling out the changes from (A) to (B). In case (B), naturally, the particular affine transformations (3.9)–(3.10) are a bit different (i.e., for case (B)), but they are of the same nature. In particular, it follows that the maximal entropy (IFS) measure for the Sierpinski carpet is also slice-singular. Moreover, the other conclusions from Lemmas 3.6 and 3.8, and Remark 3.9, carry over from case (A) to case (B), *mutatis mutandis*. As the underlying ideas and methods involved are the same, interested readers will be able to fill in details.

Moreover the above remarks, regarding extension of the conclusions for case (A) to that of (B), also apply *mutatis mutandis*, to the case of Figure 3.9, the fractal Eiffel Tower. There again, we conclude that the associated maximal entropy (IFS) measure is also slice-singular.

(A) Sierpinski gasket (B) Sierpinski carpet

FIGURE 3.10

4. The nonsymmetric case as polymorphisms

In a number of applications to mathematical models of machine learning, e.g., optimization involving statistical regression analysis, it is important to allow for a modification of the notion of polymorphism such as to allow for different measure spaces in the two factors. The present section is devoted to this analysis.

Let X, and Y be two spaces with respective σ-algebras $\mathscr{B}_X$ and $\mathscr{B}_Y$, and consider the product $Z := X \times Y$ with the associated product σ-algebra. Here we shall consider probability measures ρ on Z as follows: We shall assume that Y is contained in $\mathbb{R}$ with marginal measures

$$\mu_i := \rho \circ \pi_i^{-1}$$

and conditional measures

$$\rho\,(\cdot \mid \pi_1 = x)\,,\ \text{abbreviated}\ \rho\,(\cdot \mid x)\ \text{on}\ (Y, \mathscr{B}_Y)\,, \tag{4.1}$$

and

$$\rho\,(\cdot \mid \pi_2 = y)\,,\ \text{abbreviated}\ \rho\,(\cdot \mid y)\ \text{on}\ (X, \mathscr{B}_X)\,. \tag{4.2}$$

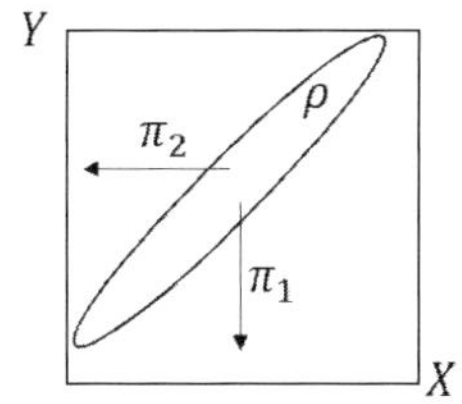

FIGURE 4.1. Nonsymmetric measures.

For measurable functions f on $X \to Y\,(\subseteq \mathbb{R})$, set

$$f_y := \int f(x)\,\rho(dx \mid y) = \mathbb{E}_\rho(f \mid y)\,. \tag{4.3}$$

It is clear from the above discussion that the nonsymmetric case also fits the general framework of polymorphism.

The application of the nonsymmetric measures ρ (i.e., on $X \times Y$) include learning theory, and in particular regression analysis.

THEOREM 4.1. *Let ρ on $Z = X \times Y$ be as above, in particular we assume that $\rho(Z) = 1$. We then obtain the following regression formula:*

$$\iint_Z |f(x) - y|^2\,\rho(dxdy) = \int_Y |y - \mathbb{E}_\rho(f \mid y)|^2\,\mu_2(dy)\,. \tag{4.4}$$

PROOF. Use the disintegration formulas from Section 2, we may carry out the integral in (4.4) as follows:

$$\begin{aligned}
&\iint_Z |f(x) - y|^2\,\rho(dxdy)\\
&= \int_X |f(x)|^2\,\mu_1(dx) - 2\int_Y \mathbb{E}_\rho(f \mid y)\,y\mu_2(dy) + \int_Y y^2\mu_2(dy)\\
&= \int_Y |y - \mathbb{E}_\rho(f \mid y)|^2\,\mu_2(dy) - \int_Y |\mathbb{E}_\rho(f \mid y)|^2\,\mu_2(dy) + \int_X |f(x)|^2\,\mu_1(dx)\\
&= \mathrm{RHS}_{(4.4)}
\end{aligned}$$

since

$$\int_Y |\mathbb{E}_\rho(f \mid y)|^2\,\mu_2(dy) = \int_X |f(x)|^2\,\mu_1(dx)\,.$$

□

5. Probability spaces: Polymorphisms in stochastic processes

The present section is devoted to a brief outline of the role played by polymorphisms in probability theory and statistical inference.

We shall adopt the standard definition of probability spaces $(\Omega, \mathscr{C}, \mathbb{P})$ where:

- Ω is a set representing sample points;
- $\mathscr{C}$ is a specified σ-algebra of subsets of Ω representing events (usually the sigma-algebra is generated from subsets of Ω, called cylinder sets, which are constructed from all choices of finite marginals);
- $\mathbb{P}$ is a probability measure defined on $\mathscr{C}$.

Let A, B be a pair of random variables from the probability space, see Figure 5.1.

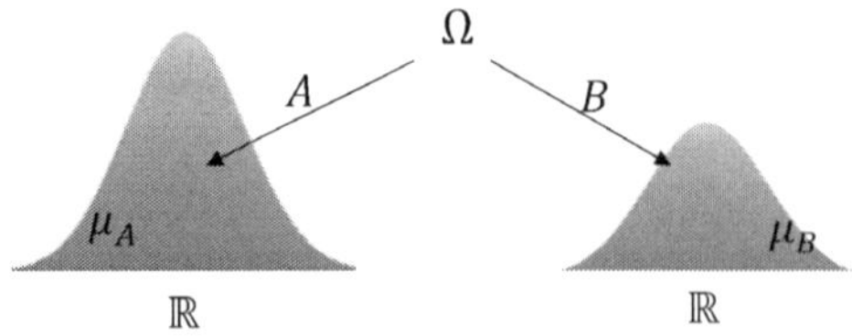

FIGURE 5.1. Pairs of random variables.

We introduce the joint distribution ν for the pair (A, B), and the marginal distributions μ_A and μ_B as follows.

For a Borel set $K \subset \mathbb{R}^2$,

$$\nu(K) := \mathbb{P}((A, B) \in K), \tag{5.1}$$

and for Borel subsets $J_i \subset \mathbb{R}$ $(i = 1, 2)$, Borel sets,

$$\mu_A(J_1) = \mathbb{P}(A \in J_1), \text{ and} \tag{5.2}$$

$$\mu_B(J_2) = \mathbb{P}(B \in J_2). \tag{5.3}$$

LEMMA 5.1. *Let $(\Omega, \mathscr{C}, \mathbb{P}, A, B)$ be as above. Then*

$$\mu_A \leftarrow \nu \rightarrow \mu_B \tag{5.4}$$

is a polymorphism, i.e.,

$$\left.\begin{aligned} \nu \circ \pi_1^{-1} &= \mu_A, \text{ and} \\ \nu \circ \pi_2^{-1} &= \mu_B. \end{aligned}\right\} \tag{5.5}$$

PROOF. For product sets $J_1 \times J_2$, where $J_i \subset \mathbb{R}$ is a Borel subset, we have

$$\nu(J_1 \times J_2) = \mathbb{P}(A \in J_1, B \in J_2) \tag{5.6}$$

and desired conclusion (5.5) follows! □

THEOREM 5.2. *For a pair of random variables (A, B) with joint distribution ν (as a Borel measure on $\mathbb{R}^2$), we introduce the polymorphism in (5.4). The corresponding conditional measures $\nu(\cdot \mid x)$ and $\nu(\cdot \mid y)$ from (2.5) are as follows: (defined via conditional probabilities.)*

$$\nu(J_2 \mid x) = \mathbb{P}(B \in J_2 \mid A = x), \text{ and} \tag{5.7}$$

$$\nu(J_1 \mid y) = \mathbb{P}(A \in J_1 \mid B = y). \tag{5.8}$$

PROOF. The conclusions follow from Lemma 5.1, and the discussion above. For the definition of conditional probabilities, we get:

$$\int_{\mathbb{R}} \mathbb{P}(B \in J_2 \mid A = x)\, \mu_A(dx) = \mathbb{P}(B \in J_2) = \mu_B(J_2), \text{ and}$$

$$\int_{J_1} \mathbb{P}(B \in J_2 \mid A = x)\, \mu_A(dx) = \mathbb{P}(A \in J_1, B \in J_2) = \nu(J_1 \times J_2),$$

and the desired conclusions (5.7)–(5.8) follow. □

REMARK 5.3. The system of isometries corresponding to (5.7)–(5.8) in Theorem 5.2 is as follows.

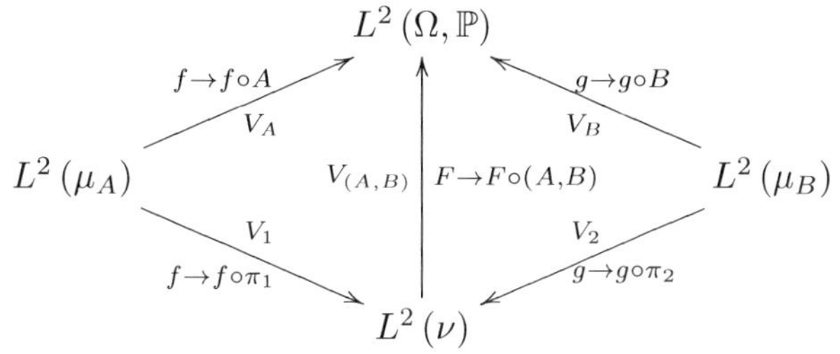

FIGURE 5.2. Isometries corresponding to pairs of random variables A and B.

With the operators V_1, V_2, V_A, V_B, and $V_{(A,B)}$ defined as in Figure 5.2, we have

$$\begin{cases} V_{(A,B)}V_1 &= V_A, \text{ and} \\ V_{(A,B)}V_2 &= V_B. \end{cases} \tag{5.9}$$

The conclusions stated in (5.9) follow from Theorem 5.2.

6. Generalized Brownian motion

We have already identified the special nature of those polymorphisms that arise from selfsimilar IFS-measures ν, specifically our focus was on the case when the system of self-similar maps are affine contractions, see Section 3. Now every choice of such an IFS polymorphism induces a generalized Brownian motion $W^{(\nu)}$ specified by the covariance (6.2). The purpose of the present section is to identify the IFS property for $W^{(\nu)}$ which is induced from a given IFS property for ν itself.

Fix an IFS measure ν subject to

$$\nu = \sum_{i=1}^{N} p_i \nu \circ \tau_i^{-1} \tag{6.1}$$

where $(p_i)_1^N$ and $(\tau_i)_1^N$ are as described above, and assume that the system $(\tau_i)_1^N$ is non-overlap.

Here, a system of maps $\{\tau_i\}$ in a measure space specified by a fixed measureμ is said to be a non-overlap system iff the intersection of the ranges of all pairs of maps τ_i and τ_j have zero μ measure for distinct indices i and j.

We now turn to the generalized Brownian motion $\{W_A^{(\nu)}\}$ indexed by $A \in \mathscr{B}_X$, determined by

$$\begin{aligned} &\mathbb{E}\left(W_A^{(\nu)}\right) = 0, \\ &\mathbb{E}\left(W_A^{(\nu)} W_B^{(\nu)}\right) = \nu(A \cap B), \ \forall A, B \in \mathscr{B}_X \text{ of finite } \nu\text{-measure.} \end{aligned} \tag{6.2}$$

REMARK. The covariance kernel $(A,B)\mapsto \nu(A\cap B)$ in (6.2) is indeed positive definite. This can be verified as follows: For all $N\in\mathbb{N}$, $(c_i)_1^N\subset\mathbb{C}$ and $(A_i)_1^N\subset\mathscr{B}_X$,

$$\sum_{i,j} c_i\overline{c_j}\nu(A_i\cap A_j)=\int_X\left|\sum c_i 1_{A_i}\right|^2 d\nu\geq 0,$$

where 1_E denotes the indicator function of $E\in\mathscr{B}_X$.

We first show how, in the general case, the polymorphism-decomposition for a given ν as in Lemma 2.3, applies to the corresponding generalized Brownian motion $W^{(\nu)}$ process from (6.2), yielding a direct Itô-integral decomposition for the $W^{(\nu)}$ Gaussian process.

PROPOSITION 6.1. *Let $\mu_1\leftarrow\nu\rightarrow\mu_2$ be a polymorphism (see Definition 2.1) and let $\nu(dxdy)=\mu_1(dx)\,\nu(dy\mid x)$ be the corresponding slice direct integral decomposition, see Lemma 2.3. For the corresponding generalized Brownian motion process, we then have the following Itô-integral decomposition, indexed by $A,B\in\mathscr{B}_X$:*

$$W_{A\times B}^{(\nu)}=\int_A W_B^{(\nu(\cdot|x))}dW_x^{(\mu_1)}. \tag{6.3}$$

PROOF. The result follows directly from the lemmas in Section 2 combined with the following fact (results from an application of Itô's isometry, see, e.g., [**Jor18, JT21, AJL17, AJ15**]):

$$\begin{aligned}\mathbb{E}\left(\left|\int_A W_B^{(\nu(\cdot|x))}dW_x^{(\mu_1)}\right|^2\right) &\underset{\text{by Itô}}{=}\int_A\mathbb{E}\left(\left|W_B^{(\nu(\cdot|x))}\right|^2\right)\mu_1(dx)\\ &=\int_A\nu(B\mid x)\,\mu_1(dx)\\ &\underset{\text{by (2.5)}}{=}\nu(A\times B).\end{aligned}$$

□

PROPOSITION 6.2 (The induced IFS property for the generalized Brownian motion of an IFS measure). *If ν satisfies (6.1), and $W^{(\nu)}$ is the generalized Brownian motion subject to (6.2), then*

$$W_A^{(\nu)}=\sum_{i=1}^N\sqrt{p_i}W_{\tau_i^{-1}(A)}^{(\nu)} \tag{6.4}$$

for all $A\in\mathscr{B}_X$ s.t. $\nu(A)<\infty$.

PROOF (SKETCH). The process $W^{(\nu)}$ is determined as a Gaussian process by $\mathbb{E}(W_A^{(\nu)})=0$ and (6.2), so we only need to verify that the RHS in (6.4) satisfies the conditions. We compute as follows:

$$\begin{aligned}&\mathbb{E}\left(\left(\sum_{i=1}^N\sqrt{p_i}W_{\tau_i^{-1}(A)}^{(\nu)}\right)\left(\sum_{j=1}^N\sqrt{p_j}W_{\tau_j^{-1}(B)}^{(\nu)}\right)\right)\\ &=\sum_i p_i\mathbb{E}\left(W_{\tau_i^{-1}(A)}^{(\nu)}W_{\tau_i^{-1}(B)}^{(\nu)}\right)\quad\text{by the nonoverlap assumption}\\ &=\sum_i p_i\nu\left(\tau_i^{-1}(A\cap B)\right)\quad\text{by (6.2)}\\ &=\nu(A\cap B)\end{aligned}$$

where we used (6.1) in the last step of the computation. □

Acknowledgments

We are pleased to acknowledge helpful discussions with many colleagues, members at the University of Iowa seminars in Analysis, Operator Theory, and in Mathematical physics, and discussions with collaborators, professors Daniel Alpay, Sergii Bezuglyi, Ilwoo Cho, Eric Weber, John Herr, and Myung-Sin Song, among others. We thank anonymous referees for constructive suggestions.

References

[AJ15] Daniel Alpay and Palle Jorgensen, *Spectral theory for Gaussian processes: reproducing kernels, boundaries, and L^2-wavelet generators with fractional scales*, Numer. Funct. Anal. Optim. **36** (2015), no. 10, 1239–1285, DOI 10.1080/01630563.2015.1062777. MR3402823

[AJL17] Daniel Alpay, Palle Jorgensen, and David Levanony, *On the equivalence of probability spaces*, J. Theoret. Probab. **30** (2017), no. 3, 813–841, DOI 10.1007/s10959-016-0667-7. MR3687240

[BHS08] Michael F. Barnsley, John E. Hutchinson, and Örjan Stenflo, *V-variable fractals: fractals with partial self similarity*, Adv. Math. **218** (2008), no. 6, 2051–2088, DOI 10.1016/j.aim.2008.04.011. MR2431670

[BJ21] Sergey Bezuglyi and Palle E. T. Jorgensen, *Symmetric measures, continuous networks, and dynamics*, New directions in function theory: from complex to hypercomplex to non-commutative, Oper. Theory Adv. Appl., vol. 286, Birkhäuser/Springer, Cham, [2021] ©2021, pp. 139–197, DOI 10.1007/978-3-030-76473-9_6. MR4368036

[BJ22] Sergey Bezuglyi and Palle E. T. Jorgensen, *Harmonic analysis on graphs via Bratteli diagrams and path-space measures*, Dissertationes Math. **574** (2022), 74. MR4388381

[BS70] James R. Brown and Ray C. Shiflett, *On extreme doubly stochastic measures*, Michigan Math. J. **17** (1970), 249–254. MR265542

[CHlQ+21] Shiping Cao, Malte S. Haßler, Hua Qiu, Ely Sandine, and Robert S. Strichartz, *Existence and uniqueness of diffusions on the Julia sets of Misiurewicz-Sierpinski maps*, Adv. Math. **389** (2021), Paper No. 107922, 41, DOI 10.1016/j.aim.2021.107922. MR4289048

[CV13] Joachim Cuntz and Anatoly Vershik, *C^*-algebras associated with endomorphisms and polymorphsims of compact abelian groups*, Comm. Math. Phys. **321** (2013), no. 1, 157–179. MR3089668

[Day73] Peter W. Day, *Decreasing rearrangements and doubly stochastic operators*, Trans. Amer. Math. Soc. **178** (1973), 383–392, DOI 10.2307/1996707. MR318962

[DJ07] Dorin Ervin Dutkay and Palle E. T. Jorgensen, *Fourier frequencies in affine iterated function systems*, J. Funct. Anal. **247** (2007), no. 1, 110–137, DOI 10.1016/j.jfa.2007.03.002. MR2319756

[GP22] François Golse and Thierry Paul, *Optimal transport pseudometrics for quantum and classical densities*, J. Funct. Anal. **282** (2022), no. 9, Paper No. 109417, 53, DOI 10.1016/j.jfa.2022.109417. MR4384100

[HJW19a] John E. Herr, Palle E. T. Jorgensen, and Eric S. Weber, *A characterization of boundary representations of positive matrices in the Hardy space via the Abel product*, Linear Algebra Appl. **576** (2019), 51–66, DOI 10.1016/j.laa.2018.02.023. MR3958137

[HJW19b] John E. Herr, Palle E. T. Jorgensen, and Eric S. Weber, *Positive matrices in the Hardy space with prescribed boundary representations via the Kaczmarz algorithm*, J. Anal. Math. **138** (2019), no. 1, 209–234, DOI 10.1007/s11854-019-0026-6. MR3996038

[Hut81] John E. Hutchinson, *Fractals and self-similarity*, Indiana Univ. Math. J. **30** (1981), no. 5, 713–747, DOI 10.1512/iumj.1981.30.30055. MR625600

[Hut95] John E. Hutchinson, *Fractals: a mathematical framework*, Complex. Int. **2** (1995), 14 HTML documents. MR1656855

[Isb55] J. R. Isbell, *Birkhoff's problem* 111, Proc. Amer. Math. Soc. **6** (1955), 217–218, DOI 10.2307/2032343. MR68507

[Jor18] Palle E. T. Jorgensen, *Harmonic analysis*, CBMS Regional Conference Series in Mathematics, vol. 128, American Mathematical Society, Providence, RI, 2018. Smooth and non-smooth; Published for the Conference Board of the Mathematical Sciences, DOI 10.1090/cbms/128. MR3838440

[JP95] Palle E. T. Jorgensen and Steen Pedersen, *Estimates on the spectrum of fractals arising from affine iterations*, Fractal geometry and stochastics (Finsterbergen, 1994), Progr. Probab., vol. 37, Birkhäuser, Basel, 1995, pp. 191–219, DOI 10.1007/978-3-0348-7755-8_10. MR1391977

[JP98] Palle E. T. Jorgensen and Steen Pedersen, *Dense analytic subspaces in fractal L^2-spaces*, J. Anal. Math. **75** (1998), 185–228, DOI 10.1007/BF02788699. MR1655831

[JP99] Palle E. T. Jorgensen and Steen Pedersen, *Spectral pairs in Cartesian coordinates*, J. Fourier Anal. Appl. **5** (1999), no. 4, 285–302, DOI 10.1007/BF01259371. MR1700084

[JS18] Palle E. T. Jorgensen and Myung-Sin Song, *Markov chains and generalized wavelet multiresolutions*, J. Anal. **26** (2018), no. 2, 259–283, DOI 10.1007/s41478-018-0139-9. MR3882025

[JST20] Palle Jorgensen, Myung-Sin Song, and James Tian, *A Kaczmarz algorithm for sequences of projections, infinite products, and applications to frames in IFS L^2 spaces*, Adv. Oper. Theory **5** (2020), no. 3, 1100–1131, DOI 10.1007/s43036-020-00079-1. MR4126821

[JT20] Palle E. T. Jorgensen and James F. Tian, *Superposition, reduction of multivariable problems, and approximation*, Anal. Appl. (Singap.) **18** (2020), no. 5, 771–801, DOI 10.1142/S021953051941001X. MR4131038

[JT21] Palle Jorgensen and James Tian, *Reproducing kernels: harmonic analysis and some of their applications*, Appl. Comput. Harmon. Anal. **52** (2021), 279–302, DOI 10.1016/j.acha.2020.03.001. MR4218424

[JT22a] Palle Jorgensen and James Tian, *Reproducing kernels and choices of associated feature spaces, in the form of L^2-spaces*, J. Math. Anal. Appl. **505** (2022), no. 2, Paper No. 125535, 31, DOI 10.1016/j.jmaa.2021.125535. MR4295177

[JT22b] Palle E. T. Jorgensen and James Tian, *Dual pairs of operators, harmonic analysis of singular nonatomic measures and Krein-Feller diffusion*, J. Operator Theory **89** (2023), no. 1, 205–248, DOI 10.1007/s43670-023-00051-z. MR4567343

[Kak43] Shizuo Kakutani, *Notes on infinite product measure spaces. II*, Proc. Imp. Acad. Tokyo **19** (1943), 184–188. MR14404

[Los82] V. Losert, *Counterexamples to some conjectures about doubly stochastic measures*, Pacific J. Math. **99** (1982), no. 2, 387–397. MR658068

[PB22] Matteo Pegoraro and Mario Beraha, *Projected statistical methods for distributional data on the real line with the Wasserstein metric*, J. Mach. Learn. Res. **23** (2022), Paper No. [37], 59. MR4420762

[Rot62] Gian-Carlo Rota, *An "Alternierende Verfahren" for general positive operators*, Bull. Amer. Math. Soc. **68** (1962), 95–102, DOI 10.1090/S0002-9904-1962-10737-X. MR133847

[RS16] Robert J. Ravier and Robert S. Strichartz, *Sampling theory with average values on the Sierpinski gasket*, Constr. Approx. **44** (2016), no. 2, 159–194, DOI 10.1007/s00365-016-9341-7. MR3543997

[Ryf68] John V. Ryff, *Majorized functions and measures*, Indag. Math. **30** (1968), 431–437. Nederl. Akad. Wetensch. Proc. Ser. A **71**. MR0234263

[SV08] Klaus Schmidt and Anatoly Vershik, *Algebraic polymorphisms*, Ergodic Theory Dynam. Systems **28** (2008), no. 2, 633–642, DOI 10.1017/S0143385707001022. MR2408396

[SV09] Klaus Schmidt and Anatoly Vershik, *Algebraic polymorphisms—erratum [MR2408396]*, Ergodic Theory Dynam. Systems **29** (2009), no. 4, 1369–1370, DOI 10.1017/S0143385709000352. MR2529654

[Ver05] A. M. Vershik, *Polymorphisms, Markov processes, and quasi-similarity*, Discrete Contin. Dyn. Syst. **13** (2005), no. 5, 1305–1324, DOI 10.3934/dcds.2005.13.1305. MR2166671

Department of Mathematics, The University of Iowa, Iowa City, Iowa 52242-1419, U.S.A.
Email address: palle-jorgensen@uiowa.edu

Mathematical Reviews, 416 4th Street Ann Arbor, Michigan 48103-4816, U.S.A.
Email address: jft@ams.org

Contemporary Mathematics
Volume **797**, 2024
https://doi.org/10.1090/conm/797/15934

Fractal interpolation over curves

Peter R. Massopust

ABSTRACT. This paper introduces the novel concept of fractal interpolation over curves in Banach spaces. The contents are based on the usual methodologies involving the fractal interpolation problem over intervals but the current approach considerably extends and generalizes them and paves the way to fractal interpolation over Banach submanifolds.

1. Introduction

Fractal interpolation techniques and methodologies have been considered extensively over the last decades in order to obtain approximation results for immensely complex and complicated geometries such as fractal sets or sets modelled as fractal sets. Common to all these techniques is the existence or the approximate assumption of an underlying self-referential structure of the (approximate) fractal set.

For many application, the (approximate) fractal set is the graph of a function defined on an appropriate subset of a, say, Banach space with values in another Banach space. In other words, one tries to solve the fractal interpolation problem: Given a bounded subset X of a Banach space E and a Banach space F, construct a *global* function

$$\psi : \mathsf{X} = \coprod_{i=1}^{n} \mathsf{X}_i \to \mathsf{F}$$

belonging to some prescribed function space $\mathscr{F} := \mathscr{F}(\mathsf{X}, \mathsf{F})$ satisfying n functional equations of the form

$$\psi(h_i(x)) = q_i(x) + s_i(x)\psi(x), \quad \text{on } \mathsf{X} \text{ and for } i \in \mathbb{N}_n.$$

The functions h_i are assumed to partition X (linearly or nonlinearly) into subsets $\mathsf{X}_i = h_i(\mathsf{X})$, and the functions q_i and s_i are appropriately chosen. We remark that, if such a global solution exists, then it is pieced together in a prescribed manner from copies of itself on the subsets X_i. This latter property then defines or expresses the self-referential nature of ψ and communicates the fact that the graph of ψ is in general an object of immense geometric complexity, i.e., a fractal set.

The aforementioned fractal interpolation problem has been investigated in numerous publications and for different function spaces $\mathscr{F}$ where the domains of $f \in \mathscr{F}$ were one-dimensional, n-dimensional, or infinite-dimensional. For specific

2020 *Mathematics Subject Classification.* Primary 28A80; Secondary 46B25, 46E15, 51F30.
Key words and phrases. Iterated function system (IFS), fractal interpolation, Read–Bajraktarević operator, fractal function, Banach algebra, Lipschitz algebra.

applications though, X is usually a compact or half-open bounded interval of the numerical Banach space $\mathbb{R}$ with values in F or even just $\mathbb{R}^m$, $m \in \mathbb{N}$.

However, in many situations especially in physics and particularly electrodynamics, complex (approximately) self-referential structures may arise over domains which are curves or more generally (Banach) submanifolds. In this paper, we initiate the investigation in this latter setting and introduce fractal interpolation where the domain X is the trace of a regular curve $\gamma : I \to \mathsf{E}$ in a Banach space. (For a proper choice of the interval I the trace is then a C^1 Banach submanifold of codimension one.) The setup presented here will then indicate the general approach to fractal interpolation over (Banach) submanifolds and will be left to a follow-up paper.

The contents of this paper are organized as follows. In Section 2, iterated function systems are defined and the notation and terminology for the remainder of the paper introduced. In Section 3, some remarks about the code space naturally associated with an IFS are presented. Section 4 reviews some concepts from the theory of curves in Banach spaces highlighting those results that are needed in this paper. In Section 5, the novel concept of fractal interpolation over curves in Banach spaces is introduced and an appropriate function space $\mathscr{F}$, namely a certain Lipschitz algebra, chosen. The main result, namely the existence of a unique solution to the fractal interpolation problem over curves, is proven in this section as well. In Section 6, a relation between the unique solution of the interpolation problem and the underlying IFS and its code space is exhibited. Section 7 lists some future research directions.

2. Iterated functions systems

Let $\mathsf{E} := (\mathsf{E}, \|\cdot\|_{\mathsf{E}})$ and $\mathsf{F} := (\mathsf{F}, \|\cdot\|_{\mathsf{F}})$ be Banach spaces (over $\mathbb{R}$ or $\mathbb{C}$). For a map $f : \mathsf{E} \to \mathsf{F}$, we define the Lipschitz constant associated with f by

$$L(f) = \sup_{x \in \mathsf{E}, x \neq y} \frac{\|f(x) - f(y)\|_{\mathsf{F}}}{\|x - y\|_{\mathsf{E}}}.$$

A map f is called *Lipschitz* or a *Lipschitz function* if $L(f) < +\infty$ and a *contraction* if $L(f) < 1$.

Definition 2.1. Let $(\mathsf{E}, \|\cdot\|_{\mathsf{E}}$ be a Banach space and $\mathcal{F}$ a finite set of functions $\mathsf{E} \to \mathsf{E}$. Then, the pair $(\mathsf{E}, \mathcal{F})$ is called an iterated function system (IFS) on E. In case all maps $f \in \mathcal{F}$ are contractions then $(\mathsf{E}, \mathcal{F})$ is called a *contractive* IFS.

For a more general definition of IFS and related topics, we refer the interested reader to [**BWL14**].

Remark 2.2. Throughout this paper, we deal with contractive IFSs and therefore we drop the adjective "contractive."

With the finite set of contractions $\mathcal{F}$ on E, one associates a set-valued operator, again denoted by $\mathcal{F}$, acting on the hyperspace $\mathscr{H}(\mathsf{E})$ of nonempty compact subsets of E:

$$\mathcal{F}(E) := \bigcup_{f \in \mathcal{F}} f(E), \qquad E \in \mathscr{H}(\mathsf{E}).$$

The hyperspace $\mathscr{H}(\mathsf{E})$ endowed with the Hausdorff-Pompeiu metric $d_{\mathscr{H}}$ defined by

$$d_{\mathscr{H}}(S_1, S_2) := \max\{d(S_1, S_2), d(S_2, S_1)\},$$

where $d(S_1, S_2) := \sup_{x \in S_1} d(x, S_2) := \sup_{x \in S_1} \inf_{y \in S_2} \|x - y\|_{\mathsf{E}}$, becomes a metric space.

It is a known fact that the completeness of E implies the completeness of $(\mathscr{H}(\mathsf{E}), d_{\mathscr{H}})$ as a metric space. Moreover, it can be shown that the set-valued operator $\mathcal{F}$ is contractive on the complete metric space $(\mathscr{H}(\mathsf{E}), d_{\mathscr{H}})$ with Lipschitz constant $L(\mathcal{F}) = \max\{L(f) : f \in \mathcal{F}\} < 1$ if all $f \in \mathcal{F}$ are contractions.

In this case and by the Banach fixed point theorem, $\mathcal{F}$ has a unique fixed point in $\mathscr{H}(\mathsf{E})$. This fixed point is called the *attractor* of or the *fractal (set)* generated by the IFS $(\mathsf{E}, \mathcal{F})$. The attractor or fractal F satisfies the self-referential equation

$$F = \mathcal{F}(F) = \bigcup_{f \in \mathcal{F}} f(F), \tag{2.1}$$

i.e., F is made up of a finite number of images of itself. Equation (2.1) reflects the fractal nature of F showing that it is as an object of immense geometric complexity.

The proof of the Banach fixed point theorem also shows that the fractal F can be obtained iteratively via the following procedure: Choose an arbitrary $F_0 \in \mathscr{H}(\mathsf{E})$ and define

$$F_n := \mathcal{F}(F_{n-1}), \qquad n \in \mathbb{N}. \tag{2.2}$$

Then $F = \lim_{n \to \infty} F_n$, where the limit is taken with respect to the Hausdorff–Pompeiu metric $d_{\mathscr{H}}$.

For more details about IFSs and fractals and their properties, we refer the interested reader to the large literature on these topics and list only two references [**B12**, **M16**] which are closely related to the present exhibition.

3. Some remarks about IFSs and their code space

In this short section, we give a brief overview of the important concept of code space associated with an IFS. As a reference, we offer [**B12**, **BBHV16**].

Let $(\mathsf{E}, \mathcal{F})$ be an IFS with $\#\mathcal{F} = n$. With an IFS one associates the code space $\mathbb{N}_n^\infty$, i.e., the compact space of all infinite sequences of the form $\boldsymbol{i} := (i_1 i_2 i_3 \ldots)$ whose elements are from $\mathbb{N}_n$. The set $\mathbb{N}_n^\infty$ is endowed with a metric δ defined by

$$\delta(\boldsymbol{i}, \boldsymbol{j}) := 2^{-k},$$

where k is the least integer such that $i_k \neq j_k$. This makes $(\mathbb{N}_n^\infty, \delta)$ into a compact metric space.

Define maps

$$s_i : \mathbb{N}_n^\infty \to \mathbb{N}_n^\infty,$$
$$s_i(\boldsymbol{j}) := i\,\boldsymbol{j}.$$

Then each s_i is a contraction with Lipschitz constant $L(s_i) = \frac{1}{2}$ and a homeomorphism onto its image. In other words, the pair $(\mathbb{N}_n^\infty, s)$ with $s := \{s_i : i \in \mathbb{N}_n\}$ is an IFS referred to as the code space IFS.

For an IFS $(\mathsf{E}, \mathcal{F})$ with maps $\mathcal{F} = \{f_1, \ldots, f_n\}$ and attractor F, define the code map

$$\pi : \mathbb{N}_n^\infty \to F,$$
$$\pi(\boldsymbol{i}) := \lim_{k \to \infty} f_{i_1} \circ f_{i_2} \circ \cdots \circ f_{i_k}(x),$$

for a fixed $x \in F$ and all $\boldsymbol{i} = i_1 i_2 \ldots \in \mathbb{N}_n^\infty$.

It is known that for contractive IFSs, $\pi(\boldsymbol{i})$ is one-elemental and independent of $x \in F$. Furthermore, π is continuous and surjective. If

$$S_i : \mathbb{N}_n^\infty \to \mathbb{N}_n^\infty,$$
$$S_i(i_1 i_2 i_3 \ldots) := (i_2 i_3 \ldots)$$

denotes the left shift operator on $\mathbb{N}_n^\infty$, then π is a conjugation between the code space IFS $(\mathbb{N}_n^\infty, s)$ and the IFS $(\mathsf{E}, \mathcal{F})$ in the sense that

$$\pi \circ s_i = f_i \circ \pi \quad \text{and} \quad (\pi \circ S)(\boldsymbol{i}) \in \mathcal{F}^{-1}(\pi(\boldsymbol{i})),$$

where $\mathcal{F}^{-1}(E) := \bigcup\limits_{i \in \mathbb{N}_n} f_i^{-1}(E)$, $E \in \mathscr{H}(\mathsf{E})$ and $\boldsymbol{i} \in \mathbb{N}_n^\infty$.

In order to introduce the concept of fractal transformation, we require one more item, namely, section of the code map π. A mapping $\xi : F \to \mathbb{N}_n^\infty$ is called a section of π if $\pi \circ \xi = \mathrm{id}_F$.

Now suppose, two IFSs $(\mathsf{E}, \mathcal{F}_1)$ and $(\mathsf{E}, \mathcal{F}_2)$ with $\#\mathcal{F}_1 = \#\mathcal{F}_2$ are given. Denote by F_1 and F_2, respectively, their attractors. The fractal transformations $\mathfrak{F}_{12} : F_1 \to F_2$ and $\mathfrak{F}_{21} : F_2 \to F_1$ are defined by

$$\mathfrak{F}_{12} := \pi_2 \circ \xi_1 \quad \text{and} \quad \mathfrak{F}_{21} := \pi_1 \circ \xi_2.$$

Here, π_i and ξ_i denotes the code map and section of the IFS $(\mathsf{E}, \mathcal{F}_i)$, $i = 1, 2$. Their respective actions are best seen by the commutative diagram below.

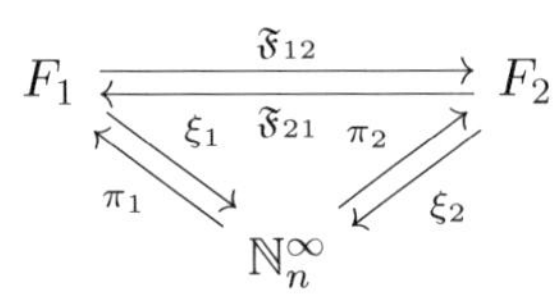

If $\mathfrak{F}_{12}$ is a homeomorphism then it is called a fractal homeomorphism and thus $\mathfrak{F}_{21} = (\mathfrak{F}_{12})^{-1}$.

4. Curves in Banach spaces

In this section, we briefly introduce the concept of a curve in a normed linear space, in particular, a Banach space.

DEFINITION 4.1. Let $(\mathsf{V}, \|\cdot\|_{\mathsf{V}})$ be a normed linear space over $\mathbb{R}$. Let $I \subset \mathbb{R}$ be an interval and $\gamma : I \to V$ a continuous mapping. Then the pair (I, γ) is called a curve in V. The family of all curves on a given Interval I is denoted by $\Gamma(I, \mathsf{V})$.

Note that if V is finite dimensional, i.e., isomorphic to $\mathbb{R}^m$, for some $m \in \mathbb{N}$, then a curve can be interpreted as a vector-valued function $\gamma : I \to \mathbb{R}^m$,

$$\gamma(t) = \begin{pmatrix} \gamma_1(t) \\ \vdots \\ \gamma_m(t) \end{pmatrix}, \tag{4.1}$$

with each $\gamma_i : I \to \mathbb{R}$.

When the interval I is understood, we simply write γ instead of (I, γ). We need to distinguish between the curve γ as a mapping $I \to \mathsf{E}$ and the trace of γ, i.e., the image set $|\gamma| := \gamma(I) \subset \mathsf{E}$.

DEFINITION 4.2. Let $(\mathsf{E}, \|\cdot\|_{\mathsf{E}})$ be a real Banach space and $I \subset \mathbb{R}$ an open interval. Suppose that $\gamma : I \to \mathsf{E}$ is a curve and $t_0 \in I$. The curve γ is called differentiable in t_0 if there exists a linear mapping $v : \mathbb{R} \to \mathsf{E}$ such that

$$\lim_{h\to 0} \frac{\|\gamma(t_0 + h) - \gamma(t_0) - v(h)\|_{\mathsf{E}}}{h} = 0.$$

We note that if a curve $\gamma : I \to \mathsf{E}$ is differentiable at a point t_0 then the derivative is uniquely determined.

The linear space

$$L(\mathbb{R}, \mathsf{E}) := \{v : \mathbb{R} \to \mathsf{E} : v \text{ is linear}\}$$

is isomorphic to E and, as every linear mapping $v : \mathbb{R} \to \mathsf{E}$ is continuous, $L(\mathbb{R}, \mathsf{E})$ is naturally a Banach space. The isomorphism between the Banach spaces E and $L(\mathbb{R}, \mathsf{E})$ allows us to identify the linear mapping $v : \mathbb{R} \to \mathsf{E}$ with the element $v(1) \in \mathsf{E}$. The vector $v(1)$ is called the derivative of the curve γ at t_0 and is denoted by $\gamma'(t_0)$.

DEFINITION 4.3. Let $(\mathsf{E}, \|\cdot\|_{\mathsf{E}})$ be a real Banach space and $I \subset \mathbb{R}$ an interval. Further, let $\gamma : I \to \mathsf{E}$ be a curve. γ is called continuously differentiable on I if

(1) γ is differentiable at every point $t \in \operatorname{int}(I)$, the interior of I;
(2) the mapping

$$I \to L(\mathbb{R}, \mathsf{E}), \quad t \mapsto \gamma'(t)$$

is continuous on $\operatorname{int}(I)$ and extends continuously to I. The family of all continuously differentiable curves $\gamma : I \to \mathsf{E}$ is denoted by $\Gamma^1(I, \mathsf{E})$.

A continuously differentiable curve $\gamma \in \Gamma^1(I, \mathsf{E})$ is called *regular* if $\gamma' \neq 0$ on I.

Now consider the case of a Banach algebra $\mathsf{A} := (\mathsf{A}, +, \cdot)$, i.e., when $(\mathsf{A}, +)$ is a vector space with a norm $\|\cdot\|$ and endowed with a product $\cdot : \mathsf{A} \times \mathsf{A}$ satisfying

(1) $(\mathsf{A}, +, \|\cdot\|)$ is a Banach space;
(2) $(\mathsf{A}, +, \cdot)$ is an associate $\mathbb{R}$-algebra;
(3) $\forall\, a_1, a_2 \in \mathsf{A} : \|a_1 \cdot a_2\| \leq \|a_1\|\|a_2\|$.

A Banach algebra A is called *unital* if it has a multiplicative identity.

EXAMPLE 4.4. For a unital Banach algebra A, we obtain for any $a \in \mathsf{A}$ a curve $\gamma : I \to \mathsf{A}$ defined by $t \mapsto \exp t \cdot a$, where exp denotes the exponential function in the Banach algebra A.

As a special case, we consider a finite dimensional Banach algebra A such as $\mathbb{K}^m$ with $\mathbb{K} \in \{\mathbb{R}, \mathbb{C}\}$, for some $m \in \mathbb{N}$, endowed with some norm. Then, $\operatorname{Hom}(\mathsf{A}, \mathsf{A})$, the set of self-maps $\mathsf{A} \to \mathsf{A}$ with multiplication $\cdot$ defined as composition of maps $\circ$, is a Banach algebra when equipped with the operator norm and for each $L \in \operatorname{Hom}(\mathsf{E}, \mathsf{E})$, we have curves $\gamma : I \to \operatorname{Hom}(\mathsf{E}, \mathsf{E})$, $t \mapsto \exp t \cdot L$ yielding curves $I \to \mathsf{E}$, $t \mapsto (\exp t \cdot L)(v)$, for each $v \in \mathsf{E}$.

For our later purposes, we require the following *generalized mean value theorem for curves.*

THEOREM 4.5. *Assume that the curve $\gamma : [a, b] \to \mathsf{E}$ is differentiable on (a, b) and $\{\gamma'(t) : t \in (a, b)\}$ is bounded. Define*

$$\|\gamma'\|_{\mathsf{E}} := \sup\{\|\gamma'(t) : t \in (a, b)\|\}.$$

Then,

$$\|\gamma(b) - \gamma(a)\|_{\mathsf{E}} \leq \|\gamma'\|_{\mathsf{E}}(b - a). \tag{4.2}$$

5. Fractal interpolation over curves

We briefly recall some basics from fractal interpolation theory and the theory of fractal functions.

To this end, let U and V be open subsets of Banach spaces E and F, respectively. A mapping $f : \mathsf{U} \to \mathsf{V}$ is called a *C^1-diffeomorphism* if f is a bijection, Fréchet differentiable on U, and the inverse of f, f^{-1}, is Fréchet differentiable on V. In a similar way, one defines higher order diffeomorphisms $\mathsf{U} \to \mathsf{V}$.

The collection of all C^α-diffeomorphisms from $\mathsf{U} \to \mathsf{V}$ with $\alpha \in \mathbb{N}_0 \cup \{\infty\}$ is denoted by $\mathrm{Diff}^\alpha(\mathsf{U}, \mathsf{V})$. In case, $\mathsf{U} := \mathsf{E} := \mathsf{F}$, we simply write $\mathrm{Diff}^\alpha(\mathsf{E})$.

If $\mathsf{X} \subset \mathsf{E}$ then $f : \mathsf{X} \to \mathsf{F}$ is called a C^α-diffeomorphism on X, in symbols $f \in \mathrm{Diff}^\alpha(\mathsf{X}, \mathsf{F})$, if there exists an open $\mathsf{U} \subset \mathsf{E}$ with $\mathsf{X} \subset \mathsf{U}$ and a C^α-diffeomorphism $g : \mathsf{U} \to \mathsf{F}$ such that $f = g|_{\mathsf{X}}$.

5.1. Fractal interpolation in general. Let X be a nonempty bounded subset of a Banach space E. Suppose we are given a finite family

$$h := \{h_i \in \mathrm{Diff}^\alpha(\mathsf{X}) : i = 1, \dots, n\}$$

of C^α-diffeomorphisms generating a partition $\Pi(h)$ of X in the sense that

$$\mathsf{X} = \coprod_{i=1}^{n} h_i(\mathsf{X}), \tag{5.1}$$

where $\coprod$ denotes the disjoint union of sets. For simplicity, we set $\mathsf{X}_i := h_i(\mathsf{X})$.

Recall that a mapping $f : \mathsf{E} \to \mathsf{F}$ is called *affine* if $f - f(0)$ is linear.

DEFINITION 5.1. A partition $\Pi(h)$ of X is called nonlinear if the maps generating $\Pi(h)$ are not affine; otherwise linear.

In case of a nonlinear partition, we may express (5.1) also as follows: The IFS (X, h) has as its attractor X which consists of finitely many disjoint *nonlinear* images of itself. Note that this extends and generalizes known methodologies. For an introduction to fractal interpolation over nonlinear partitions and related topics, we refer the interested reader to [**M22**].

In the following, we consider the partition functions h_i as either being linear or nonlinear. A more precise specification will be provided when necessary.

One of the goals of fractal interpolation is the construction of a *global* function

$$\psi : \mathsf{X} = \coprod_{i=1}^{n} \mathsf{X}_i \to \mathsf{F} \tag{5.2}$$

belonging to some prescribed function space $\mathscr{F}$ and satisfying n functional equations of the form

$$\psi(h_i(x)) = q_i(x) + s_i(x)\psi(x), \quad x \in \mathsf{X},\, i \in \mathbb{N}_n, \tag{5.3}$$

where for each $i \in \mathbb{N}_n$, $q_i \in \mathscr{F}$ and s_i is chosen such that $s_i \cdot \psi \in \mathscr{F}$. In other words, the global solution is pieced together in a prescribed manner from copies of itself on the subsets $\mathsf{X}_i = h_i(\mathsf{X})$ defined by the partition $\Pi(h)$.

We refer to (5.2) and (5.3) as the *fractal interpolation problem* and to ψ as a *fractal function.* The latter were first constructed in [**B86**] in the context of interpolating a given set of real-valued data in $\mathbb{R}^2$.

One possible way to solve the fractal interpolation problem is to consider (5.3) as the fixed point equation for an associated affine operator acting on an appropriately defined or prescribed function space.

5.2. Partitioning regular curves in Banach spaces. Let $\gamma \in \Gamma^1(I, \mathsf{E})$ be a regular curve where $I := [0,1] \subset \mathbb{R}$. The emphasis $I = [0,1]$ is only a mild restriction as any other half-open interval $[a,b]$ with $a < b$ can be obtained from I via the orientation-preserving diffeomorphism $t \mapsto (b-a)t + a$.

Note that as $\gamma \in \Gamma^1(I, \mathsf{E})$, $|\gamma| = \gamma(I)$ is compact in E and thus (totally) bounded. In particular, $\|\gamma'\|$ is bounded and Theorem 4.5 is applicable.

REMARK 5.2. Strictly speaking, a curve is an equivalence class of parametrized curves. The trace of a curve, however, is uniquely defined, as equivalently parametrized curves have the same trace. Moreover, the terms initial and terminal point of a curve are independent of its parametrization and so is its length. In what follows, we will always use the arc length parametrization of a curve

$$\sigma : I \to [0, \ell(\gamma)],$$
$$\sigma := \int_0^t \|\gamma'(\tau)\| d\tau,$$

where $\ell(\gamma)$ denotes the (finite) length of γ.

Suppose we generate a partition of $|\gamma| \in \mathscr{H}(\mathbb{R})$ by means of finitely many (prescribed) set-valued partition functions $H_i : \mathscr{H}(\mathbb{R}) \to \mathscr{H}(\mathbb{R})$ which are C^α-diffeomorphisms such that

$$H_i(|\gamma|) =: |\gamma_i|, \quad i \in \mathbb{N}_n, \tag{5.4}$$

$$H_1(\{\gamma(0)\}) = \{\gamma(0)\} =: p_0, \quad H_n(\{\gamma(1)\}) = \{\gamma(1)\} =: p_n, \tag{5.5}$$

(5.6)
$$H_i(\{\gamma(0)\}) =: p_i \in |\gamma_i|, \quad H_i(\{\gamma(1)\}) =: p_{i+1} = H_{i+1}(\{\gamma(0)\}) \in |\gamma_{i+1}|, \quad i \in \mathbb{N}_{n-1}.$$

Note that (5.6) means that the intersection between adjacent traces is one-elemental:

$$|\gamma_i| \cap |\gamma_{i+1}| = \{p_i\}, \quad i \in \mathbb{N}_{n-1}.$$

As $\gamma \in \Gamma^1(I, \mathsf{E})$ and regular, the mapping $t \mapsto \sigma = \int_0^t \|\gamma'(\tau)\| d\tau$ is a C^1-diffeomorphism. Hence, to every p_i, $i \in \mathbb{N}_{n-1}$, corresponds a unique $\sigma_i \in [0, \ell(\gamma)]$ and therefore a unique $t_i \in I$. The set $\{t_j : j \in \mathbb{N}_{n-1} \cup \{0\}\}$ defines a partition of I into subintervals $[t_j, t_{j+1}]$ with $|\gamma_{j+1}| = \gamma([t_j, t_{j+1}])$. Two adjacent subintervals of the above form have only the point $\{t_{j+1}\}$ in common thus allowing I to be written as the disjoint union

$$I = \coprod_{i=1}^{n-1} [t_{i-1}, t_i) \sqcup [t_{n-1}, t_n] =: \coprod_{i=1}^{n} I_i. \tag{5.7}$$

Conversely, we could partition the interval I via finitely many (linear or nonlinear) partition functions $h_i : I \to I_i$ into subintervals $I_i := [t_{i-1}, t_i]$ generating a partition of I of the form (5.7). As $|\gamma| = \gamma(I)$ and $|\gamma_i| = \gamma(I_i)$ we obtain

$$(\gamma \circ h_i)(I) = \gamma(h_i(I)) = \gamma(I_i) = |\gamma_i| = H_i(|\gamma|) = H_i(\gamma(I)) = (H_i \circ \gamma)(I),$$

and, thus,

$$\gamma \circ h_i = H_i \circ \gamma, \quad i \in \mathbb{N}_n. \tag{5.8}$$

Hence, for a regular $\gamma \in \Gamma^1(I, \mathsf{E})$,

$$|\gamma| = \bigcup_{i=1}^{n} |\gamma \circ h_i| =: \bigcup_{i=1}^{n} |\gamma_i|, \tag{5.9}$$

with $|\gamma_i| \cap |\gamma_{i+1}|$ being one-elemental. We can obtain a disjoint partition of $|\gamma|$ by using

$$|\gamma| = \coprod_{i=1}^{n-1} |\gamma_i| \setminus \{p_i\} \sqcup |\gamma_n|$$

Clearly, in each of the two cases, $|\gamma_i|$ is compact in E and thus bounded.

5.3. The fractal interpolation problem for curves. In this subsection, we pose and solve the fractal interpolation problem for regular curves in Banach spaces. More precisely, given the interval $I = [0, 1)$, a regular $\gamma \in \Gamma^1(I, \mathsf{E})$, and a partition $\Pi(h)$ of $|\gamma|$ of the form (5.9), we like to construct a unique function

$$\psi : |\gamma| = \coprod_{i=1}^{n} |\gamma_i| \to \mathsf{F} \tag{5.10}$$

belonging to some prescribed function space $\mathscr{F}$ and satisfying n functional equations of the form

$$\psi(\gamma_i) = q_i \circ \gamma + (s_i \circ \gamma) \cdot (\psi \circ \gamma), \quad i \in \mathbb{N}_n, \tag{5.11}$$

where for each $i \in \mathbb{N}_n$, $q_i \in \mathscr{F}$ and s_i is chosen such that $s_i \cdot \psi \in \mathscr{F}$, i.e., the global solution of (5.11) is pieced together in a prescribed manner from copies of itself on the subcurves $|\gamma_i|$ defined by the partition $\Pi(h)$. The unique solution of (5.11) will be termed a *fractal function on* γ.

Let (X, d) be a metric space and let $\mathrm{Lip}(\mathsf{X}, \mathsf{F})$ consist of all bounded Lipschitz functions $\mathsf{X} \to \mathsf{F}$ endowed with the norm

$$\|f\| := \|f\|_\infty + L(f). \tag{5.12}$$

Here, $\|\cdot\|_\infty$ denotes the supremum norm on continuous functions $\mathsf{X} \to \mathsf{F}$. Under this norm and the point-wise product, $\mathrm{Lip}(\mathsf{X}, \mathsf{F})$ becomes a unital commutative Banach algebra and is an example of a *Lipschitz algebra*. (Cf., for instance, [**J70**, **Sh63**, **W18**].)

In what follows, we regard $|\gamma|$ as a (compact) subset of the complete metric space (E, d) where d is the metric induced by the norm $\|\cdot\|_{\mathsf{E}}$. The natural function space $\mathscr{F}$ to consider here is then the Lipschitz algebra $\mathrm{Lip}(|\gamma|, \mathsf{A})$ with a Banach algebra A.

On the Lipschitz algebra $\mathrm{Lip}(|\gamma|, \mathsf{A})$, we define an affine operator T, called a Read–Bajractarević (RB) operator, as follows. Given any $f \in \mathrm{Lip}(|\gamma|, \mathsf{A})$, set

$$Tf(\gamma_i) := q_i(\gamma) + s_i(\gamma) \cdot f(\gamma), \quad i \in \mathbb{N}_n, \tag{5.13}$$

where $q_i, s_i \in \mathrm{Lip}(|\gamma|, \mathsf{A})$. Equivalently, we may express (5.13) in the form

$$T(f \circ \gamma) = \sum_{i=1}^{n} (q_i \circ \gamma) \circ h_i^{-1} \, \mathbb{1}_{I_i} + \sum_{i=1}^{n} (s_i \circ \gamma) \circ h_i^{-1} \cdot (f \circ \gamma) \circ h_i^{-1} \, \mathbb{1}_{I_i} \tag{5.14}$$

Moreover, we require that the following compatibility (join-up) conditions hold:

$$Tf(p_0) = f(p_0), \quad Tf(p_n) = f(p_n), \tag{5.15}$$

$$Tf(p_i-) = Tf(p_i+), \quad i \in \mathbb{N}_{n-1}. \tag{5.16}$$

Eqns. (5.15) and (5.16) can also be expressed in the form

$$f(p_0) = \frac{q_1(p_0)}{1 - s_1(p_0)}, \quad f(p_n) = \frac{q_n(p_n)}{1 - s_n(p_n)}, \tag{5.17}$$

$$q_i(p_n) + \frac{s_i(p_n) \cdot q_n(p_n)}{1 - s_n(p_n)} = q_{i+1}(p_0) + \frac{s_{i+1}(p_0) \cdot q_1(p_0)}{1 - s_1(p_0)}, \quad i \in \mathbb{N}_{n-1}. \tag{5.18}$$

Clearly, $Tf \in \text{Lip}(|\gamma|, \mathsf{A})$, i.e., Tf is bounded and continuous, and $\|Tf\| < \infty$.

THEOREM 5.3. *The RB operator defined in* (5.13) *and satisfying the compatibility conditions* (5.15) *and* (5.16) *is contractive on the Lipschitz algebra* $\text{Lip}(|\gamma|, \mathsf{A})$ *provided that*

$$L(T) := \max\left\{\max_{i\in\mathbb{N}_n} \|s_i\|_\infty, \max_{i\in\mathbb{N}_n} L(s_i) \cdot \|\gamma'\|_{\mathsf{E}}\right\} < 1. \tag{5.19}$$

PROOF. It suffices to consider the linear part $T - T(0)$ of T. For this purpose, we set $h := f - g$, for $f, g \in \text{Lip}(|\gamma|, \mathsf{A})$.

Now, let $t \in I$. Then, there exists an $i \in \mathbb{N}_n$ such that $t \in I_i$ and thus

$$\|Th(\gamma(t))\|_{\mathsf{A}} \leq \left\|s_i(\gamma \circ h_i^{-1}(t))\right\|_{\mathsf{A}} \cdot \left\|h(\gamma \circ h_i^{-1}(t))\right\|_{\mathsf{A}}.$$

Taking the supremum over t yields

$$\|Th\|_\infty \leq \max_{i\in\mathbb{N}_i} \|s_i\|_\infty \cdot \|h\|_\infty. \tag{5.20}$$

Furthermore, for $s, t \in I$,

$$\begin{aligned}\|Th(\gamma(t)) - Th(\gamma(s))\|_{\mathsf{A}} &\leq \|s_i(\gamma(t)) - s_i(\gamma(s))\|_{\mathsf{A}} \, \|h(\gamma(t)) - h(\gamma(s))\|_{\mathsf{A}} \\ &\leq L(s_i)\|\gamma(t) - \gamma(s)\|_{\mathsf{E}} \, L(h)\|\gamma(t) - \gamma(s)\|_{\mathsf{E}} \\ &\leq \max_{i\in\mathbb{N}_n} L(s_i) \, \|\gamma'\|_{\mathsf{E}} \, L(h) \, \|\gamma(t) - \gamma(s)\|_{\mathsf{E}},\end{aligned}$$

where we used Theorem 4.5 to obtain the last inequality. Hence,

$$L(Th) \leq \max_{i\in\mathbb{N}_n} L(s_i) \, \|\gamma'\|_{\mathsf{E}} \, L(h). \tag{5.21}$$

The inequalities (5.20) and (5.21) now yield the claim. □

Theorem 5.3 yields Corollary 5.4.

COROLLARY 5.4. *The RB operator defined in* (5.13) *and satisfying the compatibility conditions* (5.15) *and* (5.16) *has a unique fixed point* $\psi \in \text{Lip}(|\gamma|, \mathsf{A})$. *This fixed point obeys the self-referential equation*

$$\psi(\gamma_i) := q_i(\gamma) + s_i(\gamma) \cdot \psi(\gamma), \quad i \in \mathbb{N}_n,$$

and thus is the unique solution to the fractal interpolation problem (5.2) *and* (5.3) *in the present setting.*

Moreover, $\psi : |\gamma| \to \mathsf{A}$ *can be obtained by the iterative procedure*

$$\psi_0 \in \text{Lip}(|\gamma|, \mathsf{A}) \text{ chosen arbitrarily;}$$
$$\psi_k := T\psi_{k-1}, \; k \in \mathbb{N},$$

with

$$\psi = \lim_{k\to\infty} \psi_k,$$

where the limit is taken in the norm topology of $\mathrm{Lip}(|\gamma|, \mathsf{A})$.

Furthermore, one has the error estimate

$$\|\psi - \psi_k\| \leq \frac{L(T)^k}{1 - L(T)} \|T\psi_0 - \psi_0\|, \quad k \in \mathbb{N}.$$

PROOF. The statements are straightforward consequences of the Banach fixed point theorem and its constructive proof. □

DEFINITION 5.5. The unique fixed point ψ of the RB operator (5.13) is called an A-valued fractal function over the curve γ.

The next examples show the flexibility and versatility of the new concept introduced above.

EXAMPLE 5.6. In the case $\mathsf{E} := \mathbb{R}^m$, we can regard a curve $\gamma : I \to \mathbb{R}^m$ as the vector-valued function (4.1). Considering $\mathbb{R}^m$ also as a metric space $(\mathbb{R}^m, d)$ with the metric d given by

$$d(x, y) = \|x - y\| = \sqrt{\sum_{j=1}^{m} (x_j - y_j)^2}$$

and taking for A the commutative unital Banach algebra $\mathbb{R}^k$ with norm

$$\|x\|_{\mathsf{A}} := \max_{l\in\mathbb{N}_k} |x_l|, \quad x = (x_1, \ldots, x_k)^\top,$$

where $^\top$ denotes the transpose, and multiplication defined by

$$x \cdot y := (x_1 y_1, \ldots, x_k y_k)^\top,$$

the elements of the Lipschitz algebra $\mathrm{Lip}(|\gamma|, \mathbb{R}^k)$ are then vector-valued functions of m variables of the form

$$f : |\gamma| \subset \mathbb{R}^m \to \mathbb{R}^k,$$

$$f(\gamma(t)) = \begin{pmatrix} f_1(\gamma_1(t), \ldots, \gamma_m(t)) \\ \vdots \\ f_k(\gamma_1(t), \ldots, \gamma_m(t)) \end{pmatrix},$$

where $f_l : \mathbb{R}^m \to \mathbb{R}$, $l \in \mathbb{N}_k$.

EXAMPLE 5.7. Here we take again $\mathsf{E} := \mathbb{R}^m$ but for A we choose $\mathbb{R}^{k\times k}$, the collection of all $k \times k$-matrices with entries in $\mathbb{R}$ or $\mathbb{C}$. As a norm on $\mathbb{R}^{k\times k}$ we select a submultiplicative matrix norm such as, for instance, a Schatten p-norm,

$$\|A\|_p := \left(\sum_{l=1}^{k} \sigma_l^p(A)\right)^{1/p}, \quad A \in \mathbb{R}^{k\times k}, \ p \in [1, \infty],$$

where the σ_l denote the singular values of A. Multiplication is the usual multiplication of matrices. Then $\mathbb{R}^{k\times k}$ becomes a unital noncommutative Banach algebra.

In this setting, the elements of the Lipschitz algebra $\mathrm{Lip}(|\gamma|, \mathbb{R}^{k\times k})$ are matrix-valued functions of the form

$$f : |\gamma| \subset \mathbb{R}^m \to \mathbb{R}^{k\times k},$$
$$f(\gamma(t)) = \begin{pmatrix} f_{11}(\gamma(t)) & \cdots & f_{1k}(\gamma(t)) \\ \vdots & \ddots & \vdots \\ f_{k1}(\gamma(t)) & \cdots & f_{kk}(\gamma(t)) \end{pmatrix},$$

with $f_{ij} : \mathbb{R}^m \to \mathbb{R}$, $i, j \in \mathbb{N}_k$.

6. Relation to iterated function systems

In this section, the relation between the graph $G(\psi)$ of the fixed point ψ of the operator T and the attractor of the associated contractive IFS is described.

To this end, note that there are two IFSs whose attractors are the interval I and the trace $|\gamma|$ of a curve $\gamma : I \to \mathsf{E}$, respectively. The former is given by (I, h) with $h := \{h_i : i \in \mathbb{N}_n\}$ and the latter by $(|\gamma|, H)$ where $H := \{H_i : i \in \mathbb{N}_n\}$. Both IFSs have the same code space, namely, $\mathbb{N}_n^\infty$, and as (5.8) holds, we see that the fractal transformation $\mathfrak{F} : I \to |\gamma|$ is indeed a homeomorphism. In other words, for the purposes of this section, it does not matter which IFS we consider for the first variable in $G(\psi)$. In order to keep the presentation simple, we choose the IFS $(|\gamma|, H)$ and leave the other case to the interested reader.

Next, consider the product Banach space $\mathsf{E} \times \mathsf{A}$ endowed with the product norm $\|\cdot\|_\mathsf{E} + \|\cdot\|_\mathsf{A}$. For $|\gamma| \in \mathscr{H}(\mathsf{E})$ and $\mathsf{Y} \in \mathscr{H}(\mathsf{A})$, let mappings $w_i : |\gamma| \times \mathsf{Y} \to |\gamma| \times \mathsf{A}$ be defined by

$$w_i(x, y) := (H_i(x), q_i(x) + s_i(x) \cdot y), \quad i \in \mathbb{N}_n, \tag{6.1}$$

and required to map into $|\gamma| \times \mathsf{Y}$. (By [**M16**, Proposition 27], this is always possible.) Here, q_i and s_i are defined as in Section 5.

Note that the H_i are defined on the compact subset $|\gamma|$ of E and that $H_i(|\gamma|) \cap H_j(|\gamma|) \neq \emptyset$ but one-elemental for $i, j \in \mathbb{N}_n$ with $i \neq j$. However, we insist that H is *non-overlapping* with respect to the IFS $(|\gamma|, H)$ in the sense of [**BBHV16**, Definition 2.4] and that at the contact points $\{H_i(|\gamma|) \cap H_j(|\gamma|)\}$ the compatibility conditions (5.15) and (5.16) hold. This will guarantee the existence of a solution ψ (cf. also [**SB17**]).

To facilitate notation, introduce the mappings

$$v_i : \mathsf{E} \times \mathsf{Y} \to \mathsf{A},$$
$$v_i := q_i + s_i \cdot y, \quad i \in \mathbb{N}_n.$$

We remark that for a fixed $x \in |\gamma|$, arbitrary $y_1, y_2 \in \mathsf{Y}$, and $i \in \mathbb{N}_n$,

$$\begin{aligned} \|v_i(x, y_1) - v_i(x, y_2)\|_\mathsf{A} &= \|s_i(x) y_1 - s_i(x) y_2\|_\mathsf{A} \\ &\leq \|s_i\|_\mathsf{A} \|y_1 - y_2\|_\mathsf{A} \\ &\leq L(T) \|y_1 - y_2\|_\mathsf{A}, \end{aligned}$$

i.e., v_i is uniformly contractive in the second variable.

Moreover, for a fixed $y \in \mathsf{Y}$, arbitrary $x_1, x_2 \in |\gamma|$, and $i \in \mathbb{N}_n$, v_i is also uniformly Lipschitz continuous in the first variable:

$$\begin{aligned} \|v_i(x_1, y) - v_i(x_2, y)\|_\mathsf{A} &\leq \|q_i(x_1) - q_i(x_2)\|_\mathsf{A} + \|y\|_\mathsf{A} \|s_i(x_1) - s_i(x_2)\|_\mathsf{A} \\ &\leq (L(q_i) + \|y\|_\mathsf{A} L(s_i)) \|x_1 - x_2\|_\mathsf{A}. \end{aligned}$$

As $y \in \mathsf{Y} \in \mathscr{H}(\mathsf{A})$, there exists a nonnegative constant c_Y such that $\|y\|_\mathsf{A} \le c_\mathsf{Y}$. Hence,

$$\begin{aligned}\|v_i(x_1,y) - v_i(x_2,y)\|_\mathsf{A} &\le (L(q_i) + c_\mathsf{Y} L(s_i))\,\|x_1 - x_2\|_\mathsf{A}\\ &=: \lambda\|x_1 - x_2\|_\mathsf{A},\end{aligned}$$

proving the claim.

Let $\overline{L}(H) := \max\{L(H_i) : i \in \mathbb{N}_n\}$ and let $\vartheta := \frac{1-\overline{L}(H)}{2\lambda}$. Then the mapping $\|\cdot\|_\vartheta : \mathsf{E} \times \mathsf{A} \to \mathbb{R}_0^+$ given by

$$\|\cdot\|_\vartheta := \|\cdot\|_\mathsf{E} + \vartheta\,\|\cdot\|_\mathsf{F}$$

is a norm on $\mathsf{E} \times \mathsf{A}$ compatible with the product topology on $\mathsf{E} \times \mathsf{A}$.

The next theorem can also be found in [**BHM14**] in a more general setting. Here, we adopt it for our purposes.

THEOREM 6.1. *Set* $\mathcal{W} := \{w_1, \ldots, w_n\}$. *Then,* $\mathcal{F} := (|\gamma| \times \mathsf{Y}, \mathcal{W})$ *is a contractive IFS with respect to the norm* $\|\cdot\|_\vartheta$ *and the graph* $G(\psi)$ *of the solution to the fractal interpolation problem*

$$\psi(\gamma_i) = q_i(\gamma) + s_i(\gamma), \quad i \in \mathbb{N}_n, \tag{6.2}$$

is the unique attractor of the IFS $\mathcal{F}$.

Furthermore, if T *is the RB operator*

$$\begin{aligned}&T : \mathrm{Lip}(|\gamma|, \mathsf{A}) \to \mathrm{Lip}(|\gamma|, \mathsf{A}),\\ &f \mapsto q_i \circ \gamma \circ h_i^{-1} + (s_i \circ \gamma \circ h_i^{-1}) \cdot (f \circ \gamma \circ h_i^{-1}), \quad \text{on } \mathsf{X}_i,\ i \in \mathbb{N}_n,\end{aligned} \tag{6.3}$$

associated with the fractal interpolation problem (6.2) *then*

$$G(Tf) = \mathcal{F}(G(f)), \quad \forall f \in \mathrm{Lip}(|\gamma|, \mathsf{A}), \tag{6.4}$$

where $\mathcal{F}$ *denotes the set-valued operator* (2.1).

PROOF. See [**BHM14**, Theorem 3.7] adapted to the current setting. □

The above statements are paramount to the commutativity of diagram (6.5)

$$\begin{array}{ccc} |\gamma| \times \mathsf{A} & \xrightarrow{\ \mathcal{F}\ } & |\gamma| \times \mathsf{A}\\ \uparrow G & & \uparrow G\\ \mathrm{Lip}(|\gamma|, \mathsf{A}) & \xrightarrow{\ T\ } & \mathrm{Lip}(|\gamma|, \mathsf{A}) \end{array} \tag{6.5}$$

where G is the mapping $\mathrm{Lip}(|\gamma|, \mathsf{A}) \ni g \mapsto G(g) = \{(x, g(x)) : x \in |\gamma|\} \in \mathsf{E} \times \mathsf{A}$.

On the other hand, if one assumes that $\mathcal{F} = (|\gamma| \times \mathsf{A}, w_1, w_2, \ldots, w_n)$ is an IFS whose mappings w_i are of the form (6.1) where the functions H_i are linear or nonlinear partition functions $H_i : |\gamma| \to |\gamma_i|$ with nonoverlapping attractor $|\gamma|$ and contact points $\{H_i(|\gamma|) \cap H_j(|\gamma|)\}$. Further assume that the mappings v_i are uniformly Lipschitz continuous in the first variable and uniformly contractive in the second variable.

Then we can associate with the IFS $\mathcal{F}$ an RB operator T of the form (6.1) and thus a fractal interpolation problem $H_i(|\gamma|) \cap H_j(|\gamma|) \ne \emptyset$ with appropriate compatibility conditions. The attractor A of $\mathcal{F}$ is then the graph $G(\psi)$ of the solution ψ of (6.2), respectively, the fixed point of T.

7. Future research directions

The above approach may be generalized along different directions.

- Instead of considering curves in Banach spaces, one may look in general at Banach or even Fréchet submanifolds.
- The Lipschitz algebras $\mathrm{Lip}(X, \mathsf{F})$ could be replaced by the space of bounded real-valued Lipschitz functions on X which vanish at a given base point $x_0 \in X$. The corresponding Lipschitz algebras are then denoted by $\mathrm{Lip}_0(X, \mathsf{F})$.

We leave it to the interested reader to pursue these research directions.

Acknowledgment

The author would like to thank the anonymous referees for the careful reading of the manuscript and for their suggestions.

References

[B12] Michael Barnsley, *Fractals everywhere*, Academic Press, Inc., Boston, MA, 1988. MR977274

[B86] Michael F. Barnsley, *Fractal functions and interpolation*, Constr. Approx. **2** (1986), no. 4, 303–329, DOI 10.1007/BF01893434. MR892158

[BBHV16] Christoph Bandt, Michael Barnsley, Markus Hegland, and Andrew Vince, *Old wine in fractal bottles I: Orthogonal expansions on self-referential spaces via fractal transformations*, Chaos Solitons Fractals **91** (2016), 478–489, DOI 10.1016/j.chaos.2016.07.007. MR3551732

[BHM14] Michael F. Barnsley, Markus Hegland, and Peter Massopust, *Numerics and fractals*, Bull. Inst. Math. Acad. Sin. (N.S.) **9** (2014), no. 3, 389–430. MR3288909

[BHM16] M. F. Barnsley, M. Hegland, and P. R. Massopust, *Self-referential functions*, Preprint, arXiv:1610.01369, (2016).

[BWL14] Michael F. Barnsley, David C. Wilson, and Krzysztof Leśniak, *Some recent progress concerning topology of fractals*, Recent progress in general topology. III, Atlantis Press, Paris, 2014, pp. 69–92, DOI 10.2991/978-94-6239-024-9_2. MR3205482

[J70] J. A. Johnson, *Banach spaces of Lipschitz functions and vector-valued Lipschitz functions*, Trans. Amer. Math. Soc. **148** (1970), 147–169, DOI 10.2307/1995044. MR415289

[M16] Peter R. Massopust, *Fractal functions, fractal surfaces, and wavelets*, 2nd ed., Elsevier/Academic Press, London, 2016. MR3642696

[M22] Peter R. Massopust, *Fractal interpolation over nonlinear partitions*, Chaos Solitons Fractals **162** (2022), Paper No. 112503, 8, DOI 10.1016/j.chaos.2022.112503. MR4467626

[N05] M. A. Navascués, *Fractal polynomial interpolation*, Z. Anal. Anwendungen **24** (2005), no. 2, 401–418, DOI 10.4171/ZAA/1248. MR2174271

[SB17] Cristina Serpa and Jorge Buescu, *Constructive solutions for systems of iterative functional equations*, Constr. Approx. **45** (2017), no. 2, 273–299, DOI 10.1007/s00365-016-9349-z. MR3619444

[Sh63] Donald R. Sherbert, *Banach algebras of Lipschitz functions*, Pacific J. Math. **13** (1963), 1387–1399. MR156214

[W18] Nik Weaver, *Lipschitz algebras*, World Scientific Publishing Co. Pte. Ltd., Hackensack, NJ, 2018. Second edition of [MR1832645]. MR3792558

Department of Mathematics, Technical University of Munich, Boltzmannstrasse 3, 85747 Garching b. München, Germany

Email address: massopust@ma.tum.de

Contemporary Mathematics
Volume **797**, 2024
https://doi.org/10.1090/conm/797/15935

Conditional proof of the ergodic conjecture for falling ball systems

Nandor Simanyi

ABSTRACT. In this paper we present a conditional proof of Wojtkowski's ergodicity conjecture for the system of $1D$ perfectly elastic balls falling down in a half line under constant gravitational acceleration, [Ergodic Theory Dynam. Systems **5** (1985), no. 1, 145–161], [Comm. Math. Phys. **105** (1986), no. 3, 391–414], [Comm. Math. Phys. **126** (1990), no. 3, 507–533], [Comm. Math. Phys. **127** (1990), no. 2, 425–432], [Fund. Math. **157** (1998), no. 2-3, 305–341]. Namely, we prove that almost every such system is (completely hyperbolic and) ergodic, by assuming the transversality between different singularities and between singularities and stable (unstable) invariant manifolds. This transversality condition will be removed in the author's upcoming paper Simanyi, Preprint, (2023).

1. Introduction/Prerequisites

In his paper [**W90a**], Wojtkowski introduced the following Hamiltonian dynamical system with discontinuities: There is a vertical half line $\{q|\, q \geq 0\}$ given and n (≥ 2) point particles with masses $m_1 \geq m_2 \geq \cdots \geq m_n > 0$ and positions $0 \leq q_1 \leq q_2 \leq \cdots \leq q_n$ are moving on this half line so that they are subjected to a constant gravitational acceleration $a = -1$ (they fall down), they collide elastically with each other, and the first (lowest) particle also collides elastically with the hard floor $q = 0$. We fix the total energy

$$H = \sum_{i=1}^{n} \left(m_i q_i + \frac{1}{2m_i} p_i^2 \right)$$

by taking $H = 1$. The arising Hamiltonian flow with collisions $(\mathbf{M}, \{\psi^t\}, \mu)$ (μ is the Liouville measure) is the studied model of this paper.

Before formulating the result of this article, however, it is worth mentioning here three important facts:

(1) Since the phase space $\mathbf{M}$ is compact, the Liouville measure μ is finite.
(2) The phase points $x \in \mathbf{M}$ for which the trajectory $\{\psi^t(x)|, t \in \mathbb{R}\}$ hits at least one singularity (i.e., a multiple collision) are contained in a countable union of proper, smooth submanifolds of $\mathbf{M}$ and, therefore, such points form a set of μ measure zero.

2020 *Mathematics Subject Classification.* Primary 37D05.

(3) For μ-almost every phase point $x \in \mathbf{M}$ the collision times of the trajectory $\{\psi^t(x)|, t \in \mathbb{R}\}$ do not have any finite accumulation point, see [**S96**, Proposition A.1].

In [**W90a**], Wojtkowski formulated his main conjecture pertaining to the dynamical system $(\mathbf{M}, \{\psi^t\}, \mu)$:

CONJECTURE 1.1 (Wojtkowski's conjecture). *If $m_1 \geq m_2 \geq \cdots \geq m_n > 0$ and $m_1 \neq m_n$, then all but one characteristic (Lyapunov) exponents of the flow $(\mathbf{M}, \{\psi^t\}, \mu)$ are nonzero. Futhermore, the system is ergodic.*

REMARK 1.1.

(1) The only exceptional exponent zero must correspond to the flow direction.

(2) The condition of nonincreasing masses (as above) is essential for establishing the invariance of the symplectic cone field—an important condition for obtaining nonzero characteristic exponents. As Wojtkowski pointed out in [**W90a**, Proposition 4], if $n = 2$ and $m_1 < m_2$, then there exists a linearly stable periodic orbit, thus dimming the chances of proving ergodicity.

Our paper will very closely follow all the notations and definitions of the article [**W90a**], so the reader is warmly recommended to be familiar with that paper.

2. Eventually strict cone invariance

We will be working with the symplectic coordinates $(\delta h, \delta v)$ for the tangent vectors of the reduced phase space $\mathbf{M}$ satisfying the usual reduction equations $\sum_{i=1}^n \delta h_i = 0 = \sum_{i=1}^n \delta v_i$.

REMARK 2.1. The coordinates δh_i and δv_i $(i = 1, 2, \dots, n)$ serve as suitable symplectic coordinates in the codimension-one subspace $\mathcal{T}_x$ of the full tangent space $\mathcal{T}_x\mathbf{M}$ of $\mathbf{M}$ at x. Recall that the $(2n-2)$-dimensional vector space $\mathcal{T}_x$ is transversal to the flow direction, and the restriction of the canonical symplectic form

$$\omega = \sum_{i=1}^n \delta q_i \wedge \delta p_i = \sum_{i=1}^n \delta h_i \wedge \delta v_i$$

of $\mathbf{M}$ is nondegenerate on $\mathcal{T}_x$, see [**W90a**]. We also recall that

$$\delta h_i = m_i \delta q_i + m_i v_i \delta v_i = m_i \delta q_i + v_i \delta p_i.$$

Corresponding to the above choice of symplectic coordinates, the considered monotone Q-form will be

$$Q_1(\delta h, \delta v) = \sum_{i=1}^n \delta h_i \delta v_i. \tag{2.1}$$

It is clear that the evolution of $DS^t(\delta h(0), \delta v(0)) = (\delta h(t), \delta v(t))$ between colisions is

$$\frac{d}{dt}(\delta h(t), \delta v(t)) = (0, 0). \tag{2.2}$$

If a collision of type $(i, i+1)$ $(i = 1, 2, \dots, n-1)$ takes place at time t_k, then the derivative of the flow at the collision $\delta h^-(t_k) \mapsto \delta h^+(t_k)$, $\delta v^-(t_k) \mapsto \delta v^+(t_k)$ is given by the matrices

$$\begin{aligned} \delta h^+(t_k) &= R_i^* \left[\delta h^-(t_k) + S_i \delta v^-(t_k)\right] \\ \delta v^+(t_k) &= R_i \delta v^-(t_k), \end{aligned} \tag{2.3}$$

where the matrix R_i is the $n \times n$ identity matrix, except that its 2×2 submatrix at the crossings of the ith and $(i+1)$-st rows and columns is

$$R_i^{(i,i+1)} = \begin{bmatrix} \gamma_i & 1-\gamma_i \\ 1+\gamma_i & -\gamma_i \end{bmatrix}$$

with $\gamma_i = \dfrac{m_i - m_{i+1}}{m_i + m_{i+1}}$. The matrix S_i is, similarly, the $n \times n$ zero matrix, except its 2×2 submatrix at the crossings of the i-th and $(i+1)$-st rows and columns, which takes the form of

$$S_i^{(i,i+1)} = \begin{bmatrix} \alpha_i & -\alpha_i \\ -\alpha_i & \alpha_i \end{bmatrix}$$

with

$$\alpha_i = \frac{2m_i m_{i+1}(m_i - m_{i+1})}{(m_i + m_{i+1})^2}(v_i^- - v_{i+1}^-) > 0. \tag{2.4}$$

These formulas can be found, for example, in [**W90a**, Sention 4].

Concerning a floor collision $(0,1)$ at time t_k, the transformations are

$$\begin{aligned} \delta h_1^+(t_k) &= \delta h_1^-(t_k) \\ \delta v_1^+(t_k) &= \delta v_1^-(t_k) + \frac{2\delta h_1^-(t_k)}{m_1 v_1^+(t_k)}, \end{aligned} \tag{2.5}$$

see, for instance, [**W90a**, Section 4] or [**W98**].

In this section we will be studying *nonsingular* trajectory segments

$$S^{[0,T]}x_0 = \left\{x_t = S^t x_0 \middle| \, 0 \le T \le T\right\}$$

of the flow $\{S^t\}$ with the symbolic collision sequence $\Sigma = (\sigma_1, \sigma_2, \dots, \sigma_N)$, where $\sigma_k = (i_k, i_{k+1})$, $0 \le i_k \le n-1$, $k = 1, 2, \dots, N$. The collision graph $\mathcal{G} = \mathcal{G}(\Sigma)$ of Σ has the set $\{0, 1, \dots, n\}$ as its vertex set, and the unoriented edges of $\mathcal{G}$ are the unordered pairs $\{i_k, i_{k+1}\}$, $k = 1, 2, \dots, N$, counted without multiplicity.

According to Wojtkowski's arguments between the Theorem and [**W90a**, Proposition I of Section 5], in order to prove the strict invariance of the cone field $C_1 = \{Q_1 \ge 0\}$ along the considered trajectory segment

$$S^{[0,T]}x_0 = \left\{x_t = S^t x_0 \middle| \, 0 \le T \le T\right\},$$

it is enough to prove that

(*A*) for every nonzero vector $(0, \delta v) \in \mathcal{T}_{x_0}$ it is true that $Q_1(DS^T(0, \delta v)) > 0$, and

(*B*) for every nonzero vector $(\delta h, 0) \in \mathcal{T}_{x_0}$ it is true that $Q_1(DS^T(\delta h, 0)) > 0$.

Moreover, Wojtkowski's mentioned arguments in [**W90a**, Section 5] actually prove property (A) above in the case when the collision graph $\mathcal{G}(\Sigma)$ restricted to the vertex set $\{1.2.\dots.n\}$ is connected, i.e., all collisions $(i, i+1)$ with $i > 0$ occur. Here we briefly rephrase his ideas.

Formula (2.5) shows that a tangent vector of the form $(0, \delta v)$ is unchanged at any floor collision. Suppose now that a collision $(i, i+1)$ $(1 \le i \le n)$ occurs at time t_k and $\delta h^-(t_k) = 0$. Equation (2.3) shows that, after pushing the tangent vector $(0, \delta v^-(t_k))$ through the collision $(i, i+1)$, either

$$Q_1(\delta h^+(t_k), \delta v^+(t_k)) = \alpha_i(\delta v_i^-(t_k) - \delta v_{i+1}^-(t_k))^2 > 0, \tag{2.6}$$

or

$$\delta v_i^-(t_k) = \delta v_{i+1}^-(t_k) = \delta v_i^+(t_k) = \delta v_{i+1}^+(t_k). \tag{2.7}$$

Thus, as long as $Q_1(DS^T(0, \delta v(0))) = 0$ and the restriction of the collision graph $\mathcal{G}$ on the vertex set $\{1, 2, \dots, n\}$ is connected, we have that

$$\delta v_i(t) = \delta v_1(0) \tag{2.8}$$

for all t, $0 \le t \le T$ and all $i = 1, 2, \dots, n$. Therefore, $\delta v_i(t) = 0$ for all t, $0 \le t \le T$ and all $i = 1, 2, \dots, n$, thanks to the usual reduction equation $\sum_{i=1}^n \delta v_i = 0$. This finishes the proof of Property (A) in the case when the collision graph $\mathcal{G}(\Sigma)$ contains all edges $(i, i+1)$ with $1 \le i \le n-1$.

REMARK 2.2. From now on, we will always assume that the collision graph $\mathcal{G}$ contains all collisions $(i, i+1)$ with $i > 0$.

Checking Property (B) for the nonsingular trajectory segment $S^{[0,T]}x_0$ is, however, a bit harder and it requires a bit more on the side of the collision graph $\mathcal{G}(\Sigma)$ than simply containing all edges $(i, i+1)$, $i = 1, 2, \dots, n-1$.

First of all, we define the linear space of all *neutral vectors* as follows:

$$\mathcal{N}_{x_0}^T = \left\{(\delta h(0), \delta v(0)) \in \mathcal{T}_{x_0} \middle|\, Q_1(\delta h(t), \delta v(t)) = 0, \text{ for } 0 \le t \le T\right\}. \tag{2.9}$$

Recall that $(\delta h(t), \delta v(t)) = DS^t\left((\delta h(0), \delta v(0))\right)$.

It floows from the time-evolution equations (2.2), (2.3), and (2.5) that

$$DS^t\left((\delta h(0), \delta v(0))\right) = (\delta h(t), 0) \tag{2.10}$$

for $0 \le t \le T$, i.e., $\delta v(t) = 0$. Indeed, at any floor collision the form Q_1 increases by the amount of

$$\frac{2(\delta h_1^-(t_k))^2}{m_1 v_1^+(t_k)}, \tag{2.11}$$

which forces $\delta h_1^-(t_k) = \delta h_1^+(t_k) = 0$ and, as a corollary, $\delta v_1^-(t_k) = \delta v_1^+(t_k)$. According to (2.3), at an $(i, i+1)$ collision $(i \ge 1)$, occuring at time t_k, the form Q_1 increases by the amount

$$\alpha_i\left(\delta v_i^-(t_k) - \delta v_{i+1}^-(t_k)\right)^2, \tag{2.12}$$

and this forces $\delta v_i^-(t_k) = \delta v_{i+1}^-(t_k) = \delta v_i^+(t_k) = \delta v_{i+1}^+(t_k)$, thus $\delta v(t) = 0$ for $0 \le t \le T$, as claimed.

So, the neutral vector takes the form $DS^t(\delta h(0), 0) = (\delta h(t), 0)$ for $0 \le t \le T$, and $\delta h(t)$ only changes at ball-to-ball collisions $(i, i+1)$ (at time t_k) according to the law

$$\delta h^+(t_k) = R_i^* \delta h^-(t_k), \tag{2.13}$$

whereas at a floor collision, occuring at time t_k, the law

$$\delta h^-(t_k) = \delta h^+(t_k) = 0 \tag{2.14}$$

is forced by the neutrality requirement $Q_1(\delta h(t), \delta v(t)) = 0$. Accordingly, we define $R_0 = I = R_0^*$, the $n \times n$ identity matrix.

Out of the considered collisions $\sigma_1, \sigma_2, \dots, \sigma_N$, let the floor collisions be

$$\sigma_{k_1}, \dots, \sigma_{k_m},$$

$1 \le k_1 < \cdots < k_m \le N$. The above argument shows that the neutral space $\mathcal{N}_{x_0}^T$ can be defined by the following system of homogeneous linear equations:
(2.15)

$$\mathcal{N}_{x_0}^T = \mathcal{N}(\Sigma, \vec{m}) = \left\{ (\delta h(0), 0) \in \mathcal{T}_{x_0} \middle| \ \Pi_1 R_{i_{k_l}}^* \dots R_{i_2}^* R_{i_1}^* \delta h(0) = 0, \text{ for } l = 1, \dots, m \right\},$$

where $\Pi_1(\delta h_1, \delta h_2, \dots, \delta h_n) = \delta h_1$ is the projection onto the first component.

We observe that the neutral space $\mathcal{N}_{x_0}^T = \mathcal{N}(\Sigma, \vec{m})$ only depends on the n-tuple os masses $m_1 > m_2 > \cdots > m_n > 0$ and on the symbolic collision sequence $\Sigma = (\sigma_1, \sigma_2, \dots, \sigma_N)$.

DEFINITION 2.3. The nonsingular trajectory segment $S^{[0,T]}x_0$ or, equivalently, the corresponding pair $(\Sigma, \vec{m})$ is said to be *sufficient* iff

$$\mathcal{N}_{x_0}^T = \mathcal{N}(\Sigma, \vec{m}) = \{0\}.$$

Otherwise, these objects are said to be insufficient.

REMARK 2.4. A concequence of the previous definition is that the sufficiency of $(\Sigma, \vec{m})$ implies that Σ must contain all n types of collisions. Indeed, a missing $(0, 1)$ collision would mean that the system of homogeneous linear equations (2.15) contains no equations, whereas, a missing collision $(i_0, i_0 + 1)$ $(i_0 = 1, 2, \dots, n - 1)$ would establish no connection between the subsystems of particles $\{1, 2, \dots, i_0\}$ and $\{i_0 + 1, i_0 + 2, \dots, n\}$, thus preventing sufficiency.

Summarizing what we have seen so far, we have

PROPOSITION 2.16. *The sufficiency of the trajectory segment $S^{[0,T]}x_0$ (or, equivalently, the sufficiency of $(\Sigma, \vec{m})$) is equivalent to the* strict invariance *of the cone field*

$$C_1(x_t) = \left\{ (\delta h, \delta v) \in \mathcal{T}_{x_t} \middle| \ \sum_{i=1}^n \delta h_i \delta v_i \ge 0 \right\}$$

along

$$S^{[0,T]}x_0 = \left\{ x_t = S^t x_0 \middle| \ 0 \le t \le T \right\},$$

that is, this sufficiency exactly means that

$$DS^T(C_1(x_0)) \subset int(C_1(x_T)).$$

According to the time evolution equations (2.2), (2.3), (2.5), in order for the derivative DS^T of the flow to be defined on the tangent space $\mathcal{T}_{x_0}$, it is not enough to know the pair $(\Sigma, \vec{m})$, but one needs to know the relative velocities

$$\rho_k = \rho_k(\sigma_k) = v_{i_k}^-(t_k) - v_{i_k+1}^-(t_k) > 0 \tag{2.17}$$

for all ball-to-ball collisions $\sigma_k = (i_k, i_k + 1)$ (they play a role in (2.3) as a part of α_{i_k}) and the velocities

$$\rho_k = \rho_k(\sigma_k) = v_1^+(t_k) \tag{2.18}$$

for any floor collision $\sigma_k = (0, 1)$ that play a role in (2.5), $k = 1, 2, \dots, N$.

Therefore, the *strict cone invariance* formulated in (2.16) is a property of the *extended symbolic sequence*

$$\tilde{\Sigma} = ((\sigma_1, \rho_1), (\sigma_2, \rho_2), \dots (\sigma_N, \rho_N)) \tag{2.19}$$

and the vector of the masses $\vec{m}$.

The characterization of sufficiency with the strict cone invariance (2.16) has the big advantage that it appears to be *combinatorially monotone*, i.e., if one inserts an additional collision (σ^*, ρ^*) into the extended symbolic sequence

$$\tilde{\Sigma} = ((\sigma_1, \rho_1), (\sigma_2, \rho_2), \dots, (\sigma_N, \rho_N))$$

between (σ_k, ρ_k) and $(\sigma_{k+1}, \rho_{k+1})$, then the sufficiency will not be lost by this insertion:

PROPOSITION 2.20 (Combinatorial monotonicity property, CMP). *Assume that, for a given mass distribution $m_1 > m_2 > \cdots > m_n$, the extended symbolic sequence*

$$\tilde{\Sigma} = ((\sigma_1, \rho_1), (\sigma_2, \rho_2), \dots, (\sigma_N, \rho_N))$$

is sufficient. Let the extended symbolic sequence

$$\tilde{\Sigma}^* = ((\sigma_1, \rho_1), \dots, (\sigma_k, \rho_k), (\sigma^*, \rho^*), (\sigma_{k+1}, \rho_{k+1}), \dots, (\sigma_N, \rho_N))$$

be obtained from $\tilde{\Sigma}$ by the insertion of (σ^*, ρ^*), as indicated above. Then $\tilde{\Sigma}^*$ also satifies the strict cone invariance property formulated in (2.16).

PROOF. For any number l, $l = 1, 2, \dots, N$, let D_l be the derivative of the flow (in terms of $(\delta h, \delta v)$, as always) resulting from the collision (σ_l, ρ_l), and D^* be the derivative of the flow resulting from the collision (σ^*, ρ^*). We have

$$D_N D_{N-1} \dots D_1(C_1(x_0)) \subset \operatorname{int}(C_1(x_T)),$$
$$D^* D_k D_{k-1} \dots D_1(C_1(x_0)) \subset D_k D_{k-1} \dots D_1(C_1(x_0)),$$

thus

$$D_N \dots D_{k+1} D^* D_k \dots D_1(C_1(x_0)) \subset D_N D_{N-1} \dots D_1(C_1(x_0)) \subset \operatorname{int}(C_1(x_T)).$$

$\square$

Since the coefficients of the system of homogeneous linear equations (2.15) are given rational functions of the masses $m_1, m_2, \dots, m_n$, we immediately obtain

PROPOSITION 2.21 (Dichotomy). *For any given symbolic sequence*

$$\Sigma = (\sigma_1, \dots, \sigma_N)$$

either

(D_1) *for almost every n-tuple of masses $\vec{m}$ ($m_1 > m_2 > \cdots > m_n > 0$, the exceptional set being a proper algebraic subset of $\mathbb{R}^n$) it is true that $\mathcal{N}(\Sigma, \vec{m}) = \{0\}$, or*

(D_2) *for every mass vector $\vec{m}$ we have $\mathcal{N}(\Sigma, \vec{m}) \neq \{0\}$.*

DEFINITION 2.5. In the case (D_1) the symbolic sequence Σ itself is said to be sufficient, otherwise it is said to be insufficient.

Thanks to the characterization result (2.16) and the combinatorial monotonicity property (2.20) we get

COROLLARY 2.22. *For any given mass distribution $\vec{m}$, for any symbolic collision sequence Σ, and for any (not necessarily contiguous) subsequence Σ_1 of Σ, the sufficiency of the pair $(\Sigma_1, \vec{m})$ implies the sufficiency of $(\Sigma, \vec{m})$.*

Thanks to (2.21), we obtain the typical, i.e., the mass distribution-free version of (2.22):

COROLLARY 2.23. *For any symbolic collision sequence Σ and for any (not necessarily contiguous) subsequence Σ_1 of Σ, the sufficiency of Σ_1 implies the sufficiency of Σ.*

Next, we take a quick look at the limiting case $m_1 = m_2 = \cdots = m_n$. In this case, when a ball-to-ball collision $(i, i+1)$ occurs, we can simply swap the labels i and $i+1$ of the involved particles. In this way, by performing all these dynamic label changes, the entire flow becomes the independent motion of the n particles, each of them falling at unit acceleration and bouncing back elastically from the floor. (Doing so, they freely pass through each other.)

Observe that, in the case of equal masses, the transformation matrices R_i^* in (2.13) are the reflections

$$R_i = R_i^* = \begin{bmatrix} 0 & 1 \\ 1 & 0 \end{bmatrix},$$

thus the components $\delta h_i(t)$ will be independent of the time t after the dynamic relabeling.

The above makes it clear that the trajectory segment $S^{[0,T]}x_0$ is sufficient in the sense of (2.3), provided that all particles hit the floor. (After the dynamic re-labeling of the particles, of course.) This means that, for the limiting system $m_1 = m_2 = \cdots = m_n$ there exists a uniform time treshold $\tau_0 > 0$ such that $S^{[0,T]}x_0$ is sufficient, provided that $T \geq \tau_0$. By continuity, this property extends to every falling ball system $m_1 > m_2 > \cdots > m_n > 0$ with $m_n/m_1 \geq 1 - \epsilon_0$ for some fixed $\epsilon_0 > 0$. However, this property, along with the CMP (2.23), implies that there is a large enough $K \in \mathbb{Z}_+$ such that every symbolic sequence $\Sigma = (\sigma_1, \sigma_2, \ldots, \sigma_N)$ is sufficient if Σ contains at least K consecutive, contiguous, connected subsequences. (Recall that a subsequence is connected when its collision graph is connected or, equivalently, the subsequence contains all types of collisions.) However, the system of falling balls obviously possesses the property that every trajectory $S^{[0,\infty)}x_0$ contains infinitely many appearances of each collision type. Thus we obtain the main result of Section 2:

THEOREM 2.6 (Main result). *For almost every n-tuple of masses $m_1 > m_2 > \cdots > m_1 > 0$ (the exceptional set being a countable union of proper algebraic subsets of $\mathbb{R}^n$) the following is true: Every nonsingular positive orbit $S^{[0,\infty)}x_0$ is sufficient, hence possesses the property of eventually strict cone invariance.*

3. Conclusions

Here we summarize the corollaries of Theorem 2.6. These corollaries will be *conditional* on the condition that the following property is possessed by the falling ball system:

CLAIM 3.1 (Transversality condition). Singularities of different order are transversal to each other. Analogously, stable and unstable invariant manifolds are transversal to all singularities.

Our goal is to prove the ergodic conjecture of Wojtkowski [**W90a**] for almost every falling ball system, assuming the above transversality condition. The exceptional systems are the ones for which our main result (2.6) is false, so they form a countable union a proper algebraic sets in terms of the masses $\vec{m}$. For this purpose we are going to use the celebrated local ergodicity theorem (LET) of [**L-W95**], so we investigate the first return map $T: U_0 \to U_0$ of the billiard map to a suitably small, open neighborhood of a hyperbolic phase point x_0 with at most one singularity on its trajectory.

First of all, the second part of Claim 3.1 is the so-called Chernov–Sinai Ansatz, a condition of the LET of [**L-W95**].

Second, the transversality of singularites guarantees that the set of double singularities is a countable union of smooth, codimension-two (≥ 2) submanifolds, hence a slim set, negligible in dynamical considerations, see [**K-S-Sz92**, Section 2], especially [**K-S-Sz92**, Lemma 2.12] there. Thus, we may safely assume that the considered hyperbolic phase point x_0 has at most one singularity on its trajectory. Hyperbolicity means, of course, that the cone field C_1 is strictly invariant along the trajectory of x_0.

Third, the second part of 3.1 is [**Ch93**, Property 5], thus the result of [**Ch93**] guarantees the proper alignment of singularities, another condition of the LET in [**L-W95**].

Since the first return map $T: U_0 \to U_0$ enjoys strict C_1-cone invariance, according to [**L-W95**, Proposition 6.1], the minimum Q_1-expansion rate σ of T is uniformly bigger than 1. This establishes the noncontraction property and the strict unboundedness property for the first returm map $T: U_0 \to U_0$, two more conditions of the LET.

Fourth, the regularity of the singular set condition of [**L-W95**] is obtained as follows:

(1) The phase points with two singularities on their trajectory form a *slim* set, as stated before.
(2) The smooth components of different singularities do not accumulate at any point of the phase space, since there are no infinitely many collisions in finite time [**S96**, Proposition A.1 of the Appendix], and the horizon is finite.
(3) The derivatives of the singularities do not "explode", since there are no grazing (tangential) singularities in the falling ball system, only of corner type.

Finally, the local ergodicity theorem of [**L-W95**] states that the entire open neighborhood of the hyperbolic phase point x_0 (with at most one singularity on its trajectory) belongs to a single ergodic component of the falling ball system. For almost every mass vector $\vec{m}$, the set of phase points x_0 not having the two properties required above (strict cone invariance and at most one singularity along its trajectory) is a slim set. According to [**K-S-Sz92**, Lemma 2.12], such slim sets are unable to cut the phase space into different, open ergodic components. Thus, we conclude:

THEOREM 3.1 (Main corollary). *Assuming the condition* 3.1, *for almost every mass vector* $m_1 > m_2 > \cdots > m_n > 0$ *the falling ball system is ergodic.*

REMARK 3.2. In the upcoming paper [**S2023**] the author will be proving the Wojtkowski conjecture unconditionally for almost every falling ball system by employing, again, algebraic methods. Namely, the following two geometric properties will be proved there for almost every falling ball flow:

(A) For every noncollision phase point $x_0 = (q_0, v_0) \in \mathbf{M}$ of the flow it is true that if x_0 belongs to the interior (smooth part) of a singularity manifold $\mathcal{S}$, then the tangent apace $\mathcal{T}_{x_0}\mathcal{S}$ of $\mathcal{S}$ at x_0 contains the full tangent space $\mathcal{T}_{x_0}\mathbf{Q}$ of the configuration space $\mathbf{Q}$ at x_0. (The so-called "**Q**-onto property" of the singularities.)

(B) For every noncollision phase point $x_0 = (q_0, v_0) \in \mathbf{M}$ of the flow, if x_0 is the interior point of a local stable (unstable) invariant manifold γ, then the tangent space $\mathcal{T}_{x_0}\gamma$ contains the orthogonal complement of the velocity vector v_0 in $\mathcal{T}\mathbf{Q}$.

We note that these two properties together obviously imply the transversality condition 3.1 of this paper.

References

[Ch93] N. I. Chernov, *Local ergodicity of hyperbolic systems with singularities* (Russian), Funktsional. Anal. i Prilozhen. **27** (1993), no. 1, 60–64, DOI 10.1007/BF01768668; English transl., Funct. Anal. Appl. **27** (1993), no. 1, 51–54. MR1225911

[L-W95] Carlangelo Liverani and Maciej P. Wojtkowski, *Ergodicity in Hamiltonian systems*, Dynamics reported, Dynam. Report. Expositions Dynam. Systems (N.S.), vol. 4, Springer, Berlin, 1995, pp. 130–202. MR1346498

[K-S-Sz92] A. Krámli, N. Simányi, and D. Szász, *The K-property of four billiard balls*, Comm. Math. Phys. **144** (1992), no. 1, 107–148. MR1151248

[S96] Nándor Simányi, *The characteristic exponents of the falling ball model*, Comm. Math. Phys. **182** (1996), no. 2, 457–468. MR1447300

[S2023] Simányi, N.; *Finishing the Proof of Wojtkowski's Conjecture*, Under preparation (2023)

[W85] Maciej Wojtkowski, *Invariant families of cones and Lyapunov exponents*, Ergodic Theory Dynam. Systems **5** (1985), no. 1, 145–161, DOI 10.1017/S0143385700002807. MR782793

[W86] Maciej Wojtkowski, *Principles for the design of billiards with nonvanishing Lyapunov exponents*, Comm. Math. Phys. **105** (1986), no. 3, 391–414. MR848647

[W90a] Maciej P. Wojtkowski, *A system of one-dimensional balls with gravity*, Comm. Math. Phys. **126** (1990), no. 3, 507–533. MR1032871

[W90b] Maciej P. Wojtkowski, *The system of one-dimensional balls in an external field. II*, Comm. Math. Phys. **127** (1990), no. 2, 425–432. MR1037112

[W98] Maciej P. Wojtkowski, *Hamiltonian systems with linear potential and elastic constraints*, Fund. Math. **157** (1998), no. 2-3, 305–341, DOI 10.4064/fm-157-2-3-305-341. Dedicated to the memory of Wiesław Szlenk. MR1636897

Department of Mathematics, University of Alabama at Birmingham, 1402 10th Avenue South, Birmingham, Alabama 35294-1241

Email address: simanyi@uab.edu

Contemporary Mathematics
Volume **797**, 2024
https://doi.org/10.1090/conm/797/15936

Computability in dynamical systems

Michael Burr and Christian Wolf

Abstract. We present an introduction to the area of computability in dynamical systems. One of the central questions in this area is if relevant dynamical objects can be algorithmically presented by a Turing machine. After providing an overview of the relevant objects we discuss recent results concerning the computability of the entropy for symbolic systems and the computability of Julia sets as well as their Brolin–Lyubich measures.

1. Introduction

In [**16**], Milnor poses the question of whether an algorithm exists to approximate the entropy of a dynamical system up to a specified error. Such an algorithm does not exist in the most general setting, but this nonexistence leads to the following question:

For which classes of dynamical systems and invariants can these invariants be algorithmically approximated to any specified precision?

In dynamics and the sciences, computational experiments are often used to gain insight into the behavior of dynamical systems. For invariants which cannot be algorithmically approximated, the results of such computational experiments may be meaningless. We approach the generalization of Milnor's question from the point of view of computability theory.

We begin with the computational model for computability theory. From a theoretical standpoint, we may study a dynamical system $f : X \to X$ defined by an infinite amount of data. From an algorithmic standpoint, however, only a finite amount of data may be provided as input. Therefore, by necessity, we replace the dynamical system $f : X \to X$ by a convergent sequence of approximations $f_i : X_i \to X_i$, each of which may be described by a finite amount of data. Milnor's question then leads to the question

Do computations with a sequence of approximations $f_i : X_i \to X_i$ provide insight into the properties of the original dynamical system $f : X \to X$?

We apply computability theory to three case studies from dynamical systems which yield both positive and negative results.

2020 *Mathematics Subject Classification.* Primary 37A35, 37B12, 37F10, 03D15, 03D80.

The first author was partially supported by grants from the National Science Foundation (CCF-1527193 and DMS-1913119).

The second author was partially supported by grants from the Simons Foundation (#637594 to Christian Wolf) and PSC-CUNY (TRADB-51-63715 to Christian Wolf).

1.1. Dynamical systems. A dynamical system consists of a state space X (the set of attainable states) and a time evolution law

$$(\Phi_t)_{t\in T} : X \to X,$$

where T denotes the set of times when the state of the system can be observed. In other words, if the system is in state x_0 at time $t = 0$, then it is in state $\Phi_t(x_0)$ at time t. We focus on discrete dynamical systems for which $T = \mathbb{N}_0$. We assume that $\Phi_0 = \mathrm{id}_X$ and that $\Phi_{n+m} = \Phi_n \circ \Phi_m$ for all $n, m \in \mathbb{N}$. Hence

$$\Phi_n(x_0) = \underbrace{\Phi_1 \circ \ldots \circ \Phi_1}_{n-\text{times}}(x_0) =: \Phi_1^n(x_0).$$

Consequently, such a dynamical system is given by the iteration of a map $f := \Phi_1 : X \to X$.

We consider discrete dynamical systems given by continuous maps $f : X \to X$ for which (X, d) is a compact metric space. Given $x \in X$ we call $\mathcal{O}(x) := \mathcal{O}^+(x) := \{x, f(x), f^2(x), \ldots\}$ the *(forward) orbit* of x. If f is invertible, we analogously define the *backward orbit* $\mathcal{O}^-(x)$ by iterating f^{-1}. When $f^n(x) = x$ for some finite positive n, we say that x is *periodic* and define the *period* of x to be the cardinality of the orbit, i.e., a periodic point of period n has orbit $\mathcal{O}(x) = \{x, f(x), \ldots, f^{n-1}(x)\}$, where each $f^i(x)$ is distinct for $0 \leq i < n$ and $x = f^n(x)$. We denote the set of *periodic points of f with period n* by $\mathrm{Per}_n(f)$ and the set of *periodic points* of f by $\mathrm{Per}(f) = \bigcup_n \mathrm{Per}_n(f)$.

In dynamics, it is a central goal to understand the asymptotic behavior of the orbits of $f : X \to X$ for a large class of points in X. For instance, in ergodic theory, large often means a set of full measure for an f-invariant probability measure.

To motivate some of the challenges in studying families of dynamical systems computationally, the following example illustrates one of the pitfalls in the computational study of a single dynamical system.

EXAMPLE 1.1. Consider the $2x$-*mod*-1 dynamical system with $X = [0, 1)$ and $f : X \to X$ takes x to the fractional part of $2x$. It is straightforward to show that some iterate of x is periodic if and only if x is rational. Moreover, every dyadic rational is eventually mapped onto the fixed point 0. Conversely, the closure of the orbit of an irrational point is either a Cantor set or X itself.

One challenge in studying this dynamical system computationally is that the orbits of rational points are not representative of a typical orbit, even though the rationals are dense in X. Also, the most common number types on a computer are dyadic rationals, which exhibit even more atypical behavior. This example illustrates that care must be taken with both the selection of representative points and how conclusions are drawn from computational experiments.

1.2. Computability theory. A common theme in computability theory is to replace a single mathematical object by a convergent sequence, and to use the convergent sequence to study the original object, see [**4, 20**].

DEFINITION 1.2. An *oracle approximation* of a real number α is a function ψ such that for any $n \in \mathbb{N}$, $\psi(n)$ is a rational number such that $|\alpha - \psi(n)| < 2^{-n}$. A real number α is *computable* if there exists a Turing machine which is an oracle approximation for α.

For most purposes, it is sufficient to think of a Turing machine as an algorithm. Since every Turing machine is represented by a finite string of characters from a finite alphabet, we conclude that the computable numbers form a countable dense subset of the real numbers. This conclusion immediately implies that noncomputable real numbers exist.

The error estimates in the definition of a computable real number are quite strong. In some cases, there is a Turing machine that approximates α without an estimate on the convergence rate.

DEFINITION 1.3. A real number α is *computable from above* (respectively, *from below*) if there exists a Turing machine ψ such that on input n, $\psi(n)$ is a rational number such that the sequence $(\psi(n))_{n\in\mathbb{N}}$ is a decreasing (an increasing) sequence converging to α.

We observe that computability is equivalent to simultaneous computability from above and below. The ideas from computability theory can be extended to functions as well as to many other settings.

DEFINITION 1.4. A function $f : \mathbb{R} \to \mathbb{R}$ is a *computable function* if there exists a Turing machine ϕ with the following property: For any $\alpha \in \mathbb{R}$ and oracle approximation ψ for α, on input (ψ, n), $\phi(\psi, n)$ is a rational number such that $|f(\alpha) - \phi(\psi, n)| < 2^{-n}$.

The Turing machine ϕ must be able to accept every oracle approximation for any α (not merely Turing machines) and produce meaningful results. Computable functions are necessarily continuous, and there are analogous notions of computability from above and below for computable functions.

We introduce the concept of a computable metric space and refer the reader to [**13**] for more details.

DEFINITION 1.5. Let (X, d_X) be a separable metric space and let $\mathcal{S}_X = (s_i)_{i\in\mathbb{N}}$ be a dense sequence of points in X. We say that $(X, d_X, \mathcal{S}_X)$ is a *computable metric space* if there is a Turing machine $\chi : \mathbb{N}^2 \times \mathbb{N} \to \mathbb{Q}$ such that $|\chi(i,j,n) - d_X(s_i, s_j)| < 2^{-n}$.

The definition for a computable metric space is equivalent to the computability of the function $(i, j) \mapsto d_X(s_i, s_j)$. Importantly the approximation is uniform, that is, the same Turing machine χ is used for every i and j in $\mathbb{N}$. The definitions of computable numbers and functions generalize to computable metric spaces by replacing the output of Turing machines by an index $i \in \mathbb{N}$ of a point in $\mathcal{S}_X$.

2. Computability of entropy

Arguably, the most important invariant of a dynamical system f is its topological entropy $H_{\text{top}}(f)$. The topological entropy measures the complexity of the dynamical system f, and it is invariant under topological conjugation. Roughly, $H_{\text{top}}(f)$ measures the exponential growth rate of the number of orbit segments that can be distinguished up to an error of ε. Specifically, for $n \in \mathbb{N}$ and $\varepsilon > 0$, a finite set $F \subset X$ is *(n, ε)-separated* if $\max_{k=0,\dots,n-1} d(f^k(x), f^k(y)) \geq \varepsilon$ for distinct points $x, y \in F$. Let $F_n(\varepsilon)$ denote the largest cardinality of an (n, ε)-separated set. The *topological entropy* of f is

$$H_{\text{top}}(f) := \lim_{\varepsilon\to 0} \limsup_{n\to\infty} \frac{1}{n} \log F_n(\varepsilon). \tag{2.1}$$

From a computability standpoint, the limits in the formula for the topological entropy make its calculation difficult. Even if $F_n(\varepsilon)$ can be approximated for n and ε, a computation cannot decide without additional theory whether ε is small enough and n is large enough for $F_n(\varepsilon)$ to approximate the limit.

We say that f is *expansive* with *expansivity constant* $2\varepsilon_0$ if for distinct $x, y \in X$, there is some n such that $d(f^n(x), f^n(y)) > 2\varepsilon_0$. In this case, we can omit the limit on ε and compute the lim sup for ε_0. Then sequence $(h_n)_n$ defined by

$$h_n := \min_{k\leq n} \frac{1}{k} \log F_k(\varepsilon_0) \tag{2.2}$$

is nonincreasing and converges to $H_{\text{top}}(f)$.

COROLLARY 2.1. *Let $f : X \to X$ be an expansive dynamical system with expansivity constant $2\varepsilon_0$ that is given by an oracle χ. If there exists a Turing machine that takes χ and $n \in \mathbb{N}$ as input and computes $F_n(\varepsilon_0)$, then $H_{\text{top}}(f)$ is upper semi-computable.*

2.1. Symbolic dynamical systems. The class of symbolic systems form a rich family of discrete dynamical systems. These systems provide good test cases for many theories as they can be studied in a hands-on way, but have enough variability to provide challenging examples. We review the relevant material from symbolic dynamics, see [**15**].

Let $\mathcal{A} = \{0, \ldots, d-1\}$ be a finite set called the *alphabet*. Let $\mathcal{A}^*$ denote the set of all finite words in $\mathcal{A}$. The *full shift* Σ with alphabet $\mathcal{A}$ is the set of bi-infinite sequences $(x_k)_{k\in\mathbb{Z}}$ of elements of $\mathcal{A}$. With the *Tychonov product topology*, which coincides with the topology generated by the metric

$$d(x, y) := 2^{-\min\{|k|\,:\,x_k\neq y_k\}},$$

Σ is a compact metrizable topological space.

The *(left) shift map* $f : \Sigma \to \Sigma$ is defined by $f(x)_k = x_{k+1}$, and we note that f is a homeomorphism. We call a closed shift-invariant set $X \subset \Sigma$ a *shift space*, and we say that $f : X \to X$ is a *subshift*.

For a fixed subshift X, let $\mathcal{L}(X, n)$ be the set of length-n words which occur in some element of X. Moreover, we call $\mathcal{L} = \bigcup_{n\in\mathbb{N}} \mathcal{L}(X, n)$ the *language* of X. Languages and shift spaces are equivalent objects.

For a shift map $f : X \to X$, the entropy formula in Definition (2.1) reduces to

$$H_{\text{top}}(f|_X) := \lim_{n\to\infty} \frac{1}{n} \log \#\mathcal{L}(X, n). \tag{2.3}$$

Moreover, this limit converges from above.

An oracle χ of the language $\mathcal{L}(X)$ is a function such that $\chi(n)$ is a list of the words of the language of length n. In light of the simplicity of the entropy of a subshift X and Corollary 2.1, one might suspect that if the language of shift space is given by an oracle, then the entropy $H_{\text{top}}(f|_X)$ is computable. This statement, however, is not true.

The following theorem uses the notion of *computability at a point*, which is a computable version of continuity at a point, see [**8**] for more details.

THEOREM 2.2 ([**6**]). *The generalized entropy map $X \mapsto H(f|_X)$ is computable at a nonempty shift space X_0 if and only if X_0 has zero entropy. Moreover, there is one Turing machine that uniformly computes the topological entropy at all shift spaces with zero entropy.*

This fact is proved by noting that the entropy is nonnegative, and, for any value of $\mathcal{L}(X, n)$, there is a zero-entropy shift space X' which has the same language as X up to length n. For example, X' can be a union of orbits of points whose forward and backward tails are periodic. Thus, given only the language, the only possible lower bound on the entropy is zero.

2.2. SFTs of finite type and Sofic shifts. Subshifts of finite type and Sofic shifts are among the shifts for which the computability of their dynamics is best understood. Both of these shifts can be described in terms of finite data.

Let $\mathcal{F} \subset \mathcal{A}^*$ be called a *forbidden set.* Then $X_{\mathcal{F}}$ is defined to be the largest shift space such that no point $x \in X_{\mathcal{F}}$ contains any of the words in $\mathcal{F}$.

DEFINITION 2.3. A shift space X is a *subshift of finite type (SFT)* if $X = X_{\mathcal{F}}$ for some finite set of words $\mathcal{F}$.

The set $\mathcal{F}$ in this definition is not unique. A set $\mathcal{F}$ is a *minimal forbidden set* for an SFT X if for every word in $\mathcal{F}$, all of its proper subwords appear in the language of X. It follows that there is only one minimal forbidden set for an SFT. One less than the length of the largest word in the minimal forbidden set is called the *step size* of the SFT. Every SFT $X = X_{\mathcal{F}}$ is topologically conjugate to a one-step SFT $X' = X'_{\mathcal{F}'}$ on a possibly larger alphabet $\mathcal{A}'$. Moreover, given $\mathcal{F}$, the forbidden set $\mathcal{F}'$ can be computed by a Turing machine. Thus, it suffices to study one-step SFTs. One-step SFTs can also be described via a *transition matrix*, as follows: Let $A = (a_{i,j})$ be an $\mathcal{A} \times \mathcal{A}$ zero-one matrix. Then

$$X_A := \{(x_k)_k : a_{x_i, x_{i+1}} = 1\} \tag{2.4}$$

is a one-step SFT. Moreover, every one-step SFT can be written this way. Additionally, the entropy $H(f|_{X_A})$ of the SFT X_A is the logarithm of the spectral radius of the matrix A. Since there are algorithms that compute the spectral radius of A at any given precision, we obtain the following result.

PROPOSITION 2.4. *There is a Turing machine that computes the topological entropy of any SFT given by either a finite forbidden set or a transition matrix.*

DEFINITION 2.5. Let $\mathcal{T} = (G, E, L)$ be a finite labeled directed graph, where G is a graph with directed edge set E and *labeling function* $L : E \to \mathcal{A}$, which assigns a *label* $L(e)$ from the finite alphabet $\mathcal{A}$ to each edge $e \in E$. Let $\xi = \cdots e_{-1}e_0e_1\cdots$ be a bi-infinite path on G. The *label* of the path ξ is

$$L(\xi) := \cdots L(e_{-1})L(e_0)L(e_1)\cdots \in \mathcal{A}^{\mathbb{Z}}.$$

The set of all bi-infinite labels of paths is denoted by $X_{\mathcal{T}} = \{L(\xi) : \xi \in G\}$. A subset X of the full shift is a *Sofic shift* if $X = X_{\mathcal{T}}$ for a finite labeled graph $\mathcal{T}$.

A shift space X is Sofic if and only if X is a (finite-to-one) factor of some SFT Y. In particular, every SFT is Sofic, but the converse is not true. Moreover, there is a Turing machine that takes, as input, a finite labeled directed graph $\mathcal{T}$ and outputs a transition matrix A such that the Sofic shift $X_{\mathcal{T}}$ is a factor of the SFT X_A. Since finite-to-one factors preserve entropy, we obtain the following result.

PROPOSITION 2.6. *There exists a Turing machine that computes the topological entropy of any Sofic shift that is given by a finite labeled directed graph.*

2.3. Invariant measures. There is a natural connection between topological and measure-theoretic entropies, and we are interested in the computability properties of the measure-theoretic entropy. A probability measure μ is said to be *f-invariant* if $\mu(A) = \mu(f^{-1}(A))$ for all Borel sets A. For example, if $x \in \mathrm{Per}_n(f)$, then $\mu_x = \frac{1}{n}\sum_{k=0}^{n-1} \delta_{f^k(x)}$ is the unique invariant measure supported on $\mathcal{O}(x)$, where $\delta_{f^k(x)}$ denotes the Dirac measure on $f^k(x)$. Let $\mathcal{M}_f$ denote the space of all f-invariant Borel probability measures on X endowed with the weak* topology. With this topology, $\mathcal{M}_f$ is a compact convex metrizable topological space. The extreme points in this convex space are denoted by $\mathcal{M}_{f,E}$. These measures are precisely the ergodic measures, where an f-invariant measure is *ergodic* if every f-invariant set has either measure zero or one.

EXAMPLE 2.7. Let $X = [0, 1)$ and $\alpha \in [0, \infty)$. Let $f : X \to X$ be a translation by α, i.e., $f(x)$ is the fractional part of $x + \alpha$. If $\alpha \in \mathbb{Q}$, then every point in X is periodic and $\mathcal{M}_{f,E} = \{\mu_x : x \in X\} \cup \{\lambda\}$, where λ denotes the Lebesgue measure. On the other hand, if $\alpha \in \mathbb{R}^+ \setminus \mathbb{Q}$, then $\mathcal{M}_f = \mathcal{M}_{f,E} = \{\lambda\}$.

Let $H_\mu(f)$ denote the measure-theoretic entropy of f with respect to μ. Roughly, $h_\mu(f)$ is similar to the topological entropy, but ignores sets of measure zero, see Section 4 for details. The relationship between the topological and measure-theoretic entropies is the variational principle for the entropy, see Equation (4.1).

2.4. General and coded shifts. The computability properties of more general classes of shifts are more complicated. One reason why these spaces are more challenging to study is that their descriptions may be given in terms of infinite amounts of data. By Corollary 2.1, if the language is given by an oracle, then the topological entropy is upper semi-computable. Thus, we are interested in identifying when the entropy is also lower semi-computable.

We approach this problem by using a sequence $\{X_m\}_{m\in\mathbb{N}}$ of Sofic shifts which "exhausts" X in the sense that the entropy of these X_m's approaches that of X from below. The challenge with this approach is that the generalized entropy function $X \mapsto H_{\mathrm{top}}(f|_X)$ might not be continuous. Whenever such an approximation exists, one calls such a shift an *almost Sofic shift*, see [**19**]. In most cases, the language itself is not enough to compute such an approximation, we, therefore, focus on the case where X is generated from concatenations of a countable set of finite words, that is X is a coded shift.

To emphasize the generality of the class of coded shifts, we note that every shift space X can be expressed as the closure of the set of bi-infinite paths on a countable directed labeled graph, called a *representation* of X, see [**15**]. Coded shifts are precisely those shift spaces which have a representation by an irreducible countable directed labeled graph. Many classes of shifts spaces are coded shifts, including Sofic shifts, S-gap, and Beta-shifts.

DEFINITION 2.8. A two-sided shift space X over a finite alphabet $\mathcal{A} = \{0, \ldots, d-1\}$ is *coded* if there exists a *generating set* $\mathcal{G} = \{g_i\}_{i\in\mathbb{N}} \subset \mathcal{A}^*$, such that $X = X(\mathcal{G})$ is the smallest shift space that contains all bi-infinite concatenations of generators, i.e.,

$$X_{\mathrm{seq}} := \{\cdots g_{i_{-2}} g_{i_{-1}} g_{i_0} g_{i_1} g_{i_2} \cdots : i_j \in \mathbb{N}, j \in \mathbb{Z}\}.$$

In other words, X is the topological closure of X_{seq}, for more details, see [**3, 15**].

Let $X_m := X(\{g_1, \dots, g_m\})$ be the coded shift generated by the first m generators in $\mathcal{G}$. Since each X_m is a Sofic shift, its entropy is computable. We observe that $\{X_m\}_{m\in\mathbb{N}}$ is an increasing sequence of shift spaces in X_{seq}. In order to establish computability results for the entropy of coded shifts, we study when the entropy of $\{X_m\}_{m\in\mathbb{N}}$ converges to the entropy of X. For this convergence to hold, we show that the entropy on X cannot be "concentrated" on $X \setminus \bigcup_m X_m$.

To make these notions more precise, we introduce the following two definitions:

DEFINITION 2.9 ([**6**]). We say that X has *full sequential entropy* if

$$H_{\text{top}}(f|_X) = \sup_{\mu\in\mathcal{M}} \{H_\mu(f) : \mu(X_{\text{seq}}) = 1\}.$$

We say that X has *strict full sequential entropy* if

$$\sup_{\mu\in\mathcal{M}} \{H_\mu(f) : \mu(X_{\text{seq}}) = 1\} > \sup_{\mu\in\mathcal{M}} \{H_\mu(f) : \mu(X \setminus X_{\text{seq}}) = 1\}.$$

We observe that strict full sequential entropy implies full sequential entropy.

DEFINITION 2.10 (cf [**18**]). A *unique representation* of a coded shift X is a generating set $\mathcal{G}$ of X such that each $x \in X_{\text{seq}}(\mathcal{G})$ can be uniquely written as an infinite concatenation of elements of $\mathcal{G}$.

The manuscript [**1**] includes a proof that all coded shifts have a unique representation.

THEOREM 2.11 ([**6**]). *Let X be a coded shift with unique representation $\mathcal{G}$.*

(i) *If X has strict full sequential entropy, then*

$$H_{\text{top}}(f|_X) = \lim_{m\to\infty} H_{\text{top}}(f|_{X_m}).$$

(ii) *Suppose X has strict full sequential entropy. There exists a Turing machine, independent of X, that computes $H_{\text{top}}(f|_X)$ from oracles for the language and generating set $\mathcal{G}$ of X. In particular, $H_{\text{top}}(f|_X)$ is computable; and*

(iii) *If X does not have full sequential entropy, then the entropy is not continuous at X and hence is not computable at X.*

Since the entropy is not continuous as a function of the shift space, the main challenge in the proof of statement (ii) of this theorem is to show that the sequence of entropies for X_m converges to that of X. In the proof of statement (iii), it is shown that the entropy for subshifts X_m does not converge to the entropy of f on X. We refer to [**6, 19**] for examples of coded shifts without full sequential entropy. In particular, there are coded shifts for which the topological entropy is not computable.

2.5. Classification of entropies. An alternative to computing the entropy of a dynamical system is to classify the computability properties of entropies of dynamical systems. For instance, by Proposition 2.4, the entropies of all SFTs are not only computable, but also are rational logarithms of Perron numbers, where a *Perron number* is a real zero of an integer polynomial such that the zero is greater than one and is the zero of largest magnitude. The converse of this statement also holds, i.e., every rational logarithm of a Perron number is the entropy of some mixing SFT, see [**14**].

A natural generalization of one-dimensional SFTs are multidimensional SFTs which are finite type $\mathbb{Z}^d$-actions. In this case, the set of possible entropies coincides with the set of positive upper computable numbers [**12**]. Since there are upper computable numbers which are not computable, there exist symbolic systems whose entropy is not a computable number. On the other hand, the entropy of strongly irreducible higher dimensional SFTs is computable.

In more generality, for computable dynamical systems, see the definition in [**10**], the entropy is always a supremum of a computable sequence of upper-computable numbers [**10**]. Moreover, for every positive such number, there is a dynamical system over $\{0,1\}^{\mathbb{N}}$ which has this entropy and for which the time evolution function is computable.

3. Computability of Julia sets

Computer generated pictures of the Mandelbrot set and Julia sets can be quite striking, but computability theory raises the question of whether the underlying sample used to generate these sets accurately represents their true shape. Julia sets are prototypes of fractals consisting of the points with unpredictable dynamics. Since the dynamical behavior of nearby points can differ, computations with sample points may not correctly capture the dynamical behavior.

Let $\overline{\mathbb{C}} = \mathbb{C} \cup \{\infty\}$ be the Riemann Sphere endowed with the spherical metric $d_{\overline{\mathbb{C}}}$. Let

$$f_c(z) = z^d + c_{d-2}z^{d-2} + \cdots + c_0$$

be a polynomial of degree d, where $c = (c_{d-2}, \ldots, c_0) \in \mathbb{C}^{d-1}$. We denote the space of polynomials of this form by $\mathrm{Pol}(d)$.

Fatou and Julia introduced a partition of $\overline{\mathbb{C}}$ into two completely invariant sets, F_f and J_f, called the *Fatou* and *Julia* sets, respectively. The Fatou set F_f is the set of points $z \in \overline{\mathbb{C}}$ such that the family of iterates of f is equicontinuous in a neighborhood of z. The Julia set J_f is the complement of F_f in the Riemann sphere, and thus, the Fatou set is open, while the Julia set is compact. In particular, $z \in F_f$ if and only if for all $\varepsilon > 0$ there exists $\delta > 0$ such that $d_{\overline{\mathbb{C}}}(z, w) < \delta$ implies $d_{\overline{\mathbb{C}}}(f^n(z), f^n(w)) < \varepsilon$ for all $n \in \mathbb{N}$. In other words, the Fatou set is the set of points whose orbits are Lyapunov stable. The Julia set is the set of points whose orbits have sensitive dependence on the initial conditions, i.e., *chaotic dynamics*.

When $d = 2$, the polynomials are of the form $f_c(z) = z^2 + c$. The Julia set of a quadratic polynomial is either connected or totally disconnected. The connectivity-locus

$$\mathcal{M} = \{c \in \mathbb{C} : J_c \text{ is connected}\}$$

is the famous *Mandelbrot set*. The *filled-in Julia set* of f is defined by

$$K_f = \{z \in \mathbb{C} : \{f^n(z) : n \in \mathbb{N}_0\} \text{ is bounded}\}.$$

The iterates of the points in $U_f = \overline{\mathbb{C}} \setminus K_f$ converge to ∞. Moreover,

$$J_f = \partial K_f = \partial U_f \text{ and } F_f = U_f \cup \operatorname{int} K_f.$$

The sensitivity to the initial conditions makes computing the Julia set particularly challenging.

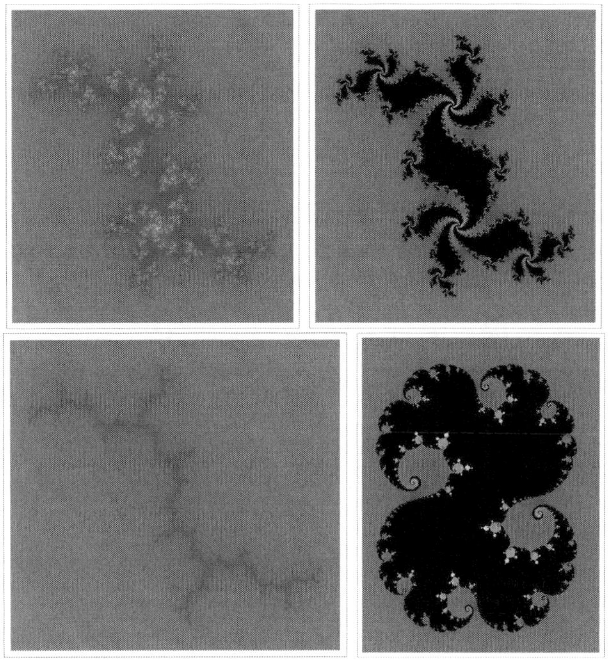

FIGURE 1. Examples of (filled-in) Julia sets for z^2+c: (a) $c \notin \mathcal{M}$, totally disconnected Julia set (b) $c \in \operatorname{int} \mathcal{M}$, f is hyperbolic (c) $c \in \partial\mathcal{M}$, Julia set is a dendrite (d) $c \in \partial\mathcal{M}$, 0 is parabolic fixed point. In all these examples the Julia set is computable. These images were created with Mathematica.

3.1. Computability of sets.

DEFINITION 3.1. Let $S \subset \mathbb{R}^m$ be an open set. S is a *recursively open set* (also called a semi-decidable set or a lower-computable set) if there exists a Turing machine ψ that produces a (possibly infinite) sequence of pairs (z_i, n_i) so that $z_i \in \mathbb{Q}^m$ is a rational vector and $n_i \in \mathbb{Z}$ so that $S = \cup_i B(z_i, 2^{-n_i})$.

There is a Turing machine, whose input is ψ, that terminates if $s \in S$, but may fail to terminate if $s \notin S$.

DEFINITION 3.2. Let $S, T \subset \mathbb{R}^m$ be compact sets. The *Hausdorff distance* between S and T is

$$d_H(S,T) := \max\{\max_{s\in S} \min_{t\in T} d(s,t), \max_{t\in T} \min_{s\in S} d(s,t)\}.$$

DEFINITION 3.3. Let $S \subset \mathbb{R}^m$ be compact. S is *computable* if there exists a Turing machine ψ such that on input n, $\psi(n)$ is a finite collection of pairs (q_i, l_i) with $q_i \in \mathbb{Q}^m$ and $l_i \in \mathbb{Z}$ such that $d_H(\cup_i \overline{B}(q_{k_i}, 2^{-l_i}), S) < 2^{-n}$.

There is a Turing machine, whose input is ψ, that terminates if $s \notin S$, but may fail to terminate if $s \in S$.

3.2. Properties of periodic points. By a theorem of Fatou and Julia, the Julia set is the closure of the set of repelling periodic orbits. A periodic point

$\alpha \in \mathrm{Per}_k(f)$ is *repelling* (respectively, *attracting*) if $|(f^k)'(\alpha)| > 1$ (if $|(f^k)'(\alpha)| < 1$). A periodic point $\alpha \in \mathrm{Per}_k(f)$ is *neutral* if $|(f^k)'(\alpha)| = 1$.

For a neutral periodic point, we replace f by its k-th iterate so that α is a fixed point. We write $f'(\alpha) = e^{2\pi i\theta}$. We say that α is *parabolic* if $f'(\alpha)$ is a root of unity, i.e., $\theta \in \mathbb{Q}$. In this case, $\alpha \in J$ and the dynamics near α is given by the Leau-Fatou Flower theorem, i.e., alternating attracting and repelling petals [**17**].

We say that α is a *Siegel point* if the dynamics near α is conjugate to the linear map $L_\alpha(z) = f'(\alpha)z$. Then $\alpha \in F_f$ and the connected component $U \subset F_f$ containing α is a *Siegel disk*, i.e., there exists a conformal map φ from U to the unit disk D such that $f(z) = \varphi^{-1} \circ L_\alpha \circ \varphi(z)$ for all $z \in U$.

By Bryuno's theorem, α is Siegel if θ is a *Bryuno number*, that is,

$$\sum_n \frac{\log(q_{n+1})}{q_n} < \infty,$$

where q_n is the denominator of the nth continued fraction expansion of θ. Moreover, the Bryuno numbers have full Lebesgue measure. We denote the set of degree d polynomials with a Siegel disk by $\mathrm{Pol}_{\mathrm{Sie}}(d)$.

3.3. Noncomputability of Julia sets. The map $c \mapsto J_c$ is not continuous at c if f_c has either a parabolic or a Siegel point, see [**5**]. In particular, the map $c \mapsto J_c$ is not computable on $\mathrm{Pol}(d)$. Furthermore, for each non-computable real number c, the Julia set J_c is not computable without oracle access to c.

Braverman and Yampolsky show that $c \mapsto J_c$ is computable on $\mathrm{Pol}(d)\backslash\mathrm{Pol}_{\mathrm{Sie}}(d)$ by one Turing machine once some non-uniform information about each parabolic point α of f_c is provided. This shows that the computability of the Juila set is problematic only for polynomials with Siegel disks. For example, consider the quadratic family

$$f_\theta(z) = z^2 + e^{2\pi i\theta}z, \quad \theta \in [0,1),$$

which has a neutral fixed point at 0. If θ is Bryuno, then f_θ has a Siegel disk centered at 0. In particular, f_θ has a Siegel disk at the origin for θ-values in a set of full Lebesgue measure. It is natural to conjecture that the non-computability of Siegel Julia sets J_θ stems from the noncomputability of θ—surprisingly, this is not the case.

Braverman and Yampolsky also construct an explicit algorithm that computes θ-values with a Siegel disk at the origin. In particular, there exist computable θ values with a noncomputable Julia set J_θ. The idea of this proof is to show that the Julia set J_θ is computable if and only if the conformal radius $r(\theta)$ of its Siegel disk is computable. In addition, they show that there are computable θ's such that $r(\theta)$ is not computable. This shows that it is impossible to compute the Julia set from θ alone since then $r(\theta)$ could be computed from θ, which is not possible.

3.4. Computability of filled Julia sets. For the filled-in Julia set the situation is better.

THEOREM 3.4 ([**5**]). *For any c, there is a Turing machine that produces a 2^{-n}-approximation of the filled-in Julia set K_c from an oracle for c and $n \in \mathbb{N}$.*

The proof of Theorem 3.4 describes a pair of Turing machines, one which terminates on points further than 2^{-n} from the filled-in Julia set, and a second which terminates on points which are near the filled-in Julia set. The second Turing machine classifies the behavior of points, e.g., near the Julia set or an iterate in the interior of a Siegel disk.

This result does not imply that $c \mapsto K_c$ is computable on Pol(d) since there would need to be *one* Turing machine that computes K_c for *all* $c \in$ Pol(d).

There are other recent results that discuss the computability of dynamically defined sets, e.g., the computability of Lorenz attractors [**11**] and the computability of generalized rotation sets [**7**].

4. Computability of measures of maximal entropy

Let $f : X \to X$ be a continuous map on a compact metric space (X, d). As in Section 2.3, there is the notion of the measure-theoretic entropy $H_\mu(f)$ of an invariant measure μ, see Section 4.1 for the definition. A measure which maximizes $H_\mu(f)$ is called a *measure of maximal entropy*.

Roughly, μ is a measure of maximal entropy if the set of its typical points carries the full topological complexity in terms of entropy. A dynamical system might not have a measure of maximal entropy, or, when such a measure exists, it might not be unique.

For several classes of dynamical systems, there exists a unique measure of maximal entropy, e.g., transitive SFTs and sofic shifts. If $f : \overline{\mathbb{C}} \to \overline{\mathbb{C}}$ is a polynomial of degree $d \geq 2$, then f has a unique measure of maximal entropy, called the *Brolin-Lyubich (B-L) measure* λ_f. The measure of maximal entropy λ_f is fully supported on the Julia set J_f.

One might suspect that since there are non-computable Julia sets, the B-L measure is also not computable. Surprisingly, this is not the case, and the B-L measure is computable.

4.1. Measure-theoretic entropy. Using a result of Katok and Brin, we provide an intuitive description of $H_\mu(f)$ as a measure-theoretic notion of complexity.

For $n \in \mathbb{N}$, define the Bowen d_n-metric on X by

$$d_n(x, y) = \max_{k=0,\dots,n-1} d(f^k(x), f^k(y)).$$

We observe that d_n induces the same topology on X as d. Let $B_n(x, \varepsilon)$ denote the Bowen-ball of radius ε with center x in the d_n-metric.

THEOREM 4.1 (Katok–Brin). *Let $\mu \in \mathcal{M}_f$. Then*

$$H_\mu(x) := \lim_{\varepsilon \to 0} \lim_{n \to \infty} -\frac{\log(\mu(B_n(x, \varepsilon))}{n}$$

exists for μ for almost every $x \in X$ and $H_\mu(\cdot)$ is μ-measurable. Moreover, if μ is ergodic, then $H_\mu(\cdot)$ is constant almost everywhere.

The quantity $H_\mu(x)$ is called the *point-wise entropy* of μ at x, and we define

$$H_\mu(f) = \int_X H_\mu(x)\, d\mu(x).$$

For $H_\mu(x)$ to be positive, the measure of the Bowen balls must decay exponentially to zero in n. Since Bowen balls are sets of points that stay close together under iteration, one infers that $H_\mu(f) > 0$ implies that μ must be "concentrated" on sets whose points are pushed apart under iteration. Thus, μ is concentrated on sets where a small change in the initial condition results in different dynamical behavior.

Bowen also defines the topology entropy $H_{\text{top}}(Y)$ for arbitrary sets $Y \subset X$. If μ is ergodic, then $H_\mu(f) = H_{\text{top}}(\mathbb{G}_\mu)$, where $\mathbb{G}_\mu$ denotes the set of μ-generic points of μ, where a point x is *μ-generic* if its orbit distribution coincides with the measure μ, i.e.,

$$\lim_{n\to\infty} \frac{1}{n}\sum_{k=0}^{n-1} \delta_{f^k(x)} = \mu,$$

in the weak* topology. As a consequence of Birkhoff's Ergodic theorem, for every ergodic measure μ, the set of μ-generic points has full μ-measure. The topological and measure-theoretic entropies are related by the *variational principle for the entropy*:

$$H_{\text{top}}(f) = \sup_{\mu\in\mathcal{M}_f} H_\mu(f). \tag{4.1}$$

4.2. Properties of B-L measures. The B-L measure λ is characterized by the following equivalent properties, see, e.g., [**2**]:

(1) $H_\lambda(f) = \log d = H_{\text{top}}(f)$;
(2) $\lambda(f(A)) = d\cdot\lambda(A)$ for all measurable sets A on which f is injective;
(3) $\lim_{n\to\infty}\frac{1}{d^n}\sum_{w\in f^{-n}(z)}\delta_w = \lambda$ for all but finitely many $z\in\mathbb{C}$ in the weak* topology;
(4) λ is the harmonic measure at infinity of the filled-in Julia set K_f.

4.3. Computability of measures. Let $\mathcal{M}(\overline{\mathbb{C}})$ denote that space of Borel probability measures on $\overline{\mathbb{C}}$ endowed with the weak* topology. We endow $\mathcal{M}(\overline{\mathbb{C}})$ with the Wasserstein-Kantorovich metric defined by

$$W_1(\mu,\nu) := \sup_{\phi\in 1-\text{Lip}(\overline{\mathbb{C}})}\left|\int\phi\,d\mu - \int\phi\,d\nu\right|,$$

where 1-Lip$(\overline{\mathbb{C}})$ is the space of real-valued Lipschitz functions on $\overline{\mathbb{C}}$ with Lipschitz constant 1. Let $\mathbb{D}$ denote the set of probability measures on $\overline{\mathbb{C}}$ that are finite rational convex combinations of Dirac measures supported on points in $\mathbb{C}$ with rational coordinates. Then $(\mathcal{M}(\overline{\mathbb{C}}), W_1, \mathbb{D})$ is a computable metric space.

4.4. Computability of B-L measures. We address the computability of the map $\text{Pol}(d)\ni c\mapsto\lambda_c$ from polynomials to B-L measures.

DEFINITION 4.2. The function $c\mapsto\lambda_c$ is computable if there exists a Turing machine ϕ with the following property: For any $c\in\text{Pol}(d)$ and any oracle approximation of ψ of c, on input ψ and n, $\phi(\psi,n)$ is a measure in $\mathbb{D}$ with $W_1(\phi(\psi,n),\lambda_c) < 2^{-n}$.

THEOREM 4.3 ([**2**]). *Let $d\geq 2$. Then $c\mapsto\lambda_c$ is computable.*

The proof of Theorem 4.3 follows from general statements about the computability of measures in $\mathcal{M}(\overline{\mathbb{C}})$ and from the characterization of the B-L measure in Property 2 of Section 4.2.

We note that Theorem 4.3 is surprising for parameters with non-computable Julia sets. To see this consider a (nonexceptional) point $z_0\in\mathbb{C}$. For any given $n\in\mathbb{N}$, the pre-images $f_c^{-n}(z_0)$ can be calculated at any pre-described accuracy. Moreover,

$$f_c^{-n}(z_0) \quad\longrightarrow\quad J_c \tag{4.2}$$

in the Hausdorff metric. If c is a parameter for which J_c is not computable, then the accuracy of the convergence in equation (4.2) can not be controlled by a Turing machine, but the convergence of

$$\frac{1}{d^n} \sum_{w \in f_c^{-n}(z_0)} \delta_w \longrightarrow \lambda_c \tag{4.3}$$

in the weak* topology can be computed at any prescribed accuracy. Since the support of the B-L measure is the entire Julia set, we interpret the convergence in equation (4.3) as covering most of the Julia set at a prescribed rate which is sufficient to compute the B-L measure.

There are additional results in the literature concerning the computability and non-computability of other dynamically relevant measures including physical measures, e.g., [**9**].

References

[1] M.-P. Béal, D. Perrin, and A. Restivo, *Unambiguously coded systems*, arXiv:2103.01012, (2021).

[2] Ilia Binder, Mark Braverman, Cristobal Rojas, and Michael Yampolsky, *Computability of Brolin-Lyubich measure*, Comm. Math. Phys. **308** (2011), no. 3, 743–771, DOI 10.1007/s00220-011-1363-1. MR2855539

[3] F. Blanchard and G. Hansel, *Systèmes codés* (French, with English summary), Theoret. Comput. Sci. **44** (1986), no. 1, 17–49, DOI 10.1016/0304-3975(86)90108-8. MR858689

[4] Vasco Brattka, Peter Hertling, and Klaus Weihrauch, *A tutorial on computable analysis*, New computational paradigms, Springer, New York, 2008, pp. 425–491, DOI 10.1007/978-0-387-68546-5_18. MR2762094

[5] Mark Braverman and Michael Yampolsky, *Computability of Julia sets* (English, with English and Russian summaries), Mosc. Math. J. **8** (2008), no. 2, 185–231, 399, DOI 10.17323/1609-4514-2008-8-2-185-231. MR2462435

[6] Michael Burr, Suddhasattwa Das, Christian Wolf, and Yun Yang, *Computability of topological pressure on compact shift spaces beyond finite type*, Nonlinearity **35** (2022), no. 8, 4250–4282, DOI 10.1088/1361-6544/ac7702. MR4452333

[7] Michael A. Burr, Martin Schmoll, and Christian Wolf, *On the computability of rotation sets and their entropies*, Ergodic Theory Dynam. Systems **40** (2020), no. 2, 367–401, DOI 10.1017/etds.2018.45. MR4048297

[8] Michael Burr and Christian Wolf, *Computability at zero temperature*, Nonlinearity **33** (2020), no. 11, 6157–6175, DOI 10.1088/1361-6544/ab9c71. MR4164675

[9] Stefano Galatolo, Mathieu Hoyrup, and Cristobal Rojas, *Dynamics and abstract computability: computing invariant measures*, Discrete Contin. Dyn. Syst. **29** (2011), no. 1, 193–212, DOI 10.3934/dcds.2011.29.193. MR2725287

[10] Silvére Gangloff, Alonso Herrera, Cristobal Rojas, and Mathieu Sablik, *Computability of topological entropy: from general systems to transformations on Cantor sets and the interval*, Discrete Contin. Dyn. Syst. **40** (2020), no. 7, 4259–4286, DOI 10.3934/dcds.2020180. MR4097543

[11] D. S. Graça, C. Rojas, and N. Zhong, *Computing geometric Lorenz attractors with arbitrary precision*, Trans. Amer. Math. Soc. **370** (2018), no. 4, 2955–2970, DOI 10.1090/tran/7228. MR3748590

[12] Michael Hochman and Tom Meyerovitch, *A characterization of the entropies of multidimensional shifts of finite type*, Ann. of Math. (2) **171** (2010), no. 3, 2011–2038, DOI 10.4007/annals.2010.171.2011. MR2680402

[13] Mathieu Hoyrup and Cristóbal Rojas, *Computability of probability measures and Martin-Löf randomness over metric spaces*, Inform. and Comput. **207** (2009), no. 7, 830–847, DOI 10.1016/j.ic.2008.12.009. MR2519075

[14] D. A. Lind, *The entropies of topological Markov shifts and a related class of algebraic integers*, Ergodic Theory Dynam. Systems **4** (1984), no. 2, 283–300, DOI 10.1017/S0143385700002443. MR766106

[15] Douglas Lind and Brian Marcus, *An introduction to symbolic dynamics and coding*, Cambridge University Press, Cambridge, 1995, DOI 10.1017/CBO9780511626302. MR1369092

[16] John Milnor. Is entropy effectively computable? `http://www.math.stonybrook.edu/~jack/comp-ent.pdf`, 2002.

[17] John Milnor, *Dynamics in one complex variable*, 3rd ed., Annals of Mathematics Studies, vol. 160, Princeton University Press, Princeton, NJ, 2006. MR2193309

[18] Ronnie Pavlov, *On entropy and intrinsic ergodicity of coded subshifts*, Proc. Amer. Math. Soc. **148** (2020), no. 11, 4717–4731, DOI 10.1090/proc/15145. MR4143389

[19] Karl Petersen, *Chains, entropy, coding*, Ergodic Theory Dynam. Systems **6** (1986), no. 3, 415–448, DOI 10.1017/S014338570000359X. MR863204

[20] Cristóbal Rojas and Michael Yampolsky, *Computable geometric complex analysis and complex dynamics*, Handbook of computability and complexity in analysis, Theory Appl. Comput., Springer, Cham, [2021] ©2021, pp. 143–172, DOI 10.1007/978-3-030-59234-9_5. MR4300755

School of Mathematical and Statistical Sciences, Clemson University, Clemson, South Carolina 29634

Email address: `burr2@clemson.edu`

Department of Mathematics, The City University of New York (CUNY) Graduate Center, 365 Fifth Ave., New York, New York

Email address: `cwolf@gc.cuny.edu`

Contemporary Mathematics
Volume **797**, 2024
https://doi.org/10.1090/conm/797/15950

Geodesics in generalizations of the Sierpinski carpet

Ethan Berkove, Elene Karangozishvili, and Derek Smith

Abstract. We analyze geodesics in a 3-parameter family of fractals that contains the Sierpinski carpet and Menger sponge, generalizing results in the literature. Between any two points in one of these fractals, we construct a geodesic path, where path lengths are induced by the 1-norm. For any of these fractals we then determine the maximum possible ratio of the geodesic metric to the Euclidean metric, and we provide examples to show that this upper bound is sharp.

1. Introduction

This paper joins a growing literature on the study of geodesic paths in fractals. When fractals are finitely ramified (post-critically finite), the analysis of geodesic paths often leans on "junction points" that the paths must pass through; for examples, see [**11**], [**12**], [**15**], [**18**], [**19**], and [**21**]. In fractals without junction points, as is the case for the members $F^{n,t,q}$ of the family studied here, there can be "a bewilderingly large set of possibilities for the paths" between two points, as Strichartz notes about the octogasket [**16**].

The geodesic paths studied here and in [**4**], [**7**], [**13**], [**17**], and [**20**] have lengths induced by a 1-norm. If $\gamma : [0,1] \to F^{n,t,q}$ is continuous, then the length $L(\gamma)$ of γ is the supremum

$$\sup\{\sum_{i=1}^{N} \|\gamma(s_i) - \gamma(s_{i-1})\|_1\}$$

over all partitions $0 = s_0 < s_1 < \cdots < s_N = 1$ (see [**5**, p. 34]), and the minimal geodesic distance between any two points x and y is the minimum of the lengths $L(\gamma)$ over all paths in $F^{n,t,q}$ from x to y. In [**4**], paths constructed to achieve this minimal distance are called "taxicab" paths since each of their internal constituent segments is parallel to one of the coordinate axes, and we retain that terminology here. While we focus exclusively on taxicab paths here, other types of paths through $F^{n,t,q}$ are possible; see [**1**], [**4**], [**6**], [**8**].

The fractal $F^{n,t,q}$ is described in several ways in Section 2, after we generalize the notions of gaps and tunnels that were initially defined in [**4**]. It is a subset of the n-dimensional cube $[0,1]^n$, where t is an integer "tunnel number" that determines which subcubes of side length q^{-1} are removed in the process of defining $F^{n,t,q}$.

2020 *Mathematics Subject Classification.* Primary 28A80, 54E35.
Key words and phrases. Sierpinski carpet, Menger sponge, geodesic, 1-norm.

The Sierpinski carpet is $F^{2,2,3}$ and the Menger sponge is $F^{3,2,3}$. The fractal family F^n studied in [**4**] coincides with $F^{n,2,3}$. The family C_s described at the end of [**7**] is $F^{s,s,3}$; it is also the subject of [**2**]. The family $S_{\mathbf{a}}$ in [**6**], [**8**], and [**9**], where $\mathbf{a} = (a^{-1}, a^{-1}, \ldots)$ for any odd integer a, is $F^{2,2,a}$. The fractals $F^{n,1,q}$ with $t = 1$ are products of "middle qth" $\frac{1}{q}$-Cantor sets and hence totally disconnected; since they contain no non-trivial paths, we will only consider $t \geq 2$.

Two possibilities for a geodesic taxicab path between two points x and y in $F^{3,3,3}$ can be seen in the relatively simple 3-dimensional example in Figure 3.1. Both paths are nonmonotonic in some coordinate, a behavior we refer to as "backtracking." Figure 4.1 and [**4**, Figure 3] offer further examples of the backtracking that may be required for a geodesic taxicab path between two points in the Menger sponge $F^{3,2,3}$. Having $t > 2$ leads to a growth of path possibilities and subtleties that are not encountered in the fractals covered in [**4**], where $t = 2$; these issues are addressed by several novel ideas in Section 3, which we outline now. Let x and y be any two points in $F^{n,t,q}$. In Section 3.1, we associate a well-defined set of "path backtracking data" to any path, taxicab or otherwise, from x to y, and identify four important properties of this set. Our next goal is to construct a taxicab path P^D of shortest length in $F^{n,t,q}$ from x to y that backtracks according to a given set D of "backtracking data" satisfying those four properties. In Section 3.2, we define two intermediate points, x^D and y^D, that appear on P^D; the construction of this path and a proof that it is indeed in $F^{n,t,q}$ are given in our main theorem on taxicab paths, Theorem 3.15 of Section 3.3. A corollary of this result (Corollary 3.17) notes the equivalence of backtracking data sets and path backtracking data sets. In Section 3.4, we introduce "containment tables" and a "backtracking tree" to help us enumerate a finite collection of all possible sets of backtracking data that could yield a geodesic taxicab path from x to y, with a comprehensive 9-dimensional example in $F^{9,4,3}$ provided to illuminate these concepts. Theorem 3.20 shows that every geodesic taxicab path is associated to some set of backtracking data in this finite collection. Finally, Theorem 3.21 in Section 3.5 justifies a greedy algorithm for choosing edge sequences in the backtracking tree that correspond to all geodesic taxicab paths from x to y.

As in [**4**], for any two distinct points x and y in $F^{n,t,q}$, define the ratio

$$\rho(x, y) = \frac{d_t(x, y)}{d_e(x, y)},$$

where $d_t(x, y)$ is the geodesic taxicab distance between x and y in $F^{n,t,q}$ and $d_e(x, y)$ is the Euclidean distance in $\mathbb{R}^n$. In [**7**], Cristea provides a sharp upper bound of $\rho(x, y) \leq 2$ for distinct points x and y in the Sierpinski carpet $F^{2,2,3}$ and an upper bound of $\rho(x, y) \leq n$ for the fractal $F^{n,n,3}$. In Section 4, we provide a sharp bound of

$$\rho(x, y) \leq \sqrt{n + 1 + \frac{1}{t - 1}}$$

for any fractal $F^{n,t,q}$, generalizing the sharp bound of $\sqrt{n+2}$ given in [**4**] for $F^{n,2,3}$.

2. Preliminary notions

In this paper, let n, t, and q be integers, where $n \geq t \geq 1$, and $q \geq 3$ is odd. The fractal $F^{n,t,q}$ is a subset of the n-dimensional unit cube $[0, 1]^n$ formed by removing "tunnels," which are square prisms of various sizes. We define these

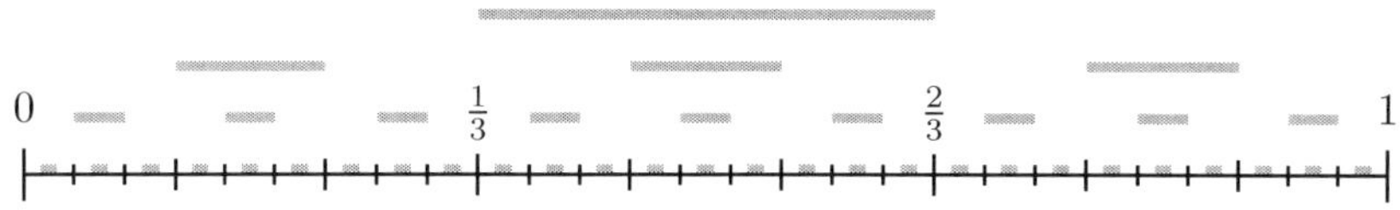

FIGURE 2.1. All k-gaps for $k = 1, 2, 3, 4$, with $q = 3$.

fractals and tunnels in Section 2.1 after defining the ideas of k-fractions and k-gaps. In Section 2.2 we then describe connections between gaps and rounding, an important concept underlying our construction of a geodesic taxicab path in Section 3.3.

2.1. Gaps, tunnels, and fractals. For a nonnegative integer k, define a *level-k fraction*, or *k-fraction*, to be a fraction in $[0, 1]$ of the form u/q^k with $u \in \mathbb{Z}$. For any positive integer k, a *level-k gap*, or *k-gap*, is any open interval in $[0, 1]$ of the form

$$\left(\frac{u}{q^k}, \frac{u+1}{q^k}\right) \text{ with } u \equiv \frac{q-1}{2} \bmod q.$$

Figure 2.1 shows k-gaps for small values of k when $q = 3$.

The collection of all k-gaps is dense in $[0, 1]$. In addition, when two distinct gaps intersect, then one is strictly contained in the other; and when a k'-fraction is contained in a k-gap, we must have $k' > k$. We note that the complement of the union of all k-gaps in $[0, 1]$ with $q = 3$ is the standard middle third $\frac{1}{3}$-Cantor set. These definitions and properties agree with those given in [**4**], which only considered the case $q = 3$.

We can use gaps to define the fractals $F^{n,t,q}$ we consider in this paper. In particular, to determine whether a point p is in a fractal, we consider how many of its n coordinates are contained in a k-gap for each value of k.

DEFINITION 2.1. For $n \geq t \geq 1$, the fractal $F^{n,t,q}$ is the set of all points p in $[0, 1]^n$ such that no t of its coordinates are in k-gaps for the same value of k.

We remark that $F^{2,2,3}$ is the Sierpinski carpet since we are excluding all points where both coordinates are in k-gaps for the same value of k. Similarly, $F^{1,1,3} = [0, 1] \setminus \{\text{all gaps}\}$ is the $\frac{1}{3}$-Cantor set. As noted in the introduction, the fractals $F^{n,1,q}$ are n-fold products of $\frac{1}{q}$-Cantor sets $F^{1,1,q} = [0, 1] \setminus \{\text{all gaps}\}$ and are totally disconnected, so we will assume $t \geq 2$ in what follows.

We will find it useful to describe $F^{n,t,q}$ in an alternate way.

DEFINITION 2.2. In $F^{n,t,q}$, a *k-tunnel* is a subset of $[0, 1]^n$ of the form

$$\prod_{j=1}^{t} (l_{i_j}, r_{i_j}) \times \prod_{j'=1}^{n-t} [0, 1]_{i_{j'}},$$

where (l_{i_j}, r_{i_j}) is a k-gap in coordinate i_j and $[0, 1]_{i_{j'}}$ is the unit interval in coordinate $i_{j'}$.

Since a point $p \in [0, 1]^n$ is in a k-tunnel if and only if it has at least t coordinates in k-gaps, Definition 2.1 tells us that the fractal $F^{n,t,q}$ is what remains in $[0, 1]^n$ after all k-tunnels have been removed, for all positive integers k. For example, the Menger sponge $F^{3,2,3}$ is the set of points in $[0, 1]^3$ such that no $t = 2$ coordinates are in a k-gap for the same value of k; the set of removed points has the appearance

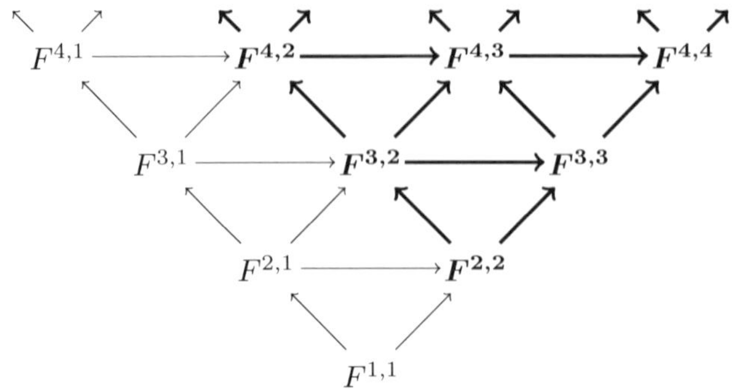

FIGURE 2.2. For any fixed q, the fractals $F^{n,t,q} = F^{n,t}$ are related by inclusion. Only fractals with $t \geq 2$ (in bold) are path connected.

of a union of tunnels that travel between opposite faces of $[0,1]^3$. We call t the *tunnel number* of $F^{n,t,q}$. We note that when $n = t$, Definition 2.2 implies that there are no k-tunnels which pass completely through $[0,1]^n$ because, for example, 0 is not in any k-gap. In this case, the sets $\prod_{j=1}^{t}(l_{i_j}, r_{i_j})$ correspond to internal cubical sets that might best be referred to as *k-cavities*, although we retain the use of the term "tunnel" to be consistent with [**4**]. For an example, we refer the reader to Figure 3.1, which shows three 2-tunnels (cavities) in part of $F^{3,3,3}$.

For each q, the fractals $F^{n,t,q}$ satisfy various inclusion relationships, some of which are displayed in Figure 2.2. Our constructions and proofs in Section 3 work uniformly for all values of q, so going forward we will often simplify our notation to $F^{n,t}$ for clarity. When $n > t$, the fractal $F^{n,t}$ is a proper subset of $F^{n,t+1}$ by the definition of the fractal (arrows pointing right denote this inclusion in Figure 2.2). For $n > t$ and any i, the subset of points in $F^{n,t}$ whose ith coordinate equals 0 or 1, or any other fixed value in the $\frac{1}{q}$-Cantor set $F^{1,1}$, is a copy of $F^{n-1,t}$ (arrows pointing up to the left), which is also the image of the orthogonal projection of $F^{n,t}$ onto any set of $n-1$ coordinates. And the subset of points in $F^{n,t}$ whose ith coordinate equals $1/2$ is a copy of $F^{n-1,t-1}$ (arrows pointing up to the right).

We find that our path constructions and proofs are more clearly stated in terms of avoiding tunnels and gaps, so our previous descriptions of the fractals in terms of these concepts are the most helpful. The fractal $F^{n,t,q}$ can also be defined in terms of an iterated function system, where the images of $[0,1]^n$ under the associated contraction maps are cubes of side length q^{-1} whose points do not have at least t coordinates that lie in the (unique) 1-gap $(\frac{q-1}{2q}, \frac{q+1}{2q})$. There are

$$R = q^n - \sum_{j=t}^{n} \binom{n}{j} (q-1)^{n-j}$$

such cubes, and the Hausdorff dimension of $F^{n,t,q}$ is $\log(R)/\log(q)$; the latter result follows from a straightforward limiting argument in conjunction with [**10**, Theorem 5.3]. It is implied in [**14**, Theorem 1.2] that the resulting fractals are path connected for $t > 1$. For any k, the union of the R^k cubes of side length q^{-k} that are the images of $[0,1]^n$ under any sequence of k contractions is called the *level-k approximation* of $F^{n,t,q}$. Equivalently, it is the set of all points not in a k'-tunnel for any $k' \leq k$.

The fractal $F^{n,t,q}$ is the intersection of its level-k approximations taken over all k and is a generalized Sierpinski carpet in the sense of [**3**].

2.2. Rounding. As seen in Figure 3.1, it is sometimes necessary for paths to move around tunnels. We use rounding to construct these paths, following [**4**] (when $q = 3$). Given $s \in [0, 1]$ and a nonnegative integer k, we define $\lfloor s \rfloor_k$ to be the value of s *rounded down* to the largest k-fraction less than or equal to s, with an analogous notion $\lceil s \rceil_k$ for rounding up:

$$\lfloor s \rfloor_k = \max\{u/q^k \mid u/q^k \leq s \text{ and } u \in \mathbb{Z}\}$$

and

$$\lceil s \rceil_k = \min\{u/q^k \mid s \leq u/q^k \text{ and } u \in \mathbb{Z}\}.$$

The following string of inequalities relates rounding up and rounding down at different levels:

$$\lfloor s \rfloor_0 \leq \lfloor s \rfloor_1 \leq \lfloor s \rfloor_2 \leq \cdots \leq s \leq \cdots \leq \lceil s \rceil_2 \leq \lceil s \rceil_1 \leq \lceil s \rceil_0.$$

We next offer two results about the relationship between rounding and gaps. The following result from [**4**] holds for any q and will be very useful; it states that when an a-fraction $\lfloor s \rfloor_a$ is in a k-gap, so is s.

LEMMA 2.3 ([**4**, Lemma 2.3]). *For any positive integers a and k, if $\lfloor s \rfloor_a$ (or $\lceil s \rceil_a$) is in a k-gap γ, then s is in γ, and thus so is $\lfloor s \rfloor_{a'}$ (or $\lceil s \rceil_{a'}$) for any $a' \geq a$.*

Our taxicab path constructions will round coordinates level-by-level, so we are interested in knowing which gaps are encountered in doing so. The next lemma provides some useful information in this regard.

LEMMA 2.4. *If the interval $[\lceil s \rceil_{a+1}, \lceil s \rceil_a]$ or $[\lfloor s \rfloor_a, \lfloor s \rfloor_{a+1}]$ contains a k-gap, then $k > a$.*

PROOF. Suppose $(\frac{u}{q^k}, \frac{u+1}{q^k}) \subset [\lceil s \rceil_{a+1}, \lceil s \rceil_a]$. Then $s \leq \lceil s \rceil_{a+1} \leq \frac{u}{q^k}$. In order for $\frac{u}{q^k} < \lceil s \rceil_a$, $\lceil s \rceil_a$ cannot be a k-fraction, so we must have $k > a$. The other case is analogous. □

Given a point $p = (p_1, p_2, \ldots, p_n) \in [0, 1]^n$, we define $\lfloor p \rfloor_k$ (resp. $\lceil p \rceil_k$) by rounding down (resp. up) to level k in each coordinate. As examples, for $q = 3$ and the point $x = \left(\frac{5}{6}, 0, \frac{14}{27}\right)$ we have

$$\begin{aligned}
\lfloor x \rfloor_0 &= (0, 0, 0) & \lceil x \rceil_0 &= (1, 0, 1)\\
\lfloor x \rfloor_1 &= \left(\tfrac{2}{3}, 0, \tfrac{1}{3}\right) & \lceil x \rceil_1 &= \left(1, 0, \tfrac{2}{3}\right)\\
\lfloor x \rfloor_2 &= \left(\tfrac{7}{9}, 0, \tfrac{4}{9}\right) & \lceil x \rceil_2 &= \left(\tfrac{8}{9}, 0, \tfrac{5}{9}\right)\\
\lfloor x \rfloor_3 &= \left(\tfrac{22}{27}, 0, \tfrac{14}{27}\right) & \lceil x \rceil_3 &= \left(\tfrac{23}{27}, 0, \tfrac{14}{27}\right).
\end{aligned}$$

3. Geodesic paths

In this section, we construct a shortest taxicab path between two distinct points $x = (x_1, \ldots, x_n)$ and $y = (y_1, \ldots, y_n)$ in $F^{n,t}$. It is clear by looking at examples of shortest paths of length n between $x = (0, 0, \ldots, 0)$ and $y = (1, 1, \ldots, 1)$ which travel along the outside edges of $[0, 1]^n$ that the shortest taxicab path between two points may not be unique. Since the fractal $F^{n,t}$ is symmetric with respect to the bisecting hyperplanes $z_i = 1/2$, we may assume that $x_i \leq y_i$ for all i. The fractal is also symmetric with respect to bisecting hyperplanes $z_i = z_j$, which allows us to reorder certain subsets of the coordinates in later arguments.

Consider the following example.

EXAMPLE 3.1. Find a shortest taxicab path in $F^{3,3,3}$ from $x = \left(\frac{5}{6}, 0, \frac{14}{27}\right)$ to $y = \left(\frac{5}{6}, \frac{1}{3}, \frac{29}{54}\right)$

As seen in Figure 3.1, any shortest path from x to y needs to avoid a 2-tunnel, which in this case is a 2-cavity since $t = n$ (see the discussion after Definition 2.2). It is also clear that either coordinate 1 or coordinate 3 will not be able to change monotonically as one moves along a path from x to y. If coordinate 1 decreases from $\frac{5}{6}$ to $\frac{7}{9}$ to allow coordinate 2 to move from 0 to $\frac{1}{3}$, and then it increases back to $\frac{5}{6}$, this will add at least $\frac{1}{18} + \frac{1}{18} = \frac{1}{9}$ to the length of a path that can otherwise change monotonically in each coordinate. In coordinate 3, the monotonic distance is $\frac{29}{54} - \frac{14}{27} = \frac{1}{54}$. However, if coordinate 3 increases from $\frac{14}{27}$ to $\frac{5}{9}$ in order for the path to avoid the 3-cavity, and then decreases to $\frac{29}{54}$, extra distance of length at least $\frac{1}{54} + \frac{1}{54} = \frac{1}{27}$ is required. We call this type of nonmonotonic behavior "backtracking" in a coordinate.

3.1. Defining backtracking data. Let x and y be two distinct points in $F^{n,t}$. The goal of this section is to define the concept of backtracking data, which occurs in two contexts: "path backtracking data" is determined by a given path from x to y, while "backtracking data" is a seemingly more general notion that depends only on x and y. We begin with a definition that applies in both contexts, motivated by considering the gaps each coordinate must move through on any possible path from x to y.

DEFINITION 3.2. For each coordinate i with $x_i < y_i$, let $q^{-\ell_i}$ be the largest width of a gap contained in the interval $[x_i, y_i]$:

$$\ell_i = \min\{k \mid \exists \text{ a } k\text{-gap } \gamma_i \text{ with } \gamma_i \subset [x_i, y_i]\}.$$

We denote the largest width of all such gaps by $q^{-\ell}$, so that $\ell = \min\{\ell_i\}$.

Since k-gaps are dense on the real line and $x \neq y$, there will always be a well-defined value of ℓ; and since k-gaps are only defined for $k > 0$, we know that $\ell > 0$. An ℓ-gap is the widest gap that any coordinate will be required to cross on any path from x to y.

We next consider backtracking that is determined by a path P from x to y. The definition of $F^{n,t}$ implies the following key observation:

LEMMA 3.3. *For any path P from x to y, if $t-1$ intervals of the form $[x_j, y_j]$ are in k-gaps, and another interval $[x_i, y_i]$ contains a k-gap γ, then at least one of these $t-1$ coordinates will need to leave its k-gap while coordinate i moves through γ.*

PROOF. Coordinate i changes continuously from x_i to y_i as P is followed continuously from x to y, so coordinate i will move through γ. If none of the $t-1$ coordinates left their respective k-gaps, there would be t coordinates simultaneously in k-gaps, and so P would leave $F^{n,t}$. □

The coordinate behavior described by Lemma 3.3 is the backtracking exhibited in both paths in Figure 3.1. For our purposes, backtracking only needs to be defined for gap levels at least as large as ℓ.

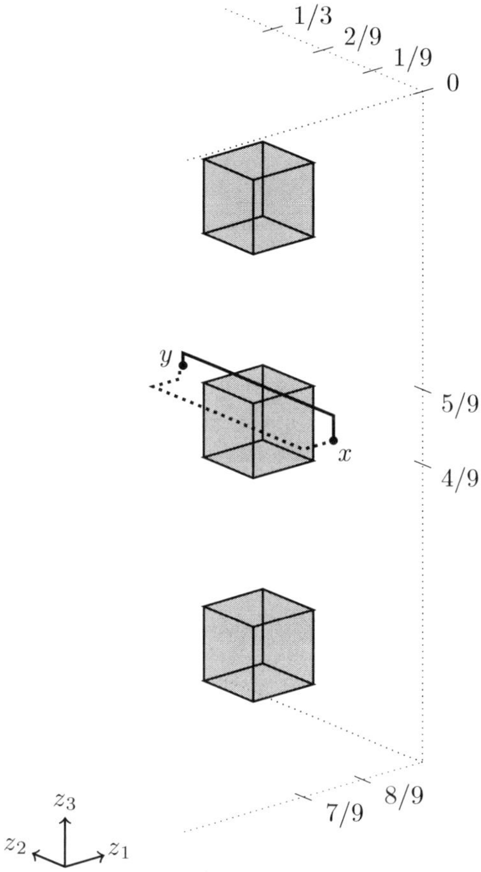

FIGURE 3.1. In $F^{3,3,3}$, the solid taxicab path from $x = \left(\frac{5}{6}, 0, \frac{14}{27}\right)$ to $y = \left(\frac{5}{6}, \frac{1}{3}, \frac{29}{54}\right)$ is shorter than the dotted path.

DEFINITION 3.4. If $k \geq \ell$, the jth coordinate of a path P is said to *backtrack at level k*, or *backtrack out of a k-gap*, if $[x_j, y_j]$ is in a k-gap and the jth coordinate leaves the gap at some point on P.

We use the existence of ℓ and the nested nature of gaps to state an important requirement for backtracking.

LEMMA 3.5. *Let x and y be points in $F^{n,t}$, and suppose that there are $t-1$ intervals $[x_{j_1}, y_{j_1}], \ldots, [x_{j_{t-1}}, y_{j_{t-1}}]$ that are in k-gaps for some $k \geq \ell$. Then in any path P from x to y, at least one of these $t-1$ coordinates must backtrack out of its k-gap.*

PROOF. By Definition 3.2, there is another interval $[x_i, y_i]$ that contains an ℓ-gap. As coordinate i moves through its ℓ-gap, it will also move through k-gaps for all $k \geq \ell$. The result then follows by Lemma 3.3. $\square$

Although it is possible for a coordinate to backtrack at many different levels, for the two notions of backtracking data that follow, we select just one of these levels.

DEFINITION 3.6. The *path backtracking data* for a path P from x to y is the set

$$\{(c_1, b_{c_1}), \ldots, (c_w, b_{c_w})\},$$

which consists of an ordered pair (c_j, b_{c_j}) for each coordinate c_j that backtracks, with b_{c_j} the smallest backtracking level for that coordinate, noting that $b_{c_j} \geq \ell$ is required by Definition 3.4.

Every path P from x to y, whether taxicab or not, determines a unique (possibly empty) set of path backtracking data. To illustrate the content of Lemma 3.3 and Definitions 3.2, 3.4, and 3.6, we return to Example 3.1, considering paths from $x = \left(\frac{5}{6}, 0, \frac{14}{27}\right)$ to $y = \left(\frac{5}{6}, \frac{1}{3}, \frac{29}{54}\right)$ in $F^{3,3,3}$.

(1) The value of ℓ_1 is not defined since $x_1 = y_1$; $\ell_2 = 2$ since $[0, \frac{1}{3}]$ contains the 2-gap $(\frac{1}{9}, \frac{2}{9})$ but no wider gap; and $\ell_3 = 5$ since $[\frac{14}{27}, \frac{29}{54}]$ contains the 5-gap $(\frac{128}{243}, \frac{129}{243})$ but no wider gap. Therefore, $\ell = \ell_2 = 2$.
(2) We note that the first and third coordinate intervals are in 2-gaps ($(\frac{7}{9}, \frac{8}{9})$ and $(\frac{4}{9}, \frac{5}{9})$ respectively), while the second coordinate interval contains the 2-gap $(\frac{1}{9}, \frac{2}{9})$. Therefore, some sort of backtracking must happen in the first or third coordinate as one moves from x to y.
(3) Figure 3.1 shows two possible paths from x to y. The path backtracking data of the solid taxicab path is $\{(3, 2)\}$, while the path backtracking data of the dotted path is $\{(1, 2)\}$.

Given points x and y in $F^{n,t}$, there are several properties that path backtracking data sets must satisfy, regardless of the specific path P that joins them, allowing us to define a notion of backtracking data that only depends on the endpoints x and y.

LEMMA 3.7. *For fixed x and y in $F^{n,t}$, every set of path backtracking data $D = \{(c_1, b_{c_1}), \ldots, (c_w, b_{c_w})\}$ satisfies four properties:*

(1) *Elements of the set $\{c_1, \ldots, c_w\}$ are distinct.*
(2) *For all i, $b_{c_i} \geq \ell$.*
(3) *For all i, the interval $[x_{c_i}, y_{c_i}]$ is contained in a b_{c_i}-gap.*
(4) *If $t-1$ intervals $[x_{j_1}, y_{j_1}], \ldots, [x_{j_{t-1}}, y_{j_{t-1}}]$ are in k-gaps for $k \geq \ell$, then there is an ordered pair $(j_h, b) \in D$ with $j_h \in \{j_1, \ldots, j_{t-1}\}$ and $b \leq k$.*

PROOF. Properties (1)–(3) follow from Definitions 3.4 and 3.6, while property (4) follows from Lemma 3.5. □

DEFINITION 3.8. Any set $D = \{(c_1, b_{c_1}), \ldots, (c_w, b_{c_w})\}$ satisfying the four properties of Lemma 3.7 is a set of *backtracking data* for x and y.

The proof that every backtracking data set is a path backtracking data set will be given in Corollary 3.17 in Section 3.4, as the taxicab path P^D constructed in Theorem 3.15 relies only on the properties of backtracking data listed in Lemma 3.7. In addition, the equivalence between these two types of sets is an important factor in Section 3.5, where we enumerate backtracking data sets as part of our identification of geodesic taxicab paths.

3.2. Defining x^D and y^D from backtracking data. In this section, we identify two important points, x^D and y^D, that we will use to build a shortest taxicab path P^D associated to a set of backtracking data D. The path P^D will go from x to x^D to y^D to y, always staying in $F^{n,t}$.

We first introduce quantities that will help us describe the distance each coordinate will travel in P^D. On any path from x to y, there is a minimal distance that each coordinate must travel:

DEFINITION 3.9. For each coordinate i, we define the *monotone distance* between x_i and y_i by $\mu_i = y_i - x_i \geq 0$.

When coordinate i backtracks at level k, there is an additional minimal distance of travel beyond the monotone distance:

DEFINITION 3.10. If $[x_i, y_i]$ is contained in a k-gap γ with $k \geq \ell$, we define ε_i^k to be the smallest distance, in addition to μ_i, that allows coordinate i to backtrack out of γ:

$$\begin{aligned}\varepsilon_i^k &= \min\{(x_i - \lfloor x_i \rfloor_k) + (y_i - \lfloor y_i \rfloor_k) - \mu_i,\ (\lceil x_i \rceil_k - x_i) + (\lceil y_i \rceil_k - y_i) - \mu_i\},\\ &= \min\{2(x_i - \lfloor x_i \rfloor_k),\ 2(\lceil y_i \rceil_k - y_i)\},\end{aligned}$$

the latter equality following from $\lfloor x_i \rfloor_k = \lfloor y_i \rfloor_k$ and $\lceil x_i \rceil_k = \lceil y_i \rceil_k$, which are the endpoints of γ.

We note that the backtracking-related value β_i given in [**4**] for fractals $F^{n,2,3}$ is $\beta_i = \mu_i + \varepsilon_i^k$.

Portions of the taxicab path P^D that we construct in Theorem 3.15 from a given set $D = \{(c_1, b_{c_1}), \ldots, (c_w, b_{c_w})\}$ of backtracking data are built by rounding the coordinates one level at a time, traveling to or from points x^D and y^D that incorporate information about all backtracking coordinates. For any coordinate i where b_i is defined, we can define x_i^D and y_i^D based on concepts already introduced. In coordinates where b_i is not defined and $x_i < y_i$, we define an appropriate rounding scale that will determine the values of x_i^D and y_i^D.

DEFINITION 3.11. When b_i is not defined and $x_i < y_i$, we define

$$m_i = \min\{k \mid \lceil x_i \rceil_k \leq \lfloor y_i \rfloor_k\}.$$

We can now determine the coordinates of x^D and y^D and describe the individual trajectory of each coordinate on P^D. As in [**4**], the coordinates are of four distinct types.

(C) If b_i is undefined and $x_i = y_i$, then $x_i = x_i^D = y_i^D = y_i$. The coordinate will remain constant.

(M) If b_i is undefined and $x_i < y_i$, then $x_i^D = \lceil x_i \rceil_{m_i}$ and $y_i^D = \lfloor y_i \rfloor_{m_i}$. The coordinate will increase monotonically a distance of μ_i from x_i to y_i.

(Bd) If b_i is defined and

$$(x_i - \lfloor x_i \rfloor_{b_i}) + (y_i - \lfloor y_i \rfloor_{b_i}) \leq (\lceil x_i \rceil_{b_i} - x_i) + (\lceil y_i \rceil_{b_i} - y_i),$$

the coordinate will backtrack out of a b_i-gap by decreasing from x_i to $\lfloor x_i \rfloor_{b_i} = x_i^D = y_i^D = \lfloor y_i \rfloor_{b_i}$, and then increasing to y_i, for a total distance of $\mu_i + \varepsilon_i^{b_i}$.

(Bi) If b_i is defined and

$$(x_i - \lfloor x_i \rfloor_{b_i}) + (y_i - \lfloor y_i \rfloor_{b_i}) > (\lceil x_i \rceil_{b_i} - x_i) + (\lceil y_i \rceil_{b_i} - y_i),$$

the coordinate will backtrack out of a b_i-gap by increasing from x_i to $\lceil x_i \rceil_{b_i} = x_i^D = y_i^D = \lceil y_i \rceil_{b_i}$, and then decreasing to y_i, for a total distance of $\mu_i + \varepsilon_i^{b_i}$.

Consider again the example paths in $F^{3,3,3}$ shown in Figure 3.1, from $x = (\frac{5}{6}, 0, \frac{14}{27})$ to $y = (\frac{5}{6}, \frac{1}{3}, \frac{29}{54})$. For these values of x and y, we have $\mu_1 = 0$, $\mu_2 = \frac{1}{3}$, and $\mu_3 = \frac{1}{54}$. Since $\frac{5}{6}$ lies in a 2-gap, as does the interval $[\frac{14}{27}, \frac{29}{54}]$, at least one of coordinate 1 and coordinate 3 will have to leave its 2-gap before coordinate 2 moves through the 2-gap $(\frac{1}{9}, \frac{2}{9})$. If coordinate 1 backtracks according to backtracking data $\{(1,2)\}$, it is of type Bd with $\varepsilon_1^2 = \frac{1}{9}$. If coordinate 3 backtracks according to $\{(3,2)\}$, it is of type Bi with $\varepsilon_3^2 = \frac{1}{27}$.

3.3. A shortest path P^D from backtracking data. In this section, we will construct a path P^D and show that it stays in $F^{n,t}$ by considering the relationships among the individual coordinate values of points on the path. The path P^D is a shortest possible taxicab path from x to y given the backtracking data D.

We will use the following two lemmas, which describe how the values x_i, y_i, x_i^D, and y_i^D interact with gaps.

LEMMA 3.12. *If x_i^D or y_i^D is in a k-gap $\gamma = (l, r)$, then $[x_i, y_i] \subset \gamma$.*

PROOF. The result is trivial for a coordinate i of type C. For a coordinate i of type Bd or Bi, $x_i^D = y_i^D$, so we can apply Lemma 2.3 with $a = b_i$ to get that x_i and y_i are in γ, and hence $[x_i, y_i] \subset \gamma$. For a coordinate i of type M, suppose $x_i^D = \lceil x_i \rceil_{m_i}$ is in γ. Then $x_i \in \gamma$ by Lemma 2.3; furthermore, $k < m_i$ as $\lceil x_i \rceil_{m_i}$ is an m_i-fraction contained in the k-gap γ. Now if $\lfloor y_i \rfloor_{m_i}$ is not in γ, then $r \leq \lfloor y_i \rfloor_{m_i} \leq y_i$. Since $k < m_i$ and r is a k-fraction, $r \leq \lfloor y_i \rfloor_k$ also holds. We conclude that $\lceil x_i \rceil_k < r \leq \lfloor y_i \rfloor_k$ which contradicts the minimality of m_i in Definition 3.11. This implies that $\lfloor y_i \rfloor_{m_i}$, like $\lceil x_i \rceil_{m_i}$, is also in γ. Hence y_i is in γ by Lemma 2.3 and thus $[x_i, y_i] \subset \gamma$. An analogous argument applies if $y_i^D = \lfloor y_i \rfloor_{m_i}$ is in γ. □

From Definition 2.2 we get an immediate corollary to Lemma 3.12, since if t coordinates of either x^D or y^D are in k-gaps then x and y are in a k-tunnel and thus not in $F^{n,t}$.

COROLLARY 3.13. *The points x^D and y^D are in $F^{n,t}$.*

LEMMA 3.14. *When any of the intervals $[x_i, x_i^D]$, $[x_i^D, x_i]$, $[x_i^D, y_i^D]$, $[y_i^D, y_i]$, or $[y_i, y_i^D]$ exists and contains a k-gap, then $k \geq \ell$.*

PROOF. The proof is by cases, depending on the coordinate type. When coordinate i is of type C, $x_i = x_i^D = y_i^D = y_i$, so all of the intervals are degenerate of length 0 and the result is vacuously true.

When coordinate i is of type M, we only need to consider the intervals $[x_i, x_i^D]$, $[x_i^D, y_i^D]$, and $[y_i^D, y_i]$, and they are all contained in the interval $[x_i, y_i]$ as $x_i \leq x_i^D \leq y_i^D \leq y_i$. By Definition 3.2, the largest gap any interval $[x_i, y_i]$ can contain is an ℓ-gap, which implies $k \geq \ell$.

When coordinate i is of type Bd, we only need to consider the intervals $[x_i^D, x_i]$ and $[y_i^D, y_i]$. By property (3) of Lemma 3.7, x_i and y_i are in a b_i-gap γ; it has left endpoint $\lfloor x_i \rfloor_{b_i} = \lfloor y_i \rfloor_{b_i}$, and hence $k > b_i$. By property (2) of Lemma 3.7, we know that $b_i \geq \ell$. Therefore, any k-gap contained in $[x_i^D, x_i]$ or $[y_i^D, y_i]$ must satisfy $k > b_i \geq \ell$. An analogous argument holds for coordinates of type Bi. □

We conclude this section with our main theorem on taxicab paths and two corollaries.

THEOREM 3.15. *Let x and y be two points in $F^{n,t}$, and let*

$$D = \{(c_1, b_{c_1}), \ldots, (c_w, b_{c_w})\}$$

be a set of backtracking data for x and y. We can construct a taxicab path P^D from x to y with backtracking data D which stays in $F^{n,t}$ and has total length

$$\mu_1 + \cdots + \mu_n + \varepsilon_{c_1}^{b_{c_1}} + \cdots + \varepsilon_{c_w}^{b_{c_w}}. \tag{3.1}$$

PROOF. As in [4], we divide the path from x to y into 3 subpaths: P_1 from x to x^D, P_2 from x^D to y^D, and P_3 from y^D to y. From Corollary 3.13, we know that x^D and y^D are in $F^{n,t}$.

We first construct the subpath P_3, which starts at y^D. There is no change in coordinates of type C, while coordinates of the other three types change according to the following rounding sequences:

$$\begin{aligned}
&(\mathrm{M}) \quad y_i^D = \lfloor y_i \rfloor_{m_i} \to \lfloor y_i \rfloor_{m_i+1} \to \lfloor y_i \rfloor_{m_i+2} \to \cdots \to y_i \\
&(\mathrm{Bd}) \quad y_i^D = \lfloor y_i \rfloor_{b_i} \to \lfloor y_i \rfloor_{b_i+1} \to \lfloor y_i \rfloor_{b_i+2} \to \cdots \to y_i \\
&(\mathrm{Bi}) \quad y_i^D = \lceil y_i \rceil_{b_i} \to \lceil y_i \rceil_{b_i+1} \to \lceil y_i \rceil_{b_i+2} \to \cdots \to y_i
\end{aligned}$$

We construct the path P_3 in steps. We start with any coordinates of y^D that are 0-fractions, and create path segments by rounding them continuously to 1-fractions according to the rounding sequences (M), (Bd), or (Bi). The rounding is done one coordinate at a time, in any order of the coordinates. We note that since $b_i \geq \ell > 0$ when b_i is defined, any coordinates that might round in the first step will be of type M with $m_i = 0$, which is only possible if $x_i = 0$ or $y_i = 1$. For the second step in the construction of P_3, we create path segments by rounding all 1-fractions to 2-fractions, again in any order of the coordinates; these coordinates have $m_i \leq 1$ or $b_i \leq 1$. We iterate this process, rounding in step $k+1$ from k-fractions to $(k+1)$-fractions, in any order of the coordinates. We remark that, if all of the coordinates of y are level-a fractions for some a, then the construction of P_3 will be complete after all coordinates have been rounded to level-a fractions. The resulting subpath P_3 will be a discrete taxicab path, comprised of only finitely many constituent line segments. Otherwise, at least one coordinate will require an infinite sequence of rounding and the limiting endpoint of P_3 is y.

We need to show that every point of the subpath P_3 is in $F^{n,t}$. Suppose that a is a nonnegative integer and we are in step $a+1$ and rounding coordinate i, so that it is moving continuously either from $\lfloor y_i \rfloor_a$ to $\lfloor y_i \rfloor_{a+1}$ or from $\lceil y_i \rceil_a$ to $\lceil y_i \rceil_{a+1}$. Let the closed interval with either pair of these endpoints be I. If there is a point $p = (p_1, \ldots, p_n)$ on the corresponding path segment of P_3 that is not in $F^{n,t}$, then p_i must be in a k-gap γ while $t-1$ other values $p_{j_1}, \ldots, p_{j_{t-1}}$ remain fixed on that segment and are all simultaneously in k-gaps. We claim that no such p exists.

Suppose otherwise, so that such a point p exists. We first note that it is not possible for one of the endpoints of the interval I to be in γ. For if so, then by Lemma 2.3, all t of the values y_i, y_{j_1}, $\ldots$, $y_{j_{t-1}}$ would be in k-gaps, implying that y is not in $F^{n,t}$. Therefore, the interval I properly contains γ, so $k > a$ by Lemma 2.4. Furthermore, since γ is in I and I is contained in either $[y_i^D, y_i]$ or $[y_i, y_i^D]$, we can also apply Lemma 3.14 to conclude that $k \geq \ell$.

We next note that, at step $a+1$ in the construction of P_3, none of the coordinates $j_1, \ldots, j_{t-1}$ have started their rounding sequences, i.e., we have $p_{j_s} = y_{j_s}^D$ for

$s = 1, \dots, t-1$. This follows from the observation that if any one of the coordinates $j_1, \dots, j_{t-1}$ has started its rounding sequence, then that coordinate has been rounded to an a-fraction or an $(a+1)$-fraction. Since that coordinate is in a k-gap, $k \le a$, which contradicts our earlier statement that $k > a$.

Therefore, by Lemma 3.12, it must be the case that all of the intervals

$$[x_{j_1}, y_{j_1}], \dots, [x_{j_{t-1}}, y_{j_{t-1}}]$$

are in k-gaps. However, this is not possible for the following reason. Using the fact that $k \ge \ell$, by property (4) of Lemma 3.7, one of these $t-1$ coordinates, j_h, is such that $(j_h, b) \in D$, with $b \le k$. Then $y^D_{j_h}$ is a k-fraction, so it is not in a k-gap, which is a contradiction. We conclude that the point p does not exist, and hence the subpath P_3 remains in $F^{n,t}$.

We remark that we can construct subpath P_1 in an analogous way. One follows the procedure described for P_3 to construct P_1^r, the *reverse* of path P_1, from x^D to x. The result is a path between x and x^D which lies in $F^{n,t}$.

We next construct subpath P_2 from x^D to y^D. Since $x^D_i = y^D_i$ for any coordinate i of type C, Bd, or Bi, we only consider coordinates of type M. The subpath is constructed by letting each coordinate i of type M increase continuously from $x^D_i = \lceil x_i \rceil_{m_i}$ to $y^D_i = \lfloor y_i \rfloor_{m_i}$, one coordinate at a time, in any order of these coordinates.

To show that P_2 stays in $F^{n,t}$, let I be the closed interval from $\lceil x_i \rceil_{m_i}$ to $\lfloor y_i \rfloor_{m_i}$. If there is a point $p = (p_1, \dots, p_n)$ on the corresponding path segment of P_2 that is not in $F^{n,t}$, then p_i and $t-1$ other values $p_{j_1}, \dots, p_{j_{t-1}}$ are all simultaneously in k-gaps. The following argument shows that this is not possible. We note that the $t-1$ values $p_{j_1}, \dots, p_{j_{t-1}}$ are all of the form x^D_j or y^D_j; for coordinates of type M, this will depend on whether the coordinate has or has not moved in P_2. Therefore, by Lemma 3.12, the intervals $[x_{j_1}, y_{j_1}], \dots, [x_{j_{t-1}}, y_{j_{t-1}}]$ are in k-gaps. Corollary 3.13 tells us that x^D and y^D are in $F^{n,t}$, so neither endpoint of I can be in a k-gap, and we conclude that I properly contains a k-gap, so $k \ge \ell$. Again, by property (4) of Lemma 3.7, one of the $t-1$ coordinates, j_h, is such that $(j_h, b) \in D$, with $b \le k$. Then, $x^D_{j_h} = y^D_{j_j}$ is a k-fraction, so it is not in a k-gap. This is a contradiction, so we conclude that the point p does not exist, and hence the subpath P_2 remains in $F^{n,t}$.

To finish the theorem we determine the length of P^D. There are n coordinates which separately contribute to the total length of the path: coordinates i of type C and M add length μ_i, while backtracking coordinates j of type Bd and Bi add length $\mu_j + \varepsilon_j^{b_j}$, as given by Definitions 3.9 and 3.10. The result follows. □

COROLLARY 3.16. *The path P^D is a minimal length taxicab path for fixed choices of x, y, and D.*

PROOF. The μ_i terms have to be part of the total path length since any path from x to y must travel in coordinate i at least the distance from x_i to y_i and only one coordinate of a taxicab path changes at a time. Further, the construction of P^D realizes the smallest additional possible lengths that come from backtracking according to D, namely the terms $\varepsilon_{c_i}^{b_{c_i}}$. □

We end this section with a result mentioned at the end of Section 3.1.

COROLLARY 3.17. *Backtracking data sets are path backtracking data sets.*

PROOF. Theorem 3.15 constructs a (taxicab) path P^D from x to y for any backtracking data set D corresponding to x and y. □

3.4. Enumerating minimal sets of backtracking data. In the previous section we produced a taxicab path P^D from x to y whose length is as short as possible for a given set D of backtracking data. The goal of this section is to enumerate all sets of backtracking data that are minimal in two important ways: no pair (c_i, b_{c_i}) can be removed from the set and still leave a valid set of backtracking data for x and y; and no pair (c_i, b_{c_i}) can be replaced by (c_i, b) with $b > b_{c_i}$ and still leave a valid set of backtracking data. The theorem at the end of this section then allows us to determine a minimal D that corresponds to a geodesic taxicab path from x to y.

Lemma 3.5 gives the conditions for backtracking: $t-1$ intervals $[x_j, y_j]$ are in k-gaps for some $k \geq \ell$. Since any $[x_j, y_j]$ can be simultaneously in gaps of many widths, and since backtracking might need to happen at multiple levels, we introduce containment tables to allow us to track this information.

A *containment table* $C(\tilde{x}, \tilde{y})$ can be defined for any pair of points $\tilde{x}$ and $\tilde{y}$ in $F^{n,t}$ with $\tilde{x}_i \leq \tilde{y}_i$ for all i. For our purposes we will only need to focus on the original pair of points x and y, as well as certain modifications of these points that are determined by choices of coordinates that backtrack.

To construct the initial table $C(x, y)$, we first identify all coordinates i such that $\ell_i = \ell$. Without loss of generality, because reflections in the bisecting hyperplanes $z_i = z_j$ preserve $F^{n,t}$ and permit all possible coordinate permutations, we may assume that these are the initial coordinates. The remaining coordinates, where either $\ell_i > \ell$ or ℓ_i does not exist, index the rows of $C(x, y)$, and there are columns for each level $k \geq \ell$. We place an asterisk in row i and column k if $[x_i, y_i]$ is in a k-gap.

The data in the containment table tell us a great deal about how backtracking on a path from x to y might proceed. When a coordinate i with $\ell_i = \ell$ moves through its ℓ-gap, it also moves through gaps of all levels k with $k > \ell$. As this happens, by Lemma 3.3 there cannot be $t-1$ other coordinates in k-gaps for any $k \geq \ell$. This motivates the next definition.

DEFINITION 3.18. The column associated to level k in a containment table is *saturated* if it contains $t-1$ asterisks, i.e., if there are $t-1$ coordinates j such that each $[x_j, y_j]$ is contained in a k-gap.

EXAMPLE 3.19. In $F^{9,4,3}$, consider the points

$$x = \left(0, \frac{1}{3}, \frac{1}{3}, \frac{29}{81}, \frac{31}{81}, \frac{148}{243}, \frac{67}{81}, \frac{33}{162}, \frac{193}{243}\right)$$

and

$$y = \left(1, \frac{2}{3}, \frac{1}{3}, \frac{33}{81}, \frac{32}{81}, \frac{148}{243}, \frac{68}{81}, \frac{33}{162}, \frac{194}{243}\right).$$

We find that $\ell = \ell_1 = \ell_2 = 1$, $\ell_4 = 3$, $\ell_5 = \ell_7 = 4$, and $\ell_9 = 5$, while ℓ_3, ℓ_6, and ℓ_8 do not exist. Since coordinates 1 and 2 have to move through an ℓ-gap, only coordinates 3 through 9 correspond to rows in Table 1(a), where the columns shown are the first five of the levels k with $k \geq \ell$. Coordinate 3, with a constant value of $\frac{1}{3}$, is in no gaps. Columns 1 through 4 are saturated.

TABLE 1. The first five columns of the containment tables for Example 3.19 in $F^{9,4,3}$, before and after accounting for backtracking in coordinate 6 at level 1.

(a) Original $C(x, y)$

		Level				
		1	2	3	4	5
Coordinate	3					
	4	$*$				
	5	$*$		$*$		
	6	$*$		$*$	$*$	
	7		$*$	$*$		
	8		$*$		$*$	$*$
	9		$*$		$*$	

(b) Updated $C(x', y')$

		Level				
		1	2	3	4	5
Coordinate	3					
	4	$*$				
	5	$*$		$*$		
	6					
	7		$*$	$*$		
	8		$*$		$*$	$*$
	9		$*$		$*$	

We note that a containment table $C(\tilde{x}, \tilde{y})$ is directly related to the q-ary expansion of the coordinates of $\tilde{x}$ and $\tilde{y}$: there is an asterisk in entry (i, k) if and only if all q-ary expansions of $\tilde{x}_i$ and $\tilde{y}_i$ have their kth digits equal to $\frac{q-1}{2}$. For example, there is an asterisk in the kth entry of row 8 of $C(x, y)$ for $k = 2$ and for all $k \geq 4$ since

$$x_8 = y_8 = 0.012111\ldots_3.$$

Containment tables are used to create the backtracking tree, which captures all possible minimal sets of backtracking data. If column k of $C(x, y)$ is saturated, on any path from x to y at least one of the coordinates must backtrack at level at most k. We also note that when a coordinate i backtracks out of a level-k gap, it is no longer in k'-gaps for $k' \geq k$. Thus if we modify the values of x and y by letting x' and y' agree with x and y, except that coordinate i backtracks by the smallest possible amount at level k (so that $x'_i = y'_i$ is equal to x_i and y_i rounded down or up to level k as in Definition 3.10), the updated containment table $C(x', y')$ will no longer contain asterisks in entries (i, k') for $k' \geq k$ but will otherwise agree with $C(x, y)$. An example is shown in Table 1(b), where we unsaturate column 1 by having x' and y' agree with x and y in Example 3.19, except that x_6 and y_6 have been rounded up to give $x'_6 = y'_6 = \lceil x_6 \rceil_1 = \lceil y_6 \rceil_1 = 2/3$. There are no longer asterisks in columns 1, 3, and 4 of the row for coordinate 6, which means that columns 3 and 4 are now also unsaturated.

The *backtracking tree* $B(x, y)$ is created by making modifications to containment tables in the manner just described. It is a labelled, rooted, and directed tree, with each vertex labelled by an ordered pair of points. The root vertex of $B(x, y)$ is $\langle x, y \rangle$. There is a directed edge labelled (i, k) leaving a vertex $\langle \tilde{x}, \tilde{y} \rangle$ if coordinate i contributes one of the $t-1$ asterisks in the saturated column of $C(\tilde{x}, \tilde{y})$ with smallest (left-most) column index k. The edge (i, k) enters a vertex $\langle \tilde{x}', \tilde{y}' \rangle$ with $\tilde{x}'_i = \tilde{y}'_i$ equal to $\tilde{x}_i$ and $\tilde{y}_i$ rounded up or down to level k and with all of the other coordinates of $\tilde{x}'$ and $\tilde{y}'$ equaling the corresponding coordinates of $\tilde{x}$ and $\tilde{y}$, respectively. The vertex $\langle \tilde{x}, \tilde{y} \rangle$ will have outdegree $t - 1$, with one edge for each of the asterisks in column k.

Thus the edges in $B(x, y)$ emanating from the root vertex $\langle x, y \rangle$ correspond to the $t - 1$ coordinates with asterisks at the smallest column index for a saturated

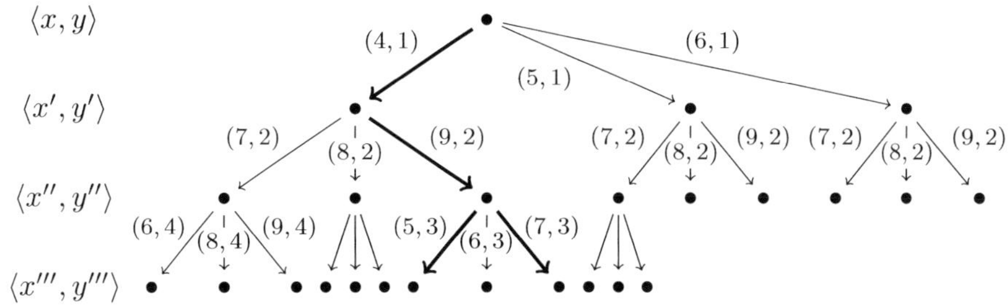

FIGURE 3.2. Directed paths in $B(x, y)$ from the root $\langle x, y\rangle$ to a leaf correspond to all minimal sets of backtracking data for Example 3.19. The two bold directed paths in $B(x, y)$ correspond to a geodesic taxicab path from x to y, as explained at the end of Section 3.5.

column in $C(x, y)$, if such a column exists; and the containment tables for leaf vertices have no saturated columns. The backtracking tree is finite because every edge (i, k) taken on a directed path from the root vertex eliminates an additional coordinate from contributing to saturated columns at levels greater than k, so there are no saturated columns remaining after at most $n - t + 1$ edges have been taken.

In the backtracking tree for Example 3.19, the root vertex $\langle x, y\rangle$ is connected to three other vertices via edges $(4, 1)$, $(5, 1)$, and $(6, 1)$; see Figure 3.2. The edge $(4, 1)$ leads to three other edges, $(7, 2)$, $(8, 2)$, and $(9, 2)$, because after coordinate 4 backtracks at level 1 and thus unsaturates column 1 in Table 1(a), column 2 of $C(x, y)$ is the lowest-level saturated column remaining. Taking the edge $(6, 1)$ instead leads to the containment table in Table 1(b) and the right-most branch of Figure 3.2. There are 25 vertices in the tree, with the 17 leaves corresponding to containment tables with no saturated columns.

After $B(x, y)$ is constructed we can create sets of backtracking data by simply collecting the edge labels along any directed path from the root to a leaf. These sets of backtracking data look like

$$D = \{(c_1, b_{c_1}), \ldots, (c_w, b_{c_w})\} \tag{3.2}$$

where c_i is the backtracking coordinate corresponding to the ith edge on a directed path from the root, and b_{c_i} is its backtracking level. Moreover, the backtracking levels strictly increase, $b_{c_1} < \ldots < b_{c_w}$, because the smallest column index for a saturated column is used at each step in creating the tree. Each set D derived from $B(x, y)$ has the four properties listed in Lemma 3.7, and thus Theorem 3.15 provides a shortest taxicab path corresponding to it.

We end this section with Theorem 3.20.

THEOREM 3.20. *Let P be a taxicab path P from x to y in $F^{n,t}$ with path backtracking data set D. Then D determines a backtracking data set D' that is derived from the backtracking tree $B(x, y)$, with the taxicab path $P^{D'}$ from x to y being at least as short as P.*

PROOF. By Definition 3.6, any (taxicab) path P from x to y determines a path backtracking data set D, which is a backtracking data set by Definition 3.8. For any $k \geq \ell$, if $t - 1$ coordinate intervals are in k-gaps, then by Lemma 3.5 at least one of these $t-1$ coordinates must backtrack out of its k-gap as the path travels from x to y. This in turn implies that property (4) of Lemma 3.7 is satisfied. As a consequence, the backtracking determined by D must unsaturate all of the columns of the containment table $C(x,y)$.

The construction of the backtracking tree implies that D contains as a subset one of the finitely many sets D' of backtracking data derived from the backtracking tree (a proper subset if a coordinate that does not need to backtrack does so anyway), after some backtracking levels b_{c_i} are possibly replaced with larger values (if a backtracking coordinate c_i backtracks out of a larger gap than necessary). By Theorem 3.15, $P^{D'}$ is a taxicab path from x to y. By Definition 3.10 and Corollary 3.16, the taxicab path $P^{D'}$ is at least as short as P. □

Thus, any geodesic taxicab path in $F^{n,t}$ from x to y is determined by some set of backtracking data derived from the backtracking tree.

3.5. Efficiently finding geodesic paths. Theorem 3.15 provides the length of a shortest possible taxicab path associated to any set of backtracking data D, and it is clear from equation (3.1) that to determine a geodesic taxicab path from x to y we should choose backtracking coordinates to minimize the backtracking sum $\varepsilon_{c_1}^{b_{c_1}} + \cdots + \varepsilon_{c_w}^{b_{c_w}}$. In this section we show that rather than calculate these sums over all possible directed paths in the backtracking tree from its root to its leaves, we can determine a (possibly nonunique) taxicab path of shortest length in a greedy fashion.

THEOREM 3.21. *A geodesic taxicab path P^D from x to y in $F^{n,t}$ corresponds to backtracking data*

$$D = \{(c_1, b_{c_1}), \ldots, (c_w, b_{c_w})\}$$

that is derived from a directed path in the backtracking tree $B(x,y)$ in the following manner: starting at the root vertex in $B(x,y)$, the ith directed edge labelled (c_i, b_{c_i}) is chosen to give a smallest value of $\varepsilon_{c_i}^{b_{c_i}}$ among the $t-1$ edges that depart from the vertex at the beginning of that edge.

PROOF. If D is empty, the backtracking tree consists of the single vertex $\langle x, y\rangle$, and the path P^D is thus a geodesic taxicab path from x to y because only the required monotonic displacement occurs in each coordinate. If D is nontrivial, suppose we have chosen the first $h \geq 0$ elements $(c_1, b_{c_1}), \ldots, (c_h, b_{c_h})$ of the backtracking data, so we are at a vertex $\langle x', y'\rangle$ at the end of a partial directed path consisting of h edges, starting from the root vertex. Assume that the next backtracking level is a, the smallest level at which $C(x', y')$ has a saturated column. We will show that an optimal choice i for the next backtracking coordinate has the smallest value ε_i^a among the $t-1$ available choices, regardless of subsequent backtracking.

We proceed by contradiction, assuming that a choice of one of the other $t-1$ coordinates, j, eventually results in a strictly shorter path, even though $\varepsilon_j^a > \varepsilon_i^a$. Let S be any set of backtracking data containing $(c_1, b_{c_1}), \ldots, (c_h, b_{c_h})$, and (j, a). We first suppose that in $C(x', y')$, whenever coordinate j contributes to a saturated column at level $a' > a$, then coordinate i does, too. Any additional elements of S provide choices of backtracking coordinates and levels greater than a which account for all remaining backtracking requirements. However, we could swap (i, a) for (j, a)

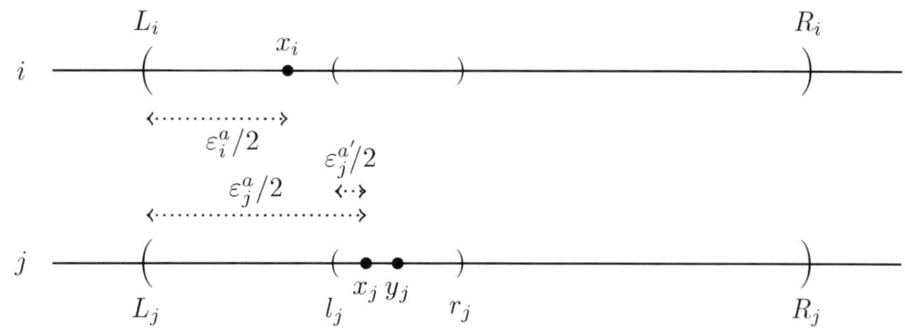

FIGURE 3.3. As described in the referencing text, the distance required for x_j to round down out of its a-gap (L_j, R_j) is at least as long as the sum of the distances for x_i to round down out of its a-gap and for x_j to round down out of its a'-gap for $a' > a$.

in S and the elements of S would still account for all backtracking requirements, with a smaller backtracking sum since $\varepsilon_j^a > \varepsilon_i^a$.

We are left to consider the case that coordinate j contributes to some saturated column at level $a' > a$, but that coordinate i does not; and we assume that a' is the smallest such level. We note that neither x_i nor y_i can be in an a'-gap because there are $t-1$ intervals $[x_{i'}, y_{i'}]$ in a'-gaps with $i' \neq i$, which would imply that either x or y would be in an a'-tunnel and hence not in $F^{n,t}$.

We remark that any two gaps at the same level have identical structure in terms of the number and placement of smaller gaps they contain. In addition, these sets of smaller gaps are centrally symmetric with respect to the containing gap, i.e., if (l, r) is a gap strictly contained in (L, R), then (L, R) also contains the gap $(l', r') = (R + L - r, R + L - l)$.

We will denote the a-gaps which contain $[x_i, y_i]$ and $[x_j, y_j]$ by (L_i, R_i) and (L_j, R_j), respectively, and (l_j, r_j) is the a'-gap containing $[x_j, y_j]$. By central symmetry, we may also assume that the backtracking in coordinate j at level a is by rounding down. Then we have

$$\varepsilon_j^a = 2(x_j - L_j) = 2(x_j - l_j) + 2(l_j - L_j). \tag{3.3}$$

See Figure 3.3. Since x_j is in the a'-gap (l_j, r_j), $2(x_j - l_j) \geq \varepsilon_j^{a'}$. If the backtracking in coordinate i happens by rounding down, then as x_i is not in an a'-gap and $\varepsilon_i^a < \varepsilon_j^a$, we know that x_i is no further from L_i than l_j is from L_j, so we have $\varepsilon_i^a = 2(x_i - L_i) \leq 2(l_j - L_j)$. Next, assume that the backtracking in coordinate i happens by rounding up, and recall that the gaps (l, r) and (l', r') are centrally symmetric in (L, R). Then as coordinate j is in the a'-gap (l_j, r_j), by central symmetry and the fact that $\varepsilon_i^a < \varepsilon_j^a$, we must have $2(l_j - L_j) = 2(R_j - r_j') \geq 2(R_i - y_i) = \varepsilon_i^a$. In both cases, we substitute into equation (3.3) to find

$$\varepsilon_j^a \geq \varepsilon_j^{a'} + 2(l_j - L_j) \geq \varepsilon_j^{a'} + \varepsilon_i^a. \tag{3.4}$$

Equation (3.4) shows that we could replace (j, a) in S with the pairs (i, a) and (j, a') and not increase the backtracking sum, so long as these replacements unsaturate any columns in $C(x', y')$ that j had unsaturated at levels greater than or equal to a. Indeed, including (i, a) in S unsaturates relevant columns from a up to $a'-1$ due to the minimality of a', while including (j, a') unsaturates any relevant higher-level columns. □

Returning to Example 3.19, the set of backtracking data $\{(4,1),(9,2),(5,3)\}$ corresponds to a geodesic taxicab path from x to y with

$$\varepsilon_4^1 + \varepsilon_9^2 + \varepsilon_5^3 = \frac{26}{243},$$

where $\varepsilon_4^1 = 4/81$ and $\varepsilon_9^2 = 8/243$ are minimal values among the options at levels 1 and 2. The backtracking data $\{(4,1),(9,2),(7,3)\}$ labels the other bold directed path on the backtracking tree in Figure 3.2, corresponding to a different taxicab path with the same shortest length, as $\varepsilon_5^3 = \varepsilon_7^3 = 2/81 < 8/243 = \varepsilon_6^3$.

4. Comparing taxicab and Euclidean metrics

In Section 3 we constructed a geodesic taxicab path P between any two points x and y in $F^{n,t}$. In this section we determine a sharp upper bound for

$$\rho(x,y) = \frac{d_t(x,y)}{d_e(x,y)},$$

where $d_t(x,y)$ is the length of a geodesic taxicab path from x to y and $d_e(x,y)$ is the standard Euclidean distance from x to y.

First we consider the case when none of the coordinates require backtracking, so the backtracking data set for P is empty.

THEOREM 4.1. *If a geodesic taxicab path in $F^{n,t}$ from x to y does not backtrack, then*

$$\rho(x,y) \leq \sqrt{n},$$

and there are points x and y which give equality.

PROOF. When there is no backtracking, there are no $\varepsilon_{c_i}^{b_{c_i}}$ terms in equation (3.1), so the ratio of metrics has the form

$$\rho(x,y) = \frac{\mu_1 + \mu_2 + \cdots + \mu_n}{\sqrt{\mu_1^2 + \mu_2^2 + \cdots + \mu_n^2}}.$$

The result then follows from a straightforward Cauchy–Schwarz argument given for [**4**, Lemma 4.1], with $x = (0,\ldots,0)$ and $y = (1,\ldots,1)$ realizing the bound. □

Next, we assume that backtracking is required in at least one coordinate. From the discussion following equation (3.2) we know that the backtracking levels for P are distinct; we can label them as $b_1 < b_2 < \cdots < b_w$. Since b_w is one of these levels, there must be $t-1$ coordinates i with $[x_i, y_i]$ in a b_w-gap, only one of which will backtrack.

Consider the $t-1$ coordinates that comprise the saturated b_w column. By Theorem 3.21, there is a shortest path from x to y that will backtrack in the coordinate whose backtracking distance at the b_w level is minimal. However, we can choose to backtrack in a different coordinate, paying the potential penalty of increasing the path length slightly from the minimum. We elect to backtrack in the coordinate i whose value μ_i is the largest among those $t-1$ potential backtracking coordinates at level b_w. The resulting ratio of this taxicab distance to the Euclidean distance will be greater than or equal to $\rho(x,y)$. In fact, we will later see that when this ratio is maximized, all of these μ_i values are equal, so this ratio equals $\rho(x,y)$.

Without loss of generality in what follows, we may assume that the coordinates with the potential to backtrack at level b_w are the first $t-1$ coordinates, and that the first coordinate is the one with the largest associated value of μ_i among these $t-1$ coordinates, i.e., $\mu_1 \geq \mu_j$ for all $1 \leq j \leq t-1$. We also assume that $\mu_n \geq \mu_j$ for all $1 \leq j \leq n$. We can do this as b_w is one of the backtracking levels, which means that there is a coordinate i such that $[x_i, y_i]$ contains a b_w-gap, namely any coordinate associated to ℓ. This coordinate is not one of the first $t-1$ coordinates, so we may set $i = n$. By this discussion, $\mu_n \geq q^{-\ell}$.

Associated to each backtracking coordinate i is its additional backtracking distance $\varepsilon_i^{b_i}$. To simplify notation below, we will denote these additional distances as ε^{b_i}, recalling that the additional backtracking distance associated with coordinate 1 is ε^{b_w}. We get a first bound on

$$\rho(x,y) = \frac{d_t(x,y)}{d_e(x,y)} = \frac{\mu_1 + \cdots + \mu_n + \varepsilon^{b_1} + \cdots + \varepsilon^{b_w}}{\sqrt{\mu_1^2 + \cdots + \mu_n^2}}$$

by bounding the sizes of the ε^{b_i} terms. To start, $\mu_1 + \varepsilon^{b_w} \leq q^{-b_w}$, since the total distance traveled by a backtracking coordinate is at most the length of the gap it backtracks out of. Next, consider a backtracking coordinate which backtracks out of a b_i-gap with $i < w$. We claim that

$$\varepsilon^{b_i} \leq q^{-b_i} - q^{-b_{i+1}}. \tag{4.1}$$

This is straightforward when $\mu_i \geq q^{-b_{i+1}}$, since $\varepsilon^{b_i} \leq q^{-b_i} - \mu_i \leq q^{-b_i} - q^{-b_{i+1}}$. If $\mu_i < q^{-b_{i+1}}$, then we claim that neither x_i nor y_i can be in a b_{i+1}-gap. We note that, after backtracking at level b_i, level b_{i+1} remains saturated, with $t-1$ coordinates in b_{i+1}-gaps. If either x_i or y_i were also in such a gap, then x or y would be in a b_{i+1}-tunnel and not in $F^{n,t}$. Since $\mu_i < q^{-b_{i+1}}$, x_i and y_i must lie on the same side of the central b_{i+1}-gap inside of the b_i-gap. This implies that $\varepsilon^{b_i} \leq q^{-b_i} - q^{-b_{i+1}}$ in this case as well.

We apply inequality (4.1) to the sum

$$\varepsilon^{b_1} + \varepsilon^{b_2} + \cdots + \varepsilon^{b_{w-1}} + \mu_1 + \varepsilon^{b_w}$$

and find that it is no larger than

$$(q^{-b_1} - q^{-b_2}) + \cdots + (q^{-b_{w-1}} - q^{-b_w}) + q^{-b_w} = q^{-b_1} \leq q^{-\ell}.$$

Hence

$$\rho(x,y) \leq \frac{q^{-\ell} + \mu_2 + \cdots + \mu_n}{\sqrt{\mu_1^2 + \cdots + \mu_n^2}}. \tag{4.2}$$

The following two lemmas will lead us to Theorem 4.4 and the sharp bound on $\rho(x,y)$ we are seeking. The lemmas are proved in the cases when $2 < t < n$; the cases when $t = 2$ or $t = n$ involve some simplified expressions in the proofs and lead to the same bound. We note that the case when $t = 2$ and $q = 3$ is covered in [**4**].

First, we show that we can bound $\rho(x,y)$ using expressions that combine most of the μ_i terms.

LEMMA 4.2. *Let $2 < t < n$, with $\mu_1 \geq \mu_j$ for $j \leq t-1$. Then*

$$\rho(x,y) \leq \frac{q^{-\ell} + \mu_2 + \cdots + \mu_{t-1} + \mu_t + \cdots + \mu_{n-1} + \mu_n}{\sqrt{\mu_1^2 + \mu_2^2 + \cdots + \mu_{t-1}^2 + \mu_t^2 + \cdots + \mu_{n-1}^2 + \mu_n^2}}$$
$$\leq \frac{q^{-\ell} + (t-2)A + (n-t)B + \mu_n}{\sqrt{(t-1)A^2 + (n-t)B^2 + \mu_n^2}}$$

where $A = \sqrt{\frac{\mu_2^2 + \cdots + \mu_{t-1}^2}{t-2}}$ *and* $B = \sqrt{\frac{\mu_t^2 + \cdots + \mu_{n-1}^2}{n-t}}$.

PROOF. The first line of the lemma is inequality (4.2). To get the second inequality, note that by the Cauchy–Schwarz inequality we have

$$\mu_2 + \cdots + \mu_{t-1} \leq \sqrt{(t-2)(\mu_2^2 + \cdots + \mu_{t-1}^2)} = (t-2)A.$$

Similarly,

$$\mu_t + \cdots + \mu_{n-1} \leq \sqrt{(n-t)(\mu_t^2 + \cdots + \mu_{n-1}^2)} = (n-t)B.$$

Also, since $\mu_1^2 \geq \mu_j^2$ for $2 \leq j \leq t-1$, we have

$$\mu_1^2 \geq \frac{\mu_2^2 + \cdots + \mu_{t-1}^2}{t-2} = A^2.$$

By applying the first two displayed inequalities in the numerator and the third in the denominator, we arrive at our stated bound. □

We note that the cases $t = 2$ and $t = n$ correspond to setting $A = 0$ and $B = 0$, respectively, in Lemma 4.2. The inequality of the lemma holds in these cases as well.

Our next lemma uses the fact that $[x_n, y_n]$ contains an ℓ-gap.

LEMMA 4.3. *Let $2 < t < n$. With $\mu_n \geq q^{-\ell}$, the expression*

$$\frac{q^{-\ell} + (t-2)A + (n-t)B + \mu_n}{\sqrt{(t-1)A^2 + (n-t)B^2 + \mu_n^2}} \tag{4.3}$$

obtains its maximum possible value of

$$\sqrt{n + 1 + \frac{1}{t-1}}.$$

when

$$\mu_n = q^{-\ell}, A = \frac{q^{-\ell}(t-2)}{2(t-1)}, \text{ and } B = \frac{q^{-\ell}}{2}.$$

PROOF. We write $\langle A, \ldots, A, B, \ldots, B, \mu_n \rangle = s\langle v_1, \ldots, v_1, v_2, \ldots, v_2, v_3 \rangle$, where A and B have multiplicities $t-1$ and $n-t$, respectively, and $\langle v_1, \ldots, v_1, v_2, \ldots, v_2, v_3 \rangle$ is a unit vector. Expression (4.3) then takes the form

$$\frac{q^{-\ell} + (t-2)v_1 s + (n-t)v_2 s + v_3 s}{s\sqrt{(t-1)v_1^2 + (n-t)v_2^2 + v_3^2}} = \frac{q^{-\ell}}{s} + (t-2)v_1 + (n-t)v_2 + v_3.$$

The restriction that $q^{-\ell} \leq \mu_n$ is equivalent to $q^{-\ell} \leq v_3 s$, or $\frac{q^{-\ell}}{s} \leq v_3$. Therefore, we have the inequality

$$\frac{q^{-\ell}}{s} + (t-2)v_1 + (n-t)v_2 + v_3 \leq (t-2)v_1 + (n-t)v_2 + 2v_3 = h(v_1, v_2, v_3).$$

Let $h(v_1, v_2, v_3)$ be our objective function, with the variables v_1, v_2, and v_3 constrained by the equation

$$(t-1)v_1^2 + (n-t)v_2^2 + v_3^2 = 1.$$

We apply the method of Lagrange multipliers and note that the resulting solution will be a maximum since the objective function is a plane whose normal points towards the first octant and the constraint is an ellipsoid centered at the origin.

The method of Lagrange multipliers generates the following system of equations:

$$\begin{aligned} t-2 &= 2\lambda(t-1)v_1 \\ n-t &= 2\lambda(n-t)v_2 \\ 2 &= 2\lambda v_3 \\ (t-1)v_1^2 + (n-t)v_2^2 + v_3^2 &= 1. \end{aligned}$$

From the first three equations, we find that

$$\frac{1}{2\lambda} = \frac{(t-1)v_1}{t-2} = v_2 = \frac{v_3}{2},$$

and the fourth equation then leads to

$$v_3 = 2\sqrt{\frac{t-1}{nt+t-n}}.$$

Therefore,

$$v_1 = \frac{t-2}{\sqrt{(t-1)(nt+t-n)}} \quad \text{and} \quad v_2 = \sqrt{\frac{t-1}{nt+t-n}},$$

and it can be checked that h attains the maximum value given in the statement of the lemma for these values of the v_i. Expression (4.3) also equals the stated maximum when $\mu_n = v_3 s = q^{-\ell}$ and

$$A = \frac{q^{-\ell}(t-2)}{2(t-1)} \quad \text{and} \quad B = \frac{q^{-\ell}}{2}. \qquad \square$$

In the boundary cases of $t = 2$ or $t = n$ we apply similar methods and it is straightforward to check that we get the same bounds.

The main result of this section follows directly from Lemmas 4.2 and 4.3:

THEOREM 4.4. *For any distinct points x and y in $F^{n,t,q}$, we have*

$$\rho(x,y) \leq \sqrt{n+1+\frac{1}{t-1}},$$

and there are points x and y which give equality.

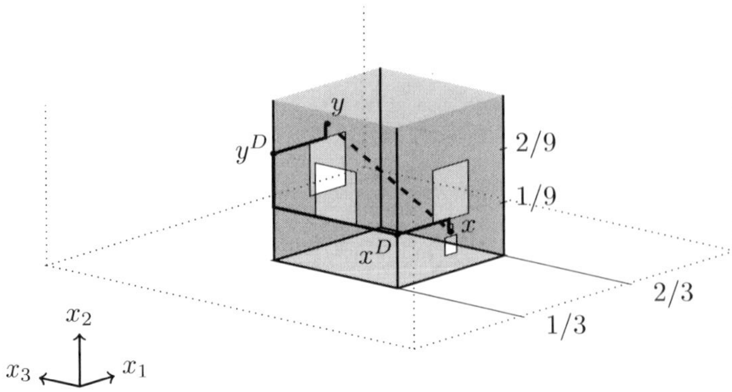

FIGURE 4.1. A geodesic path (solid) and the straight-line Euclidean path (dashed) are displayed from $x = (1/2, 1/12, 1/3)$ to $y = (1/2, 3/12, 2/3)$ in the Menger sponge $F^{3,2,3}$, with $\rho(x, y) = \sqrt{5}$. The two points are on opposite sides of a vertical level-1 tunnel, with square windows signaling several higher-level tunnels that the geodesic path must also avoid.

PROOF. To show that the bound is sharp, let

$$x = \left(\underbrace{\frac{1}{2} - \frac{t-2}{4q(t-1)}, \ldots, \frac{1}{2} - \frac{t-2}{4q(t-1)}}_{t-1}, \underbrace{\frac{1}{4q}, \ldots, \frac{1}{4q}}_{n-t}, \frac{q-1}{2q} \right),$$

$$y = \left(\underbrace{\frac{1}{2} + \frac{t-2}{4q(t-1)}, \ldots, \frac{1}{2} + \frac{t-2}{4q(t-1)}}_{t-1}, \underbrace{\frac{3}{4q}, \ldots, \frac{3}{4q}}_{n-t}, \frac{q+1}{2q} \right).$$

Any one of the first $t-1$ coordinates can be taken as the single backtracking coordinate, and we have $\ell = 1$, $\varepsilon^1 = \frac{t}{2q(t-1)}$,

$$\mu_1 = \cdots = \mu_{t-1} = \frac{t-2}{2q(t-1)}$$
$$\mu_t = \cdots = \mu_{n-1} = \frac{1}{2q}$$

and $\mu_n = \frac{1}{q}$. □

Figure 4.1 shows the example of Theorem 4.4 hitting the sharp bound of $\rho(x, y) = \sqrt{5}$ in the case of the Menger sponge $F^{3,2,3}$.

Acknowledgment

We thank Robert Niemeyer and the reviewers for thoughtful comments that significantly improved this paper.

References

[1] Christoph Bandt and Mohamed Mubarak, *Distribution of distances and interior distances for certain self-similar measures* (English, with English and Arabic summaries), Arab. J. Sci. Eng. Sect. C Theme Issues **29** (2004), no. 2, 111–124. MR2126592

[2] M. T. Barlow, *Analysis on the Sierpinski carpet*, Analysis and geometry of metric measure spaces, CRM Proc. Lecture Notes, vol. 56, Amer. Math. Soc., Providence, RI, 2013, pp. 27–53, DOI 10.1090/crmp/056/02. MR3060498

[3] Martin T. Barlow and Richard F. Bass, *Brownian motion and harmonic analysis on Sierpinski carpets*, Canad. J. Math. **51** (1999), no. 4, 673–744, DOI 10.4153/CJM-1999-031-4. MR1701339

[4] Ethan Berkove and Derek Smith, *Geodesics in the Sierpinski carpet and Menger sponge*, Fractals **28** (2020), no. 7, 2050120, DOI: 10.1142/S0218348X20501200.

[5] Dmitri Burago, Yuri Burago, and Sergei Ivanov, *A course in metric geometry*, Graduate Studies in Mathematics, vol. 33, American Mathematical Society, Providence, RI, 2001, DOI 10.1090/gsm/033. MR1835418

[6] Joe P. Chen and Robert G. Niemeyer, *Periodic billiard orbits of self-similar Sierpiński carpets*, J. Math. Anal. Appl. **416** (2014), no. 2, 969–994, DOI 10.1016/j.jmaa.2014.03.001. MR3188750

[7] Ligia L. Cristea, *A geometric property of the Sierpiński carpet*, Quaest. Math. **28** (2005), no. 2, 251–262, DOI 10.2989/16073600509486126. MR2146391

[8] Estibalitz Durand-Cartagena and Jeremy T. Tyson, *Rectifiable curves in Sierpiński carpets*, Indiana Univ. Math. J. **60** (2011), no. 1, 285–309, DOI 10.1512/iumj.2011.60.4382. MR2952419

[9] Estibalitz Durand-Cartagena and Jeremy T. Tyson, *Erratum to: Rectifiable curves in Sierpiński carpets, Volume 60:1 (2011), 285–309 [MR2952419]*, Indiana Univ. Math. J. **62** (2013), no. 1, 355–356, DOI 10.1512/iumj.2013.62.5346. MR3158512

[10] Kenneth Falconer, *Fractal geometry*, 3rd ed., John Wiley & Sons, Ltd., Chichester, 2014. Mathematical foundations and applications. MR3236784

[11] Jiangwen Gu, Jiaqi Fan, Qianqian Ye, and Lifeng Xi, *Mean geodesic distance of the level-n Sierpinski gasket*, J. Math. Anal. Appl. **508** (2022), no. 1, Paper No. 125853, 31, DOI 10.1016/j.jmaa.2021.125853. MR4345992

[12] Jiangwen Gu, Qianqian Ye, and Lifeng Xi, *Geodesics of higher-dimensional Sierpinski gasket*, Fractals **27** (2019), no. 4, 1950049, 10, DOI 10.1142/S0218348X1950049X. MR3980704

[13] Yiming Li and Lifeng Xi, *Manhattan property of geodesic paths on self-affine carpets*, Arch. Math. (Basel) **111** (2018), no. 3, 279–285, DOI 10.1007/s00013-018-1199-4. MR3834922

[14] Alexandru Mihail, *On the connectivity of attractors of iterated function systems*, Rocky Mountain J. Math. **40** (2010), no. 6, 1949–1964, DOI 10.1216/RMJ-2010-40-6-1949. MR2764231

[15] Mustafa Saltan, Yunus Özdemir, and Bünyamin Demir, *Geodesics of the Sierpinski gasket*, Fractals **26** (2018), no. 3, 1850024, 8, DOI 10.1142/S0218348X1850024X. MR3817232

[16] Robert S. Strichartz, *Isoperimetric estimates on Sierpinski gasket type fractals*, Trans. Amer. Math. Soc. **351** (1999), no. 5, 1705–1752, DOI 10.1090/S0002-9947-99-01999-6. MR1433127

[17] Xuemin Wang and Jiaqi Fan, *Mean geodesic distance of n-level Sierpinski carpet*, Fractals **29** (2021), no. 4, 2150096, DOI: 10.1142/S0218348X21500961.

[18] Songjing Wang, Zhouyu Yu, and Lifeng Xi, *Average geodesic distance of Sierpinski gasket and Sierpinski networks*, Fractals **25** (2017), no. 5, 1750044, 8, DOI 10.1142/S0218348X1750044X. MR3693010

[19] Jinjin Yang, Songjing Wang, Lifeng Xi, and Yongchao Ye, *Average geodesic distance of skeleton networks of Sierpinski tetrahedron*, Phys. A **495** (2018), 269–277, DOI 10.1016/j.physa.2017.12.051. MR3778874

[20] Luming Zhao, Songjing Wang, and Lifeng Xi, *Average geodesic distance of Sierpinski carpet*, Fractals **25** (2017), no. 6, 1750061, 8, DOI 10.1142/S0218348X1750061X. MR3725300

[21] Jiali Zhu, Li Tian, and Qin Wang, *Average geodesic distance on Sierpinski hexagon and Sierpinski hexagon networks*, Fractals **27** (2019), no. 5, 1950077, 9, DOI 10.1142/S0218348X19500774. MR3999126

Department of Mathematics, Lafayette College, Easton, Pennsylvania 18042

Department of Mathematics, Lafayette College, Easton, Pennsylvania 18042

Department of Mathematics, Lafayette College, Easton, Pennsylvania 18042

Contemporary Mathematics
Volume **797**, 2024
https://doi.org/10.1090/conm/797/15951

IFS measures on generalized Bratteli diagrams

Sergey Bezuglyi and Palle E. T. Jorgensen

Dedicated to the memory of Robert Strichartz

ABSTRACT. The purpose of or paper is a general analysis of path space measures. Our focus is a certain path space analysis on generalized Bratteli diagrams. We use this in a systematic study of systems of self-similar measures (the term "IFS measures" is used in the paper) for both types of such diagrams, discrete and continuous. In special cases, such measures arise in the study of iterated function systems (IFS). In the literature, similarity may be defined by, e.g., systems of affine maps (Sierpinski), or systems of conformal maps (Julia). We study new classes of semi-branching function systems related to stationary Bratteli diagrams. The latter play a big role in our understanding of new forms of harmonic analysis on fractals. The measures considered here arise in classes of discrete-time, multi-level dynamical systems where a similarity is specified between levels. These structures are made precise by prescribed systems of functions which in turn serve to define self-similarity, i.e., similarity of large scales, and small scales. For path space systems, in our main result, we give a necessary and sufficient condition for existence of such generalized IFS measures. For the corresponding semi-branching function systems, we further identify the measures which are also shift-invariant.

CONTENTS

1. Introduction

We present new results in the study of self-similar measures on path space for infinite graph models,here called generalized Bratteli diagrams. By this, we mean particular graph systems B with the property that the corresponding sets V of vertices, and E of edges, admit discrete level-structures, defined below. This means

2020 *Mathematics Subject Classification.* Primary 28A33, 28A80, 37B10, 46G12, 60J22.
Key words and phrases. Bratteli diagrams, invariant path space measures, self-similar measures, iterated and semibranching function systems, IFS measures.

that V is a disjoint union of the sets V_n and E is a disjoint union of the sets E_n. We emphasize that the class of generalized Bratteli diagrams include two cases: (1) all the sets V_n, E_n are countable (if all V_n's are finite, then we have a standard Bratteli diagram); (2) all the set V_n, E_n are standard Borel spaces. In case (2), we will say that B is a measurable (or continuous) Bratteli diagram.

Our present framework is motivated by, but more versatile than, the more familiar setting in earlier studies of Bratteli diagrams. We use the discrete levels in order to identify a class of self-similar path space measures, called iterated function system (IFS) measures. The latter in turn are inspired by earlier studies of IFS systems arising the analysis of fractals, such as various Sierpinski constructs, or conformal attractors. However, by contrast our analysis centers around a new path space analysis for discrete-time random walk models in generalized Bratteli diagrams. We also discuss IFS measures for the case when the system of levels making up B are instead standard measure spaces, so nondiscrete.

We recall that a measure μ on a standard Borel space X is called *self-similar* with respect to an iterated function system $f_i : X \to X, i \in I$, if

$$\mu = \sum_{i\in I} p_i \mu \circ f_i^{-1},$$

where $p = (p_i)_{i\in I}$ is a probability vector and I is finite or countable.

A main aim of our paper is an identification of, and an analysis of, iterated function system (IFS) measures on the path space X_B of generalized Bratteli diagrams. This entails two tasks, (a) an analysis of IFS structure of the path spaces, and (b) a study of the particular IFS measures on them.

Our main results consist of finding an explicit construction of IFS measures for both discrete and measurable generalized Bratteli diagrams.

In order to motivate, and to place this in context, we add the following three comments here:

(i) Previously, IFS measures have been considered in special geometries of self-similar fractals (see, e.g., [**35**] and the papers cited there). These standard self-similar fractals may be realized in finite-dimensional Euclidean space. By contrast, there are no previous studies of IFS measures in path space, i.e., IFS measures realized in the class of path space structures considered here, standard and generalized Bratteli diagrams.

(ii) While there are earlier results on other, different but related, classes of measures on path space of generalized Bratteli diagrams, e.g., tail invariant measures, and Markov measures, our present identification of IFS measures on path space is new.

(iii) In order to prepare the reader for the IFS path space measures, it will be necessary for us to begin with an account of tail invariant measures, and Markov measures. The tools involved there for generalized Bratteli diagrams are also needed in our introduction of the new IFS measures. But this means our main results for IFS measures will be postponed to Sections 3 and 4 below, after the necessary preliminaries have been presented.

Generalized Bratteli diagrams considered here arise in various areas and have many applications. We mention Cantor and Borel dynamics where they are used to construct models of transformations, see [**30**], [**29**], [**20**], [**4**], [**6**]. Measurable Bratteli diagrams can be met in the theory of Markov chains, we refer to [**42**] as an example of such literature. Several other applications come to mind, (1) models from financial mathematics, and (2) neural networks. In each case, the role of the IFS measures must be specified. In some applications, there are important limit theorems, for example for financial derivative models both discrete and continuous pricing formulas are important. And it may be stated in the setting of generalized Bratteli diagrams, measurable setting. We mean first of all financial derivative models, e.g., binomial models; vs continuous/measurable (cm), e.g., pricing of options via Ito calculus, see, e.g., [**15**], [**37**]. In these applications, we also have theorems to the effect that the continuous models are limits of discrete counterparts. Typically, the limit arguments involve the central limit theorem from probability. Another case of discrete models take the form of deep neural network models, deep means a "large number of levels", so many steps through the levels in our diagrams.

The organization of the paper is the following. Section 2 contains the basic definitions related to the concept of generalized Bratteli diagram (discrete case), and the description of various classes of measures on the path space of a generalized Bratteli diagrams. We consider tail invariant, shift-invariant, and Markov measures and their relations to a semibranching function system generated by a stationary generalized Bratteli diagram. In Section 3, we consider a semi-branching function system $\{\tau_e\}$ defined on the path space of a stationary generalized Bratteli diagram and indexed by the edge set E. We prove there one of the main results by giving necessary and sufficient conditions on the existence of an IFS measure for $\{\tau_e\}$, see Theorem 3.4. Section 4 focuses on measurable Bratteli diagrams. Since this notion is relatively new, we give detailed definitions and discuss the properties of such diagrams. Then we prove a measurable version of the main theorem about the existence of an IFS measure, Theorem 4.7.

At the end of this introduction, we mention the literature that may be interesting for the reader. The literature on standard Bratteli diagrams their applications in dynamics is very extensive. We mention here the following pioneering papers [**30**], [**29**], [**19**], a recent book [**43**], and surveys [**20**], [**4**], [**11**]. Generalized Bratteli diagrams are less studied. The stationary case uses the Perron–Frobenius theory for infinite matrices. We refer to the book [**38**] and the literature there. These diagrams are discussed in [**8**] and [**10**]. More references and numerous connections with other areas can be found therein. In particular, the following papers on IFS measures and fractals are related to the current paper [**14**], [**13**], [**22**], [**23**], [**25**], [**34**], [**35**], [**44**], [**1**], [**16**].

2. Basics on generalized Bratteli diagrams

We consider the fundamentals of path space for generalized Bratteli diagrams in this section. This notion was first introduced in [**6**] under the name of Borel-Bratteli diagram. More detailed exposition of this concept can be found in [**8**]. For the reader's convenience we give a concise version here.

2.1. Main definitions.

In the introduction, we described the notion of a Bratteli diagram as an infinite graded graph. A natural extension of this concept consists of consideration of diagrams with countably infinite levels.

DEFINITION 2.1 (Generalized Bratteli diagrams, vertices, edges, incidence matrices). Let V_0 be a countable set (which can be identified with either $\mathbb{N}$ or $\mathbb{Z}$ if necessary). Set $V_i = V_0$ for all $i \geq 1$, and $V = \bigsqcup_{i=0}^{\infty} V_i$. A countable graded graph $B = (\mathcal{V}, \mathcal{E})$ is called a *generalized Bratteli diagram* if it has the following properties.

(i) The set of edges $\mathcal{E}$ of B is represented as $\bigsqcup_{i=0}^{\infty} E_i$ where E_i is the set of edges between the vertices of levels V_i and $V_{i+1}, i \geq 0$.

(ii) The set $E(w, v)$ of edges e between the vertices $w \in V_i$ and $v \in V_{i+1}$ is finite (or empty). Let $f^{(i)}_{v,w} = |E(w, v)|$ where $|\cdot|$ denotes the cardinality of a set. It defines a sequence of infinite (countable-by-countable) *incidence matrices* $(F_i : i \in \mathbb{N}_0)$ with entries $F_i = (f^{(i)}_{v,w} : v \in V_{i+1}, w \in V_i)$, $f^{(i)}_{v,w} \in \mathbb{N}_0$.

(iii) It is required that the matrices F_i have at most *finitely many non-zero entries* in each row. In general, we do not impose any restrictions on the columns of F_i.

(iv) The maps $r, s : E \to V$ are defined on the diagram B: for every $e \in \mathcal{E}$, there are w, v such that $e \in E(w, v)$; then $s(e) = w$ and $r(e) = v$. They are called the *range* (r) and *source* (s) maps.

(v) For every $w \in V_i,\ i \geq 0$, there exist an edge $e \in E_i$ such that $s(e) = w$; for every $v \in V_i,\ i > 1$, there exists an edge $e' \in E_{i-1}$ such that $r(e') = w$. In other words, every row and every column of the incidence matrix F_i has non-zero entries.

(vi) If all entries of incidence matrices F_n are zero or ones, the corresponding generalized Bratteli diagram is called a 0-1 diagram.

REMARK 2.2.

(1) It follows from Definition 2.1 that the structure of every generalized Bratteli diagram is completely determined by a sequence of matrices (F_n) such that every matrix F_n satisfies (iii) and (iv). Indeed, the entry $f^{(n)}_{v,w}$ indicates the number of edges between the vertex $w \in V_n$ and vertex $v \in V_{n+1}$. It defines the set $E(w, v)$; then one takes

$$E_n = \bigcup_{w \in V_n, v \in V_{n+1}} E(w, v).$$

In this case, we write $B = B(F_n)$. If all $F_n = F$, the corresponding generalized Bratteli diagram $B(F)$ is called *stationary*.

(2) If V_0 is a singleton, and each V_n is a finite set, then we obtain the standard definition of a Bratteli diagram originated in [**12**]. Later it was used in the theory of C^*-algebras and dynamical systems for solving important classification problems of Cantor dynamics and construction of various models of homeomorphisms of a Cantor set (for references, see Introduction).

DEFINITION 2.3 (Path space and cylinder sets). A finite or infinite *path* in a generalized Bratteli diagram $B = (V, E)$ is a sequence of edges $(e_i : i \geq 0)$ such that $r(e_i) = s(e_{i+1})$ for all $i \geq 0$. Denote by X_B the set of all infinite paths. Every finite path $\overline{e} = (e_0, \ldots, e_n)$ determines a *cylinder subset* $[\overline{e}]$ of X_B:

$$[\overline{e}] := \{x = (x_i) \in X_B : x_0 = e_0, \ldots, x_n = e_n\}.$$

The collection of all cylinder subsets forms a base of neighborhoods for a topology on X_B. In this topology, X_B is a Polish zero-dimensional space, and every cylinder set is clopen. In general, X_B is not locally compact. But if the set $s^{-1}(v)$ is finite for every vertex $v \in \mathcal{V}$, then the path space X_B is locally compact.

In the following remark, we formulate several statements about properties of generalized Bratteli diagrams and their path spaces.

REMARK 2.4.

(1) If $x = (x_i)$ is a point in X_B, then it is obviously represented as the intersection of clopen sets:

$$\{x\} = \bigcap_{n \geq 0} [\overline{e}]_n \tag{2.1}$$

where $[\overline{e}]_n = [x_0, \ldots, x_n]$.

(2) Define a metric on X_B compatible with the clopen topology: for $x = (x_i)$ and $y = (y_i)$ from X_B, we set

$$\text{dist}(x, y) = \frac{1}{2^N}, \quad N = \min\{i \in \mathbb{N}_0 : x_i \neq y_i\}.$$

(3) We will *assume* that the diagram B is chosen so that the space X_B *has no isolated points*. This means that every column of the incidence matrix $F_n, n \in \mathbb{N}_0$, has more than one nonzero entry.

(4) Let $\mathbb{1} = (\ldots, 1, 1, \ldots.)$ be the vector indexed by $v \in V_0$ such that every entry equals 1. Define $H^{(n)} := F_{n-1} \cdots F_0 \mathbb{1}$. Let $E(V_0, v), v \in V_n$, denote the set of all finite paths between V_0 and a fixed vertex $v \in V_n$. Then $H_v^{(n)} = |E(V_0, v)|$.

(5) Let $X_v^{(n)}$ be a subset of X_B such that

$$X_v^{(n)} = \bigcup_{v_0 \in V_0} \bigcup_{\overline{e} \in E(v_0, v)} [\overline{e}]. \tag{2.2}$$

For every level V_n, the collection $\{X_v^{(n)} : v \in V_n\}$ forms a partition ζ_n of X_B into disjoint clopen sets. Every set $X_v^{(n)}$ is a *finite union* of cylinder sets. The number of the cylinder sets here is exactly $H_v^{(n)}$. The sequence of partitions (ζ_n) is refining. According to (2.2), the cylinder sets from all $X_v^{(n)}$ generate the topology (and Borel σ-algebra) on X_B.

For a generalized Bratteli diagram B, define the *tail equivalence relation* $\mathcal{R}$ on the path space X_B.

DEFINITION 2.5 (Tail equivalence relation). It is said that two infinite paths $x = (x_i)$ and $y = (y_i)$ are *tail equivalent* if there exists $m \in \mathbb{N}$ such that $x_i = y_i$ for all $i \geq m$. Let $[x]_{\mathcal{R}} := \{y \in X_B : (x, y) \in \mathcal{R}\}$ be the set of points tail equivalent to x. We say that a point x is *periodic* if $|[x]_{\mathcal{R}}| < \infty$. If there is no periodic points, then the tail equivalence relation is called *aperiodic*.

Without loss of generality, we will consider generalized Bratteli diagrams with aperiodic $\mathcal{R}$. Clearly, $\mathcal{R}$ is a hyperfinite countable Borel equivalence relation, see [**18**] for definitions.

DEFINITION 2.6. It is said that a generalized Bratteli diagram B is *irreducible* if, for any two vertices v and w and any $n \in \mathbb{N}_0$, there exists a level $V_m (m > n)$ such that $w \in V_n$ and $v \in V_m$ are connected by a finite path. This is equivalent to the property that, for any fixed v, w, there exists $m \in \mathbb{N}$ such that the product of matrices $F_{m-1} \cdots F_n$ has a nonzero (v, w)-entry.

2.2. Measures on the path space of a Bratteli diagram. In this subsection, we will consider two classes of Borel measures on the path space X_B of a generalized Bratteli diagram. They are tail invariant measures and Markov measures.

DEFINITION 2.7 (Tail equivalent measures). Let $B = (\mathcal{V}, \mathcal{E})$ be a generalized Bratteli diagram, and X_B the path space of B. A Borel measure μ on X_B (finite or σ-finite) is called *tail invariant* if, for any two finite paths $\overline{e}$ and $\overline{e}'$ such that $r(\overline{e}) = r(\overline{e}')$, one has

$$\mu([\overline{e}]) = \mu([\overline{e}']), \tag{2.3}$$

where $[e]$ and $[e']$ denote the corresponding cylinder sets.

If $\mu(X_B) = 1$, then the property of tail invariance means that the probability to arrive at a vertex $v \in V_n$ does not depend on a starting point $w \in V_0$ and does not depend on the path connecting w and v.

Let μ be a Borel tail invariant measure on X_B. Relation (2.3) defines a sequence of nonnegative vectors $(\mu^{(n)})$ where $\mu^{(n)} = (\mu_v^{(n)} : v \in V_n)$:

$$\mu_v^{(n)} = \mu([\overline{e}]), \quad \overline{e} \in E(V_0, v), \ v \in V_n. \tag{2.4}$$

Because μ is tail invariant the value $\mu_v^{(n)}$ does not depend on the choice of $\overline{e} \in E(V_0, v)$.

The following theorem is a key tool in the study of tail invariant measures, see [**5**], [**20**], [**4**], [**8**].

THEOREM 2.8. *Let $B = (\mathcal{V}, \mathcal{E})$ be a generalized Bratteli diagram defined by a sequence (F_n) of incidence matrices. Let μ be a Borel probability measure on the path space X_B of B which is tail invariant. Then the corresponding sequence of vectors $\mu^{(n)}$ (defined as in (2.4)) satisfies the property*

$$A_n \mu^{(n+1)} = \mu^{(n)}, \tag{2.5}$$

where $A_n = F_n^T$ is the transpose of F_n.

Conversely, if a sequence of vectors $\mu^{(n)}$ satisfies (2.5), then it defines a unique tail invariant measure μ.

The theorem remains true for σ-finite measures ν satisfying the property $\nu([\overline{e}]) < \infty$ for every cylinder set $[\overline{e}]$.

The other interesting class of measures on Bratteli diagrams is Markov measures. In the context of Bratteli diagrams, these measures were considered in [**17**], [**45**], [**8**], and some other papers.

DEFINITION 2.9 (Markov measures). Let $B = (\mathcal{V}, \mathcal{E})$ be a generalized Bratteli diagram constructed by a sequence of incidence matrices (F_n). Let $q = (q_v)$ be a strictly positive vector, $q_v > 0, v \in V_0$, and let (P_n) be a sequence of non-negative infinite matrices with entries $(p_{v,e}^{(n)})$ where $v \in V_n, e \in E_n, n = 0, 1, 2, \ldots$ To define a *Markov measure* m, we require that the sequence (P_n) satisfies the following properties:

$$\text{(a)}\ p_{v,e}^{(n)} > 0 \iff (s(e) = v); \qquad \text{(b)} \sum_{e:s(e)=v} p_{v,e}^{(n)} = 1. \tag{2.6}$$

Condition (2.6)(a) shows that $p_{v,e}^{(n)}$ is positive only on the edges outgoing from the vertex v, and therefore the matrices P_n and $A_n = F_n^T$ share the same set of zero entries. For any cylinder set $[\overline{e}] = [(e_0, e_1, \dots, e_n)]$ generated by the path $\overline{e}$ with $v = s(e_0) \in V_0$, we set

$$m([\overline{e}]) = q_{s(e_0)} p^{(0)}_{s(e_0),e_0} \cdots p^{(n)}_{s(e_n),e_n}. \tag{2.7}$$

Relation (2.7) defines the value of the measure m of the set $[\overline{e}]$. By (2.6)(b), this measure satisfies the *Kolmogorov consistency condition* and can be extended to the σ-algebra of Borel sets. To emphasize that m is generated by a sequence of *stochastic matrices*, we will also write $m = m(P_n)$.

If all stochastic matrices P_n are equal to a matrix P, then the corresponding measure $m(P)$ is *called stationary Markov measure.*

We refer to [**8**] for a detailed study of Markov measures. We mention here only the following result.

THEOREM 2.10. *Let ν be a tail invariant probability measure on the path space X_B of a generalized Bratteli diagram $B = (\mathcal{V}, \mathcal{E})$. Then there exists a sequence of Markov matrices (P_n) such that $\nu = m(P_n)$.*

For stationary generalized Bratteli diagrams, we can find explicit formulas for tail invariant measures. In the following statement, we use the terminology from the Perron–Frobenius theory, see [**38**] for details.

THEOREM 2.11 ([**8**]).

(1) *Let $B = B(F)$ be a stationary Bratteli diagram such that the incidence matrix F (and therefore $A = F^T$) is irreducible, aperiodic, and recurrent. Let $t = (t_v : v \in V_0)$ be a right eigenvector corresponding to the Perron eigenvalue λ for A, $At = \lambda t$. Then there exists a tail invariant measure μ on the path space X_B whose values on cylinder sets are determined by the following rule: for every finite path $\overline{e}(w, v)$ that begins at $w \in V_0$ and terminates at $v \in V_n$, $n \in \mathbb{N}_0$, we set*

$$\mu^{(n)}_v = \mu([\overline{e}(w, v)]) = \frac{t_v}{\lambda^n}. \tag{2.8}$$

(2) *The measure μ is finite if and only if the right eigenvector $t = (t_v)$ has the property $\sum_v t_v < \infty$.*

In particular, $\mu(X^{(0)}_w) = t_w, w \in V_0$, and

$$\mu(X^{(n)}_v) = H^{(n)}_v \frac{t_v}{\lambda^n}. \tag{2.9}$$

2.3. Semi-branching function systems on Bratteli diagrams. Here we give the definition of a semibranching function system following [**39**] and [**3**]. This notion was used in the literature, in particular, for the construction of representations of Cuntz–Krieger algebras, see [**3**], [**27**], [**28**].

DEFINITION 2.12 (Semibranching function systems and coding maps).

(1) Let (X, μ) be a probability measure space with non-atomic measure μ. We consider a family $\{\sigma_i : i \in \Lambda\}$ of one-to-one μ-measurable maps indexed by a finite (or countable) set Λ. The family $\{\sigma_i\}$ is called a *semibranching function system (s.f.s.)* if the following conditions hold:

  (i) σ_i is defined on a subset D_i of X and takes values in $R_i = \sigma_i(D_i)$ such that $\mu(R_i \cap R_j) = 0$ for $i \neq j$ and $\mu(X \setminus \bigcup_{i\in\Lambda} R_i) = 0$;

(ii) the measure $\mu \circ \sigma_i$ is equivalent to μ and, i.e.,

$$\rho_\mu(x, \sigma_i) := \frac{d\mu \circ \sigma_i}{d\mu}(x) > 0 \text{ for } \mu\text{-a.e. } x \in D_i;$$

(iii) there exists an endomorphism $\sigma : X \to X$ (called a *coding map*) such that $\sigma \circ \sigma_i(x) = x$ for μ-a.e. $x \in D_i$, $i \in \Lambda$.
If, additionally to properties (i)–(iii), we have $\bigcup_{i\in\Lambda} D_i = X$ (μ-a.e.), then the s.f.s. $\{\sigma_i : i \in \Lambda\}$ is called *saturated.*

(2) It is said that a saturated s.f.s. satisfies *condition C-K*[1] if for any $i \in \Lambda$ there exists a subset $\Lambda_i \subset \Lambda$ such that up to a set of measure zero

$$D_i = \bigcup_{j\in\Lambda_i} R_j.$$

In this case, condition C-K defines a 0-1 matrix $\widetilde{A}$ by the rule:

$$\widetilde{a}_{i,j} = 1 \iff j \in \Lambda_i, \quad i \in \Lambda. \tag{2.10}$$

Then the matrix $\widetilde{A}$ is of the size $|\Lambda| \times |\Lambda|$.

Semi-branching function system associated with a generalized stationary Bratteli diagram.

Let B be a generalized stationary 0-1 Bratteli diagram. We construct an s.f.s. Σ which is defined on the path space X_B endowed with a Markov measure m. As we will see below, this measure must have some additional properties to satisfy Definition 2.12. The role of the index set Λ for this s.f.s. is played by the edge set E which is the set of edges between any two consecutive levels of B. For any $e \in E$, we denote

$$D_e = \{y = (y_i) \in X_B : s(y) = s(y_0) = r(e)\}, \tag{2.11}$$

$$R_e = \{y = (y_i) \in X_B : y_0 = e\}. \tag{2.12}$$

We see that D_e depends on $r(e)$ only so that $D_e = D_{e'}$ if and only if $r(e) = r(e')$.

The collection of maps $\{\tau_e : e \in E\}, \tau_e : D_e \to R_e$, is defined by the formula

$$\tau_e(y) := (e, y_0, y_1, \dots), \quad y = (y_i). \tag{2.13}$$

Since $s(y_0) = r(e)$, the map τ_e is well defined on D_e.

REMARK 2.13. We defined the metric dist in Remark 2.4. It follows from the definition of τ_e that

$$\mathrm{dist}(\tau_e(x), \tau_e(y)) = \frac{1}{2}\mathrm{dist}(x, y), \ e \in E,$$

that is τ_e is a contractive map for every e.

PROPOSITION 2.14. *The system $\{D_e, R_e, \tau_e : e \in E\}$ is a saturated s.f.s. on the path space X_B of a generalized stationary Bratteli diagram satisfying conditions* (i), (iii) *of Definition* 2.12 *and the C-K condition.*

PROOF. Let $\sigma : X_B \to X_B$ be defined as follows: for any $x = (x_i)_{i\geq 0} \in X_B$,

$$\sigma(x) := (x_1, x_2, \dots) \tag{2.14}$$

It follows from (2.13) and (2.14) that the map σ is onto and

$$\sigma \circ \tau_e(x) = x, \quad x \in D_e;$$

[1] C-K stands for Cuntz–Krieger.

Hence σ is a coding map.

We deduce from (2.12) that $\{R_e : e \in E\}$ constitutes a partition of X_B into clopen sets. Relation (2.11) implies that $\{\tau_e : e \in E\}$ is a saturated s.f.s. Moreover, we claim that it satisfies condition C-K. Indeed,

$$D_e = \bigcup_{f:s(f)=r(e)} R_f, \quad e \in E, \tag{2.15}$$

because $y = (y_i) \in D_e \iff s(y_0) = r(e) \iff \exists f = y_0$ such that $y = (f, y_1, \dots) \iff y \in \bigcup_{f:s(f)=r(e)} R_f$. Thus, $\Lambda_e = \{f : s(f) = r(e)\}$, see (2.10).

Relation (2.15) shows that the nonzero entries of the 0-1 matrix $\widetilde{A}$ from Definition 2.12 are defined by the rule:

$$(\widetilde{a}_{e,f} = 1) \iff (s(f) = r(e)). \tag{2.16}$$

We observe that the matrix $\widetilde{A}$ has the following property: there are finitely many nonzero entries in every column of $\widetilde{A}$, but the rows of $\widetilde{A}$ may contain infinitely many nonzero entries.

Next, we observe that $\sigma : X_B \to X_B$ is a finite-to-one continuous map. Indeed,

$$|\sigma^{-1}(x)| = |r^{-1}(r(x_1))| = \sum_{u \in V_0} f_{v,u}.$$

The latter is finite by Definition 2.1. □

Thus, it remains to find out under what conditions property (ii) of Definition 2.12 holds. We consider here two classes: tail invariant measures and Markov measures.

Let m be a Borel probability measure on X_B. Since X_B is naturally partitioned into a refining sequence of clopen partitions $\mathcal{Q}_n$ formed by cylinder sets of length n, we can apply de Possel's theorem (see, for instance, [**47**]). We have that for m-a.a x,

$$\rho_m(x, \tau_e) = \lim_{n\to\infty} \frac{m(\tau_e([\overline{e}(n)])}{m[\overline{e}(n)]} \tag{2.17}$$

where $\{x\} = \bigcap_n [\overline{e}(n)]$.

Tail invariant measure. We first consider the case when m is the tail invariant measure μ determined in Theorem 2.11. Let $At = \lambda t$ where $t = (t_v)$. If $\overline{f} = (f_0, f_1, \dots, f_n) \in D_e$, then $r(\overline{f}) = r(f_n)$ and $\tau_e(\overline{f}) = (e, f_0, \dots, f_n)$. By (2.8), we have

$$\mu([\overline{f}]) = \frac{t_{r(\overline{f})}}{\lambda^n}, \quad \mu(\tau_e([\overline{f}])) = \frac{t_{r(\overline{f})}}{\lambda^{n+1}},$$

and therefore

$$\rho_\mu(x, \tau_e) = \lambda^{-1}. \tag{2.18}$$

Stationary Markov measure. Let m be a stationary Markov measure determined by a stochastic matrix P, $m = m(P)$ as in (2.7). If $\{x\} = \bigcap_n [f(n)] \in D_e$, then

$$m([\overline{f}(n)]) = q_{s(f_0)} p_{s(f_0),f_0} \cdots p_{s(f_n),f_n},$$
$$m(\tau_e([\overline{f}(n)])) = q_{s(e)} p_{s(e),e} p_{s(f_0),f_0} \cdots p_{s(f_n),f_n},$$

and the Radon–Nikodym derivative can be found by

$$\rho_m(x, \tau_e) - \frac{q_{s(e)} p_{s(e),e}}{q_{s(f_0)}}. \tag{2.19}$$

It follows from (2.18) and (2.19) that ρ_m is positive on D_e if and only if all entries of the vector $q = (q_v)$ are positive. The latter means that the support of m is the entire space X_B.

Nonstationary Markov measure. In this case, we need some additional conditions to guarantee that the Radon–Nikodym derivative $\rho_m(x, \tau_e)$ is positive on D_e. As above, we represent $x = (x_i) \in D_e$ by means of the sequence $[\overline{f}(n)] = [(f_0, f_1, \dots, f_n)]$ such that $x_i = f_i, i = 0, 1, \dots, n$, and find

$$m([f(n)]) = q_{s(f_0)} p^{(0)}_{s(f_0),f_0} \cdots p^{(n)}_{s(f_n),f_n}$$

and

$$m(\tau_e([f(n)])) = q_{s(e)} p^{(0)}_{s(e),e} p^{(1)}_{s(f_0),f_0} \cdots p^{(n+1)}_{s(f_n),f_n}.$$

A direct computation gives the following result.

LEMMA 2.15. *Let m be a Markov measure on the path space of a generalized stationary 0-1 Bratteli diagram. Then $\rho_m(x, \tau_e) > 0$ on D_e if and only if*

$$0 < \prod_{i=1}^{\infty} \frac{p^{(i+1)}_{s(f_i),f_i}}{p^{(i)}_{s(f_i),f_i}} < \infty \tag{2.20}$$

for any $x = \bigcap_n [\overline{f}(n)] \in D_e$

A Markov measure satisfying (2.20) is called a *quasi-stationary measure.* Condition (2.20) appeared first in [**24**], [**26**] in a different context.

We summarize the above results in the following theorem.

THEOREM 2.16. *Given a generalized stationary 0-1 Bratteli diagram B with the edge set E, the collection of maps $\{\tau_e : D_e \to R_e\}$, $e \in E$, defined in (2.13) on the space (X_B, m), forms a saturated s.f.s. Σ satisfying C-K condition where the Markov measure m is either the tail invariant measure μ, or a stationary Markov measure $m(P)$ of full support, or a quasi-stationary measure Markov measure of full support.*

REMARK 2.17. Let B be a generalized stationary Bratteli diagram of *bounded size*, see [**9**] for the definition. In particular, if the incidence matrix A is banded, then B is of bounded size. In this case, the definition of the Cuntz–Krieger algebra $\mathcal{O}_A$ can be given similar to the case of finite matrices. Then we can use the methods developed in [**3**, Theorem 4.12] to construct a representation of the Cuntz–Krieger algebra $\mathcal{O}_A$ generated by the s.f.s. defined in Theorem 2.16. We omit the details.

2.4. Shift invariant measures on stationary Bratteli diagrams. Let $B = (\mathcal{V}, \mathcal{E})$ be a generalized stationary Bratteli diagram defined by the incidence infinite matrix F and $A = F^T$. In this subsection, we discuss σ-invariant measures on X_B. We will consider two cases, tail invariant measures and Markov measures.

Recall that, for every $v \in V_1$, the row sum $H^{(1)}_v = \sum_{w \in V_0} f_{v,w}$ is finite, see Remark 2.4 for notation. In (2.14), we defined a finite-to-one endomorphism σ acting on the path space X_B. For $x = (e_0.e_1, \dots)$, we have

$$\sigma^{-1}(x) = \{y = (y_i) \in X_B : r(y_0) = s(e_0), y_i = e_{i-1}, i \geq 1\} \tag{2.21}$$

and $|\sigma^{-1}(x)| = H^{(1)}_{s(e_0)}$.

Tail invariant measures. There is a special case of a generalized stationary Bratteli diagram such that the tail invariant measure is also shift-invariant.

PROPOSITION 2.18. *Let the matrix A has a Perron-Frobenius eigenpair (ξ, λ). Let μ be the tail invariant measure on X_B defined as in Theorem 2.11. Then μ is σ-invariant if and only if $H_v^{(1)} = \lambda$ for all $v \in V_0$.*

PROOF. For a cylinder set $[\overline{e}] = [e_0, e_1, \dots, e_n]$, calculate $m([\overline{e}])$ and $m(\sigma^{-1}([\overline{e}]))$. it follows from (2.21) that

$$\sigma^{-1}([\overline{e}]) = \bigcup_{f : r(f) = s(e_0)} [f, e_0, \dots, e_n] = \bigcup_{f : r(f) = s(e_0)} [f, \overline{e}] \tag{2.22}$$

and this union is disjoint. Let λ and ξ be a Peron-Frobenius eigenpair, $A\xi = \lambda\xi$. By Theorem 2.11, we have

$$\mu([\overline{e}]) = \frac{\xi_v}{\lambda^n}, \qquad \mu([f, \overline{e}]) = \frac{\xi_v}{\lambda^{n+1}}, \ v = r(\overline{e})$$

Then

$$\mu(\sigma^{-1}[\overline{e}]) = \sum_{f : r(f) = s(e_0)} \mu([f, \overline{e}]) = \sum_{f : r(f) = s(e_0)} \frac{\xi_v}{\lambda^{n+1}} = |\sigma^{-1}([\overline{e}])| \frac{\xi_v}{\lambda^{n+1}} = H_w^{(1)} \frac{\xi_v}{\lambda^{n+1}}$$

where $w = s(e_0) \in V_1$. By de Possel's theorem (2.17),

$$\frac{d\mu \circ \sigma^{-1}}{d\mu}(x) = \lim_{n \to \infty} \frac{\mu \circ \sigma^{-1}(\overline{e}]_n)}{\mu([\overline{e}]_n)} = H_w^{(1)} \lambda^{-1}$$

where $x = \bigcap_n [\overline{e}]_n$. Therefore,

$$\mu \circ \sigma^{-1} = \mu \iff H_w^{(1)} = \sum_{u \in V_0} f_{w,u} = \lambda, \ \forall w \in V_0.$$

□

REMARK 2.19. If $\max\{H_v^{(1)} : v \in V_1\} = M < \infty$, then

$$\lambda^{-1} \le \frac{d\mu \circ \sigma^{-1}}{d\mu}(x) \le M\lambda^{-1}.$$

This means that μ is equivalent to a σ-invariant measure μ'. In particular, this is true for bounded size diagrams.

Markov measures. Consider first a stationary Markov measure on the path space of a generalized stationary Bratteli diagram, see Definition 2.9. For simplicity we will work with a $0 - 1$ diagram; the general case is considered similarly.

Let $q = (q_v : v \in V_0)$ be a positive probability vector (called initial distribution). Let P be a Markov matrix. For a cylinder set $[\overline{e}] = [e_0, e_1, \dots, e_n]$, we use (2.7) to determine the value of the corresponding Markov measure $m(P)$.

THEOREM 2.20. *For $B, q, P, m(P)$ as above, the measure $m = m(P)$ is σ-invariant if and only if $qP = P$.*

PROOF. Suppose that $qP = q$. We know that

$$m([\overline{e}]) = q_{s(e_0)} p_{s(e_0), r(e_0)} \cdots p_{s(e_n), r(e_n)}$$

and

$$
\begin{aligned}
m(\sigma^{-1}[\overline{e}]) &= \sum_{f:r(f)=s(e_0)} m([f,\overline{e}]) \\
&= \sum_{f:r(f)=s(e_0)} q_{s(f)}p_{s(f),r(f)}p_{s(e_0),r(e_0)}\cdots p_{s(e_n),r(e_n)} \\
&= q_{r(f)}p_{s(e_0),r(e_0)}\cdots p_{s(e_n),r(e_n)} \\
&= m([\overline{e}])
\end{aligned}
\tag{2.23}
$$

because $q_{r(f)} = q_{s(e_0)}$ and q is P-invariant. Since $\overline{e}$ is arbitrary, we see that m is σ-invariant.

The converse statement follows from (2.23): the equality $m([\overline{e}]) = m(\sigma^{-1}[\overline{e}])$ implies $qP = q$. □

It remains to consider a non-stationary Markov measure m defined on the path space of a generalized stationary Bratteli diagram B. In this case, the measure m is defined by a sequence of Markov matrices (P_n) and an initial distribution $q = (q_v)$:

$$
m([e_0,\dots,e_n]) = q_{s(e_0)}p^{(0)}_{s(e_0),r(e_0)}\cdots p^{(n)}_{s(e_n),r(e_n)}.
$$

PROPOSITION 2.21. *Suppose that $qP_0 = q$. Then the measures $m = m(P_n)$ is σ-invariant if and only if*

$$
\prod_{i=1}^{\infty}\frac{p^{(i+1)}_{s(e_i),r(e_i)}}{p^{(i)}_{s(i_i),r(e_i)}} = 1.
$$

PROOF. Using the same technique, we can show that, under the condition $qP = q$,

$$
\frac{d\mu\circ\sigma^{-1}}{d\mu}(x) = \prod_{i=1}^{\infty}\frac{p^{(i+1)}_{s(e_i),r(e_i)}}{p^{(i)}_{s(i_i),r(e_i)}}
$$

where $x = \bigcap_n [e_0,\dots,e_n$. We omit the details. □

3. IFS measures on discrete generalized Bratteli diagrams

In this section, we consider the notion of *iterated function system (IFS)*. This concept have been discussed in many books and papers. We mention here only several related references such as **[32]**, **[33]**, **[34]**, **[2]**, **[35]**, **[21]**, **[22]**, **[36]**, **[41]**. Our goal is to describe IFS measures defined on the path space of a generalized Bratteli diagram.

3.1. Iterated function systems and measures. By an *endomorphism* σ we mean a finite-to-one (or countable-to-one) Borel map of a standard Borel space $(X,\mathcal{B})$ onto itself. In this case, there exists a family of one-to-one maps $\{\tau_i\}_{i\in\Lambda}$ such that $\tau_i : X \to X$ and $\sigma\circ\tau_i = \mathrm{id}_X$ where Λ is at most countable. The maps τ_i are called the *inverse branches* for σ. The collection of maps $(\tau_i : 1 \in \Lambda)$ gives an example of *iterated function system (IFS)*. In general, an IFS is defined by a collection of maps $\{\tau_i : i \in \Lambda\}$ of a complete metric space (or a compact space) such that τ_i's are continuous (or even contractions). In this work, we focus on the following features of the family $\{\tau_i : i \in \Lambda\}$: (i) each τ_i is a one-to-one map defined on a subset D_i of X, and (ii) there exists an endomorphism σ of X such that $\sigma\circ\tau_i = \mathrm{id}_X$ for each i.

The general theory of *infinite iterated function systems* is more detailed and requires additional assumptions (see, for example, the expository article [**40**]). As an example, one can consider the Gauss map $x \mapsto \{1/x\}$ of the unit interval with the piecewise monotone inverse branches $\tau_i : 1/(i+x)$ defined on $(1/(i+1), 1/i)$.

DEFINITION 3.1 (IFS measures). Suppose that $(\tau_i : i \in \Lambda)$ is a given IFS on a Borel space $(X, \mathcal{B})$. Let $p = (p_i : i \in \Lambda)$ be a strictly positive vector indexed by a countable set Λ. A measure μ_p on $(X, \mathcal{B})$ is called an *IFS measure* for (τ_i) if

$$\mu_p = \sum_{i\in\Lambda} p_i \, \mu_p \circ \tau_i^{-1}, \tag{3.1}$$

or, equivalently,

$$\int_X f(x)\, d\mu_p(x) = \sum_{i\in\Lambda} p_i \int_X f(\tau_i(x))\, d\mu_p(x), \qquad f \in L^1(\mu_p).$$

The main tool in the study of an IFS $(\tau_i : i \in \Lambda)$ on a space X is a realization of the IFS as the full one-sided shift S on a symbolic product space Ω. Then any S-invariant and ergodic product-measure $\mathbb{P}$ on Ω can be pulled back to X. This construction gives ergodic invariant measures for IFSs.

In Subsection 3.2, we will find an explicit method for construction of an IFS measure on the path space of a generalized stationary Bratteli diagram.

We discuss here a method that leads to construction of IFS measures. This approach works perfectly for many specific applications under some additional conditions on X and maps τ_i.

Suppose that $(X; \tau_i, i \in \lambda)$ is a given IFS. Let Ω be the infinite direct product

$$\Omega = \prod_{i\in\mathbb{N}_0} \Lambda_i, \qquad \Lambda_i = \Lambda.$$

For $\omega = (\omega_1, \omega_2, \dots) \in \Omega$, let $\omega|_n$ denote the finite word $(\omega_0, \dots, \omega_n)$. Then, $\omega|_n$ defines a map $\tau_{\omega|_n}$ acting on X by the formula:

$$\tau_{\omega|_n}(x) := \tau_{\omega_0} \cdots \tau_{\omega_n}(x), \qquad x \in X, \;\; \omega \in \Omega, \; n \in \mathbb{N}_0.$$

It is said that Ω is an *encoding space* if, for every $\omega \in \Omega$,

$$F(\omega) = \bigcap_{n\geq 1} \tau_{\omega|_n}(X) \tag{3.2}$$

is a singleton. Relation (3.2) defines a Borel map $F : \Omega \to F(\Omega)$. If each τ_i is a contraction and X is a complete metric, then a coding map $F : \Omega \to X$ always exists.

Let S be the left shift on Ω:

$$S(\omega_0, \omega_1, \dots) = (\omega_1, \omega_2, \dots).$$

The inverse branches $s_i, i \in \Lambda$, of S are

$$s_i(\omega_0, \omega_1, \dots) = (i, \omega_0, \omega_1, \dots).$$

Clearly,

$$s_i(\Omega) = C(i) = \{\omega \in \Omega : \omega_0 = i\},$$

and the space Ω is partitioned into the sets $C(i), \in \Lambda$.

Let $p = (p_i : i \in \Lambda)$ be a positive probability vector. It defines the product measure $\mathbb{P} = p \times p \times \cdots$ on Ω. We observe that the maps $(s_i : i \in \Lambda)$ constitute an IFS on Ω such that $\mathbb{P}$ is an IFS measure:

$$\mathbb{P} = \sum_{i \in \Lambda} p_i \, \mathbb{P} \circ s_i^{-1}. \tag{3.3}$$

Proposition 3.2 shows how IFS measures arise.

PROPOSITION 3.2. *Suppose that $(X; \tau_i, i \in \Lambda)$ is an IFS that admits a coding map $F : \Omega \to X$. Let $p = (p_i)$ be a probability vector generating the product measure $\mathbb{P} = p \times p \times \cdots$. Then the measure $\mu := \mathbb{P} \circ F^{-1}$ is an IFS measure satisfying*

$$\mu = \sum_{i=1}^{N} p_i \mu \circ \tau_i^{-1}.$$

Moreover, if F is continuous, then μ has full support.

The proof can be found in [**7**].

In Section 4, we will consider a analogue of the above construction for measurable Bratteli diagrams.

3.2. IFS measures on generalized stationary Bratteli diagrams. In this subsection we discuss an explicit formula for an IFS measure on the path space of a generalized stationary 0-1 Bratteli diagram $B = (\mathcal{V}, \mathcal{E})$.

In the following remark, we show that the requirement to be a 0-1 Bratteli diagram is not restrictive.

REMARK 3.3. Every generalized stationary Bratteli diagram $B = (\mathcal{V}, \mathcal{E})$ can be represented as a 0-1 Bratteli diagram. Indeed, we can use the matrix $\widetilde{A}$ defined in (2.16) to built a new generalized Bratteli diagram $\widetilde{B} = (\widetilde{V}, \widetilde{E})$. Note that $\widetilde{A}$ is a 0-1 matrix such that every row of $\widetilde{F} = \widetilde{A}^T$ has finitely many non-zero entries so that $\widetilde{F}$ can be viewed as an incidence matrix of $\widetilde{B}$. In this case, the set of vertices $\widetilde{V}$ for each level is coincides with E and two vertices $\widetilde{v} = f \in E$ and $\widetilde{w} = f \in E$ are connected by a singe edge if and only if $r(e) = s(f)$. Moreover, the path spaces $X_{\widetilde{B}}$ coincides with X_B.

The above argument can be applied to any non-stationary generalized Bratteli diagram.

We can now apply the results of Subsection 2.3 and construct an s.f.s. $\Psi = (X_B; \tau_e, e \in E)$ where $\tau_e : D_e \to R_e$ is defined in (2.13).

THEOREM 3.4. *Let $B = (\mathcal{V}, \mathcal{E})$ be a generalized stationary 0-1 Bratteli diagram, $p = (p_e : e \in E)$ a probability vector, and Ψ an s.f.s. defined by $\tau_e : D_e \to R_e$, $e \in E$. We identify every edge $e \in E$ with the pair of vertices $(s(e), r(e))$. Let ν be a Borel probability full measure on X_B; we define a probability positive vector $q = (q_v : v \in V_0)$ by setting $q_v = \nu([w])$ where $[w] = \{xinX_B : s(x) = w\}$. Then*

$$\nu = \sum_{e \in E} p_e \, \nu \circ \tau_e^{-1} \tag{3.4}$$

if and only if

$$q_w = \sum_{w \in V_0} p_{w,v} q_v = \sum_{e : s(e) = w} p_{s(e),r(e)} q_{r(e)}, \tag{3.5}$$

that is $Pq = q$ where the matrix P has the entries $(p_{s(e),r(e)} : e \in E)$.

PROOF. We use in the proof notation from (2.11)–(2.15). We note that relation (3.4) will be proved if we show that it holds for any cylinder set $C \subset X_B$. Recall that it follows from the definition of τ_e that τ_e^{-1} is uniquely determined on R_e and $\tau_e^{-1}(e, y_1, y_2, \dots) = (y_1, y_2, \dots)$.

Let $q = (q_w) > 0$ be a solution to $q = Pq$. Define $\nu([w]) = q_w$, $w \in V_0$. For any edge $f \in E$ and the corresponding cylinder set $[f]$, we set

$$\nu([f]) = q_{r(f)} p_f. \tag{3.6}$$

Check that this definition of ν satisfies (3.4) for the cylinder sets $[f]$. Indeed, since τ_e^{-1} is defined only on $R_e = [e]$ and $\tau_f^{-1}([f]) = [r(f)]$, we have

$$\sum_{e \in E} p_e \nu \circ \tau_e^{-1}([f]) = p_f \nu \circ \tau_f^{-1}([f]) = p_f \nu([r(f)]) = q_{r(f)} p_f = \nu([f]).$$

Then, by induction, we define the values of ν on all cylinder sets of length n:

$$\nu([f_0, \dots, f_{n-1}]) := p_{f_0} \cdots p_{f_{n-1}} q_{r(f_{n-1})}.$$

Verify that this definition satisfies the Kolmogorov extension theorem. Because

$$[f_0, \dots, f_{n-1}] = \bigcup_{e : s(e) = r(f_{n-1})} [f_0, \dots, f_{n-1}, e],$$

we find that

$$\begin{aligned}
\sum_{e : s(e) = r(f_{n-1})} \nu([f_0, \dots, f_{n-1}, e]) = & \sum_{e : s(e) = r(f_{n-1})} p_{f_0} \cdots p_{f_{n-1}} p_e q_{r(e)} \\
= & p_{f_0} \cdots p_{f_{n-1}} \sum_{e : s(e) = r(f_{n-1})} p_e q_{r(e)} \\
= & p_{f_0} \cdots p_{f_{n-1}} q_{r(f_{n-1})} \\
= & \nu([f_0, \dots, f_{n-1}])
\end{aligned}$$

(we used (3.5) here).

It remains to show that (3.4) holds for any cylinder $[f_0, \dots, f_n]$. Indeed,

$$\begin{aligned}
\sum_{e \in E} p_e \nu \circ \tau_e^{-1}([f_0, \dots, f_n]) = & p_{f_0} \nu([f_1, \dots, f_n]) \\
= & p_{f_0} p_{f_1} \cdots p_{f_n} q_{r(f_n)} \\
= & \nu([f_0, \dots, f_n]).
\end{aligned}$$

Remark that the condition $Pq = q$ is used to check that the measure ν defined inductively on cylinder sets can be extended to all Borel sets. □

THEOREM 3.5. *Let B be an irreducible generalized stationary Bratteli diagram, and let ν be the IFS measure defined in Theorem 3.4. Then:*

(i) *ν is not tail invariant;*
(ii) *ν is shift-invariant.*

PROOF.

(i) Suppose that ν is tail invariant. Let $f, e \in E$ be two edges with $r(f) = r(e)$, i.e., f, e are tail-equivalent. Then it follows from (3.6) that $\nu([f]) = \nu([e])$ or

$$q_{r(f)} p_f = q_{r(e)} p_e.$$

This means that $p_f = p_e$. In other words, the matrix

$$P = (p_{w,v}) : e = (w, v) \in E)$$

has constant columns.

Consider two cylinder sets $[f_0, f_1]$ and $[e_0, e_1]$ such that $r(f_1) = r(e_1)$ and there exist edges g, h satisfying the properties: $s(g) = s(h)$, $r(g) = r_{f_0}$, and $r(h) = r_{e_0}$. From the tail invariance of ν and the case considered above, we obtain that $p_{f_1} = p_{e_1}$ and

$$q_{r(f_1)} p_{f_0} p_{f_1} = q_{r(e_1)} p_{e_0} p_{e_1}$$

Hence, $p_{f_0} = p_{e_0}$ and therefore $p_g = p_h$. This proves that the rows of the matrix P are constant. But the vector $(p_e : e \in E)$ is probability, contradiction.

(ii) Let $\varphi(x)$ be a bounded positive Borel function on the path space X_B. For the left shift σ on X_B, compute

$$\begin{aligned}
\int_{X_B} \varphi(x)\, d\nu \circ \sigma^{-1}(x) &= \int_{X_B} \varphi(\sigma x)\, d\nu(x) \\
&= \sum_{e \subset E} \int_{X_B} p_e \varphi(\sigma x)\, d\nu \circ \tau_e^{-1}(x) \\
&= \sum_{e \in E} \int_{X_B} p_e \varphi(\sigma(\tau_e x))\, d\nu(x) \qquad (3.7)\\
&= \sum_{e \in E} \int_{X_B} p_e \varphi(x)\, d\nu(x) \\
&= \int_{X_B} \varphi(x)\, d\nu(x).
\end{aligned}$$

This proves that $\nu \circ \sigma^{-1} = \nu$.

□

4. Measurable Bratteli diagrams and IFS measures

In this section, we discuss a measurable analogue of generalized Bratteli diagrams. The principal differences between discrete and measurable generalized Bratteli diagrams are: (a) the levels V_n of a measurable Bratteli diagram are formed by standard Borel spaces $(X_n, \mathcal{A}_n)$, and (b) the sets of edges E_n are Borel subsets of $X_n \times X_{n+1}$. Because these objects are non-discrete, we need to use new methods and techniques.

4.1. Measurable Bratteli diagrams and path space measures.

We give the definitions of main objects in this subsection.

DEFINITION 4.1. (*Measurable Bratteli diagrams*) Let $\{(X_n, \mathcal{A}_n) : n \in \mathbb{N}_0\}$ be a sequence of standard Borel spaces. Let $\{E_n : n \in N_0\}$ be a sequence of Borel subsets such that $E_n \subset X_n \times X_{n+1}$. Denote by $\mathcal{E} = \bigsqcup_{i \geq 0} E_i$ and $\mathcal{X} = \bigsqcup_{i \geq 0} X_i$ the sets of "edges" and "vertices", respectively. For $e = (x, y) \in E_i$, the maps $s_i(x, y) = x$ and $r_i(x, y) = y$ are onto projections of E_i to X_i and X_{i+1}. Define s, r on $\mathcal{E}$ by setting $s = s_i, r = r_i$ on E_i. Then we call $\mathcal{B} = (\mathcal{X}, \mathcal{E})$ a *measurable Bratteli diagram*. The pair (X_n, E_n) is called the n-th level of the measurable Bratteli diagram $\mathcal{B}$.

If all $E_n = E$, then the measurable Bratteli diagram $\mathcal{B}$ is called *stationary.*

REMARK 4.2.

(1) We will identify the standard Borel spaces $\{(X_n, \mathcal{A}_n) : n \in \mathbb{N}_0\}$ with an uncountable standard Borel space $(X, \mathcal{A})$. This means that a paint x can be seen in all levels $(X_n, \mathcal{A}_n)$. This fact explains why we do not require that all levels $X_n = X_0$ in the definition of a stationary Bratteli diagram. Nevertheless, we will keep using subindeces to indicate the level of a measurable Bratteli diagram.

(2) The set of edges $E_i, i \geq 0$, can be represented as follows:

$$E_i = \bigcup_{x \in X_i} s^{-1}(x) = \bigcup_{x \in X_{i+1}} r^{-1}(x).$$

This means that E_i can be seen as the union of "vertical" and "horizontal" sections. The fact that r is an onto map says that $\forall y \in X_{i+1}\ \exists x \in X_i$ such that $(x, y) \in E_i$. A similar property holds for the map s.

(3) To define the *path space* $\mathcal{X}_\mathcal{B}$ of a measurable Bratteli diagram $\mathcal{B} = (\mathcal{X}, \mathcal{E})$, we take a sequence $\overline{x} = \{e_i = (x_i, x_{i+1})\}, e_i \in E_i$, such that $r(e_i) = s(e_{i+1})$ for all i. Equivalently, $\overline{x} = (x_0, x_1, x_2, \dots)$ where every pair (x_i, x_{i+1}) is in E_i, $i \in \mathbb{N}_0$. Then $s(\overline{x}) = x_0$.

(4) We denote

$$\mathcal{X}_\mathcal{B}(w) = \{\overline{x} \in \mathcal{X}_\mathcal{B} : s(\overline{x}) = w\}. \tag{4.1}$$

Clearly, the sets $\mathcal{X}_\mathcal{B}(w)$ form a partition η of $\mathcal{X}_\mathcal{B}$, and X_0 is the quotient space with respect to this partition.

We will consider below *measurable stationary Bratteli diagrams*, i.e., $E_i = E$. For every $e = (w, v) \in E$, we set

$$D_e = \{\overline{x} \in \mathcal{X}_\mathcal{B} : s(\overline{x}) = r(e) = v\},$$

$$R_e = \{\overline{x} \in \mathcal{X}_\mathcal{B} : (x_0, x_1) = e\}.$$

For $\overline{x} = (x_0, x_1, \dots) \in D_e$, define the map $\tau_e : D_e \to R_e$ by setting

$$\tau_e(\overline{x}) = (w, v, x_1, x_2, \dots).$$

We note that $\overline{x} \in D_e$ means that $x_0 = v$ so that $\tau_e(\overline{x})$ is well defined and belongs to R_e. Moreover, the map τ_e is one-to-one on its domain and the map $\tau_e^{-1} : R_e \to D_e$ is defined by

$$\tau_e^{-1}(e, \overline{y}) = \overline{y}.$$

We can consider the metric dist on the path space $\mathcal{X}_\mathcal{B}$ similarly to the case of generalized Bratteli diagrams. As shown in Remark 2.13, the maps τ_e are contractive for all $e \in E$.

Let $\mathcal{B}$ be a measurable Bratteli diagram with the path space $\mathcal{X}_\mathcal{B}$. By a measure μ on $X_\mathcal{B}$, we mean a Borel positive (finite or sigma-finite) measure.

Consider a sequence of Borel sets $C_i \subset X_i : i \in \mathbb{N}_0$. Then $(C_0 \times C_1 \times \cdots \times C_N) \cap \mathcal{X}_\mathcal{B}$ is called a *cylinder subset* of $X_\mathcal{B}$ of length N and denoted by $[C_0, \dots, C_N]$. In particular, we can consider the set R_e as a cylinder set $[e]$ defined by the edge e, $e \in E$. It defines the partition

$$\xi = \{[e] : e \in E\} \tag{4.2}$$

of $\mathcal{X}_\mathcal{B}$. We remark that the quotient $\mathcal{X}_\mathcal{B}/\xi$ is isomorphic to the set E. Since $D_e = \mathcal{X}_\mathcal{B}(r(e))$ for every $e \in E$, the partition η (see Remark 4.2(4)) is an enlargement of ξ because

$$D_e = \bigcup_{f:s(f)=r(e)} [f].$$

Observation. The collection of cylinder sets $[C_0, \dots, C_N]$ of any finite length generates the Borel sigma-algebra of $X_\mathcal{B}$. A measure μ on $X_\mathcal{B}$ is completely determined by its values on cylinder sets. These facts are obvious.

4.2. IFS measures on measurable Bratteli diagrams. Let $\mathcal{B} = (\mathcal{X}, \mathcal{E})$ be a stationary measurable Bratteli diagram. Recall that E denotes a Borel subset of $X_0 \times X_1$ (in fact, E can be viewed as a subset of $X_0 \times X_0$). Suppose that p is a Borel measure on E, then (E, p) becomes a standard measure space. We consider both cases, finite and sigma-finite measures p.

The key tool of this subsection is the disintegration theorem for a finite or sigma-finite measure with respect to a measurable partition of a measure space. The literature on this subject is very extensive. We refer to the original paper by Rokhlin [**46**] and more recent paper [**48**] (see [**7**] also).

DEFINITION 4.3. Suppose that $(Z, \mathcal{C}, \mu)$ and $(Y, \mathcal{D}, \nu)$ are standard σ-finite measure spaces. Let $\pi : Z \to Y$ be a measurable function. A system of *conditional measures* for μ with respect to π is a collection of measures $\{\mu_y : y \in Y\}$ such that

(i) μ_y is a Borel measure on $\pi^{-1}(y)$;
(ii) for every $B \in \mathcal{C}$, the function $y \mapsto \mu_y(B)$ is measurable and $\mu(B) = \int_Y \mu_y(B)\, d\nu(y)$.

THEOREM 4.4 ([**48**]). *Let $(Z, \mathcal{C}, \mu)$ and $(Y, \mathcal{D}, \nu)$ be as above. For any measurable function $\pi : Z \to Y$ such that $\mu \circ \pi^{-1} \ll \nu$ there exists a uniquely determined system of conditional measures $(\mu_y)_{y \in Y}$ which disintegrates the measure μ, i.e.*

$$\mu = \int_Y \mu_y d\nu(y).$$

REMARK 4.5. For the edge set E of a stationary measurable Bratteli diagram $\mathcal{B}$ and a Borel measure p on E, consider the partition of E into the sets $r(s^{-1}(x)), x \in X_0$. This partition is measurable in the sense of [**46**]. We can apply the disintegration theorem cited above. More precisely, the sets $r^{-1}(x) = \{(x, y) \in E : y \in X_1\}$, where $x \in X_0$, are vertical sections of the set E. Setting $\widehat{p}(\cdot) = p(s^{-1}(\cdot))$, we have the projection of the measure p onto X_0. Denoting by $x \mapsto p(x, \cdot)$ the corresponding system of conditional measures, we have

$$p = \int_{X_0} p(x, dy)\, d\widehat{p}(x)$$

where the measure $p(x, dy)$ is supported by the set $r(s^{-1}(x))$.

We will consider several versions of disintegration theorem related to measures on the path space $\mathcal{X}_\mathcal{B}$ of a measurable Bratteli diagram $\mathcal{B}$. We assume that the measures will satisfy conditions (i) and (ii) formulated below.

(i) For a given measure m on $\mathcal{X}_\mathcal{B}$ and the set $\mathcal{X}_\mathcal{B}(y)$ defined in (4.1), the measurable function

$$q_m(y) = m(\mathcal{X}_\mathcal{B}(y)),\ y \in X_0,$$

takes finite values.

(ii) The partition ξ defined in (4.2) is obviously measurable, and the measure m on $\mathcal{X}_{\mathcal{B}}$ can be disintegrated with respect to ξ. Denote by $\widehat{m}$ the projection of m onto $E = X/\xi$. Let p be a measure on E. Assuming that $\widehat{m} \ll p$ and applying Theorem 4.4, we have

$$m = \int_E m_e \, dp(e), \tag{4.3}$$

where $e \mapsto m_e$ is the system of conditional measures of m with respect to (E, p). We note that the conditional measure m_e is supported by the set $[e]$, $e \in E$.

The class of IFS measures is determined by a special form of measures m_e defined in (4.3). We will assume that the measures considered on the path space $\mathcal{X}_{\mathcal{B}}$ satisfy conditions (i) and (ii).

DEFINITION 4.6. (*IFS measures and disintegration*) Let $\mathcal{B}$ be a stationary measurable Bratteli diagram, and let $\{\tau_e, e \in E\}$ be the system of contractive maps defined in subsection 4.1. Suppose p is a fixed probability measure on the set E. A Borel measure μ is called an *IFS measure* with respect to p if

$$\mu = \int_E \mu \circ \tau_e^{-1} \, dp(e). \tag{4.4}$$

Our goal is to find conditions under which a measure m defined on the path space of a stationary measurable Bratteli diagram is an IFS measure.

THEOREM 4.7. *Let p be a Borel probability measure on E where E is the edge set of a stationary measurable Bratteli diagram $\mathcal{B} = (\mathcal{X}, \mathcal{E})$. Let μ be a measure on $\mathcal{X}_{\mathcal{B}}$ and $q(x) = \mu(\mathcal{X}_{\mathcal{B}}(x))$. Then μ is and IFS measure if and only if the following condition holds:*

$$\int_{X_1} p(x, dy) q(y) = q(x). \tag{4.5}$$

PROOF. In the proof of the theorem, we apply the idea used in Theorem 3.4.

We will construct an IFS measure μ by defining its values on cylinder sets of consequently increasing length. This construction will be based on the application of (4.4) so that the measure μ will be an IFS measure automatically. Relation (4.5) is used to satisfy the Kolmogorov extension theorem.

As mentioned in Remark 4.5, the projection of p onto X_0 defines the measure $\widehat{p}$. This allows us to define μ on cylinder subsets of $\mathcal{X}_{\mathcal{B}}$ generated by Borel sets on X_0:

$$\mu([C_0]) = \int_{C_0} q(x_0) \, d\widehat{p}(x_0),$$

where C_0 is a Borel subset of X_0.

For a cylinder set $[C_0, C_1] = \{x = (x_i) \in \mathcal{X}_{\mathcal{B}} : x_0 \in C_0, x_1 \in C_1\}$ of length two, we define

$$\mu([C_0, C_1]) = \int_{C_0} \left(\int_{C_1 \cap r(s^{-1}(x_0))} p(x_0, dy) q(y) \right) d\widehat{p}(x_0). \tag{4.6}$$

Show that this definition of the measure μ satisfies (4.4):

$$\begin{aligned}\int_E \mu\circ\tau_e^{-1}([C_0, C_1])\, dp(e) &= \int_E \mu\circ\tau_e^{-1}([C_0, C_1]\cap[e])\, dp(e)\\ &= \int_{C_0}\int_{C_1\cap r(s^{-1}(x_0))} \mu(\mathcal{X}_{\mathcal{B}}(y)\, p(x_0, dy)\, d\widehat{p}(x_0)\\ &= \int_{C_0}\int_{C_1\cap r(s^{-1}(x_0))} q(y)\, p(x_0, dy)\, d\widehat{p}(x_0)\\ &=\mu([C_0, C_1])\end{aligned}$$

In general, if we defined the measure μ on cylinder sets of the form $[C_0, \dots, C_k]$, $k = 1, \dots, n-1$,, then we set

$$\mu([C_0, \dots, C_n]) = \int_E \mu\circ\tau_e^{-1}([C_0, \dots, C_{n-1}])\, dp(e).$$

This means that this definition of μ shows that relation (4.4) holds automatically.

It remains to show that the measure μ is well defined, that is it satisfies the Kolmogorov extension theorem. We check this property for the cylinder sets of length two. Indeed, in this case we take $C_1 = X_1$ and compute

$$\begin{aligned}\mu([C_0, X_1]) &= \int_{C_0}\left(\int_{X_1\cap r(s^{-1}(x_0))} q(y)p(x_0, dy)\right) d\widehat{p}(y)\\ &= \int_{C_0}\left(\int_{r(s^{-1}(x_0))} q(y)p(x_0, dy)\right) d\widehat{p}(y)\\ &= \int_{C_0} q(x_0)\, d\widehat{p}(x_0)\\ &=\mu([C_0]).\end{aligned}$$

The general case is proved analogously. □

REMARK 4.8. We note that the existence of an IFS measure on $\mathcal{X}_{\mathcal{B}}$ can be proved by using a fixed point theorem following [**31**]. We consider the metric dist on $\mathcal{X}_{\mathcal{B}}$ defined in Remark 2.4. Let

$$\mathrm{Lip}_1 = \{f : \mathcal{X}_{\mathcal{B}} \to \mathbb{R} : |f(x) - f(y) \le \ \mathrm{dist}(x, y)\}.$$

For a measure ν on $\mathcal{X}_{\mathcal{B}}$, set

$$L(\nu) = \int_E \nu\circ\tau_e^{-1}\, dp(e).$$

Since τ_e is a contractive map for every e, one can show that $\rho(L(\nu), L(\mu)) \le \rho(\nu, \mu)$ where the metric ρ is defined by

$$\rho(\nu, \mu) = \sup\Big\{|\int f\, d\nu - \int f\, d\mu\,| : f\in \mathrm{Lip}_1\Big\}.$$

Then we conclude that there exists a measure μ_0 such that $L(\mu_0) = \mu_0$.

REMARK 4.9. The IFS measure μ defined in Theorem 4.7 is shift invariant. The proof of this fact is similar to that giving in Theorem 3.5.

Acknowledgments

The authors are pleased to thank our colleagues and collaborators, especially, R. Curto, H. Karpel, P. Muhly, W. Polyzou, and S. Sanadhya. We are thankful to the members of the seminars in Mathematical Physics and Operator Theory at the University of Iowa for many helpful conversations.

Declaration

The authors declare that they have no conflict of interest.

References

[1] P. Alonso Ruiz, Y. Chen, H. Gu, R. S. Strichartz, and Z. Zhou, *Analysis on hybrid fractals*, Commun. Pure Appl. Anal. **19** (2020), no. 1, 47–84, `https://doi-org.proxy.lib.uiowa.edu/10.3934/cpaa.2020004`. MR4025934

[2] Michael Fielding Barnsley, *Superfractals*, Cambridge University Press, Cambridge, 2006, DOI 10.1017/CBO9781107590168. MR2254477

[3] S. Bezuglyi and Palle E. T. Jorgensen, *Representations of Cuntz-Krieger relations, dynamics on Bratteli diagrams, and path-space measures*, Trends in harmonic analysis and its applications, Contemp. Math., vol. 650, Amer. Math. Soc., Providence, RI, 2015, pp. 57–88, DOI 10.1090/conm/650/13008. MR3441734

[4] S. Bezuglyi and O. Karpel, *Bratteli diagrams: structure, measures, dynamics*, Dynamics and numbers, Contemp. Math., vol. 669, Amer. Math. Soc., Providence, RI, 2016, pp. 1–36, DOI 10.1090/conm/669/13421. MR3546661

[5] S. Bezuglyi, J. Kwiatkowski, K. Medynets, and B. Solomyak, *Invariant measures on stationary Bratteli diagrams*, Ergodic Theory Dynam. Systems **30** (2010), no. 4, 973–1007, DOI 10.1017/S0143385709000443. MR2669408

[6] Sergey Bezuglyi, Anthony H. Dooley, and Jan Kwiatkowski, *Topologies on the group of Borel automorphisms of a standard Borel space*, Topol. Methods Nonlinear Anal. **27** (2006), no. 2, 333–385. MR2237460

[7] Sergey Bezuglyi and Palle E. T. Jorgensen, *Transfer operators, endomorphisms, and measurable partitions*, Lecture Notes in Mathematics, vol. 2217, Springer, Cham, 2018, DOI 10.1007/978-3-319-92417-5. MR3793614

[8] Sergey Bezuglyi and Palle E. T. Jorgensen, *Harmonic analysis on graphs via Bratteli diagrams and path-space measures*, Dissertationes Math. **574** (2022), 74, DOI 10.4064/dm826-12-2021. MR4388381

[9] S. Bezuglyi and O. Karpel, *Homeomorphic measures on stationary Bratteli diagrams*, J. Funct. Anal. **261** (2011), no. 12, 3519–3548, DOI 10.1016/j.jfa.2011.08.009. MR2838033

[10] Sergey Bezuglyi, Palle E. T. Jorgensen, and Shrey Sanadhya, *Invariant measures and generalized bratteli diagrams for substitutions on infinite alphabets*, Preprint, `arXiv:2203.14127v2`, 2022.

[11] Sergey Bezuglyi and Olena Karpel, *Invariant measures for Cantor dynamical systems*, Dynamics: topology and numbers, Contemp. Math., vol. 744, Amer. Math. Soc., [Providence], RI, [2020] ©2020, pp. 259–295, DOI 10.1090/conm/744/14988. MR4062569

[12] Ola Bratteli, *Inductive limits of finite dimensional C^*-algebras*, Trans. Amer. Math. Soc. **171** (1972), 195–234, DOI 10.2307/1996380. MR312282

[13] Ola Bratteli and Palle Jorgensen, *Wavelets through a looking glass*, Applied and Numerical Harmonic Analysis, Birkhäuser Boston, Inc., Boston, MA, 2002. The world of the spectrum, DOI 10.1007/978-0-8176-8144-9. MR1913212

[14] Ola Bratteli and Palle E. T. Jorgensen, *Iterated function systems and permutation representations of the Cuntz algebra*, Mem. Amer. Math. Soc. **139** (1999), no. 663, x+89, DOI 10.1090/memo/0663. MR1469149

[15] Gerard Brunick and Steven Shreve, *Mimicking an Itô process by a solution of a stochastic differential equation*, Ann. Appl. Probab. **23** (2013), no. 4, 1584–1628, DOI 10.1214/12-aap881. MR3098443

[16] Shiping Cao, Malte S. Haßler, Hua Qiu, Ely Sandine, and Robert S. Strichartz, *Existence and uniqueness of diffusions on the Julia sets of Misiurewicz-Sierpinski maps*, Adv. Math. **389** (2021), Paper No. 107922, 41, DOI 10.1016/j.aim.2021.107922. MR4289048

[17] A. H. Dooley and Toshihiro Hamachi, *Nonsingular dynamical systems, Bratteli diagrams and Markov odometers*, Israel J. Math. **138** (2003), 93–123, DOI 10.1007/BF02783421. MR2031952

[18] R. Dougherty, S. Jackson, and A. S. Kechris, *The structure of hyperfinite Borel equivalence relations*, Trans. Amer. Math. Soc. **341** (1994), no. 1, 193–225, `https://doi-org.proxy.lib.uiowa.edu/10.2307/2154620`. MR1149121

[19] F. Durand, B. Host, and C. Skau, *Substitutional dynamical systems, Bratteli diagrams and dimension groups*, Ergodic Theory Dynam. Systems **19** (1999), no. 4, 953–993, DOI 10.1017/S0143385799133947. MR1709427

[20] Fabien Durand, *Combinatorics on Bratteli diagrams and dynamical systems*, Combinatorics, automata and number theory, Encyclopedia Math. Appl., vol. 135, Cambridge Univ. Press, Cambridge, 2010, pp. 324–372. MR2759109

[21] Dorin Ervin Dutkay and Palle E. T. Jorgensen, *Harmonic analysis and dynamics for affine iterated function systems*, Houston J. Math. **33** (2007), no. 3, 877–905. MR2335741

[22] Dorin Ervin Dutkay and Palle E. T. Jorgensen, *Probability and Fourier duality for affine iterated function systems*, Acta Appl. Math. **107** (2009), no. 1-3, 293–311, `https://doi.org/10.1007/s10440-008-9384-2`. MR2520021

[23] Dorin Ervin Dutkay and Palle E. T. Jorgensen, *Spectral theory for discrete Laplacians*, Complex Anal. Oper. Theory **4** (2010), no. 1, 1–38, DOI 10.1007/s11785-008-0098-2. MR2643786

[24] Dorin Ervin Dutkay and Palle E. T. Jorgensen, *Monic representations of the Cuntz algebra and Markov measures*, J. Funct. Anal. **267** (2014), no. 4, 1011–1034, DOI 10.1016/j.jfa.2014.05.016. MR3217056

[25] Dorin Ervin Dutkay and Palle E. T. Jorgensen, *The role of transfer operators and shifts in the study of fractals: encoding-models, analysis and geometry, commutative and non-commutative*, Geometry and analysis of fractals, Springer Proc. Math. Stat., vol. 88, Springer, Heidelberg, 2014, pp. 65–95, DOI 10.1007/978-3-662-43920-3_3. MR3275999

[26] Dorin Ervin Dutkay and Palle E. T. Jorgensen, *Representations of Cuntz algebras associated to quasi-stationary Markov measures*, Ergodic Theory Dynam. Systems **35** (2015), no. 7, 2080–2093, DOI 10.1017/etds.2014.37. MR3394108

[27] Carla Farsi, Elizabeth Gillaspy, Palle Jorgensen, Sooran Kang, and Judith Packer, *Purely atomic representations of higher-rank graph C^*-algebras*, Integral Equations Operator Theory **90** (2018), no. 6, Paper No. 67, 26, DOI 10.1007/s00020-018-2493-z. MR3859437

[28] Carla Farsi, Elizabeth Gillaspy, Palle Jorgensen, Sooran Kang, and Judith Packer, *Representations of higher-rank graph C^*-algebras associated to Λ-semibranching function systems*, J. Math. Anal. Appl. **468** (2018), no. 2, 766–798, DOI 10.1016/j.jmaa.2018.08.051. MR3852552

[29] Thierry Giordano, Ian F. Putnam, and Christian F. Skau, *Topological orbit equivalence and C^*-crossed products*, J. Reine Angew. Math. **469** (1995), 51–111. MR1363826

[30] Richard H. Herman, Ian F. Putnam, and Christian F. Skau, *Ordered Bratteli diagrams, dimension groups and topological dynamics*, Internat. J. Math. **3** (1992), no. 6, 827–864, DOI 10.1142/S0129167X92000382. MR1194074

[31] John E. Hutchinson, *Fractals and self-similarity*, Indiana Univ. Math. J. **30** (1981), no. 5, 713–747, `http://dx.doi.org/10.1512/iumj.1981.30.30055`. MR625600

[32] John E. Hutchinson, *Elliptic systems*, Instructional Workshop on Analysis and Geometry, Part I (Canberra, 1995), Proc. Centre Math. Appl. Austral. Nat. Univ., vol. 34, Austral. Nat. Univ., Canberra, 1996, pp. 111–120. MR1394679

[33] John E. Hutchinson and Ludger Rüschendorf, *Selfsimilar fractals and selfsimilar random fractals*, Fractal geometry and stochastics, II (Greifswald/Koserow, 1998), Progr. Probab., vol. 46, Birkhäuser, Basel, 2000, pp. 109–123. MR1785623

[34] Palle E. T. Jorgensen, *Analysis and probability: wavelets, signals, fractals*, Graduate Texts in Mathematics, vol. 234, Springer, New York, 2006. MR2254502

[35] Palle E. T. Jorgensen, *Harmonic analysis*, CBMS Regional Conference Series in Mathematics, vol. 128, American Mathematical Society, Providence, RI, 2018. Smooth and non-smooth; Published for the Conference Board of the Mathematical Sciences, DOI 10.1090/cbms/128. MR3838440

[36] Palle E. T. Jorgensen, Keri A. Kornelson, and Karen L. Shuman, *Iterated function systems, moments, and transformations of infinite matrices*, Mem. Amer. Math. Soc. **213** (2011), no. 1003, x+105, DOI 10.1090/S0065-9266-2011-00627-1. MR2858536

[37] Ioannis Karatzas and Steven E. Shreve, *Methods of mathematical finance*, Applications of Mathematics (New York), vol. 39, Springer-Verlag, New York, 1998, DOI 10.1007/b98840. MR1640352

[38] Bruce P. Kitchens, *Symbolic dynamics*, Universitext, Springer-Verlag, Berlin, 1998. One-sided, two-sided and countable state Markov shifts, DOI 10.1007/978-3-642-58822-8. MR1484730

[39] Matilde Marcolli and Anna Maria Paolucci, *Cuntz-Krieger algebras and wavelets on fractals*, Complex Anal. Oper. Theory **5** (2011), no. 1, 41–81, DOI 10.1007/s11785-009-0044-y. MR2773056

[40] R. Daniel Mauldin, *Infinite iterated function systems: theory and applications*, Fractal geometry and stochastics (Finsterbergen, 1994), 1995, pp. 91–110, `http://dx.doi.org/10.1007/978-3-0348-7755-8_5`. MR1391972

[41] Ian D. Morris and Cagri Sert, *A strongly irreducible affine iterated function system with two invariant measures of maximal dimension*, Ergodic Theory Dynam. Systems **41** (2021), no. 11, 3417–3438, `https://doi-org.proxy.lib.uiowa.edu/10.1017/etds.2020.107`. MR4321717

[42] Esa Nummelin, *General irreducible Markov chains and nonnegative operators*, Cambridge Tracts in Mathematics, vol. 83, Cambridge University Press, Cambridge, 1984, DOI 10.1017/CBO9780511526237. MR776608

[43] Ian F. Putnam, *Cantor minimal systems*, University Lecture Series, vol. 70, American Mathematical Society, Providence, RI, 2018, DOI 10.1090/ulect/070. MR3791491

[44] Robert J. Ravier and Robert S. Strichartz, *Sampling theory with average values on the Sierpinski gasket*, Constr. Approx. **44** (2016), no. 2, 159–194, DOI 10.1007/s00365-016-9341-7. MR3543997

[45] Jean Renault, *Random walks on Bratteli diagrams*, Operator theory: themes and variations, Theta Ser. Adv. Math., vol. 20, Theta, Bucharest, 2018, pp. 187–204. MR3839554

[46] V. A. Rohlin, *On the fundamental ideas of measure theory* (Russian), Mat. Sbornik N.S. **25(67)** (1949), 107–150. MR0030584

[47] G. E. Shilov and B. L. Gurevich, *Integral, measure and derivative: a unified approach*, Revised English edition, Dover Books on Advanced Mathematics, Dover Publications, Inc., New York, 1977. Translated from the Russian and edited by Richard A. Silverman. MR0466463

[48] David Simmons, *Conditional measures and conditional expectation; Rohlin's disintegration theorem*, Discrete Contin. Dyn. Syst. **32** (2012), no. 7, 2565–2582, DOI 10.3934/dcds.2012.32.2565. MR2900561

Department of Mathematics, University of Iowa, Iowa City, Iowa 52242
Email address: `sergii-bezuglyi@uiowa.edu`

Department of Mathematics, University of Iowa, Iowa City, Iowa 52242
Email address: `palle-jorgenseni@uiowa.edu`

Contemporary Mathematics
Volume **797**, 2024
https://doi.org/10.1090/conm/797/15952

Unicritical and maximally critical laminations

Brittany E. Burdette, Caleb S. Falcione, Cameron G. Hale, and John C. Mayer

Abstract. We study the correspondence between unicritical laminations and maximally critical laminations with rotational and identity return polygons. Laminations are a combinatorial and topological way to study Julia sets. Laminations give information about the structure of parameter space of degree d polynomials with connected Julia sets.

1. Introduction

1.1. Motivation. Definitions and facts in this section are adapted from Milnor [**12**]. Let $P : \mathbb{C} \to \mathbb{C}$ be a polynomial of degree $d \geq 2$:

$$P(z) = a_d z^d + a_{d-1} z^{d-1} + \cdots + a_1 z + a_0$$

For $z_0 \in \mathbb{C}$, we define the *iterates* of z_0 to be:

$$P^n(z_0) := P(P^{n-1}(z_0)).$$

We define the *forward orbit* of z_0 to be the set:

$$\{z_0, P(z_0), P^2(z_0), \ldots, P^n(z_0), \ldots\}.$$

Compactify $\mathbb{C}$ to the Riemann sphere $\mathbb{C}_\infty$ by adding the point at infinity. For a polynomial, P, ∞ is an attracting fixed point: for $z \in \mathbb{C}$ with $|z|$ sufficiently large, $\lim_{n\to\infty} P^n(z) = \infty$. The *basin of attraction* of ∞ is:

$$B_\infty := \{z \in \mathbb{C} \mid \lim_{n\to\infty} P^n(z) = \infty\}$$

By continuity, B_∞ is an open set. Let P be a polynomial of degree $d \geq 2$. We define the following sets in $\mathbb{C}$ associated with P:

(1) The *Julia set* $J(P)$:=boundary of B_∞.
(2) The *Fatou set* $F(P)$:=$\mathbb{C}_\infty \setminus J(P)$.
(3) The *filled Julia set* $K(P) := \mathbb{C} \setminus B_\infty$.

It is known that $J(P)$ is nonempty, compact, and perfect; $K(P)$ is full (does not separate $\mathbb{C}$), but $J(P)$ may separate $\mathbb{C}$. We will only consider cases when $J(P)$ is connected. Periodic points of P can be attracting, repelling, or indifferent. Attractive periodic orbits are in the Fatou set. Repelling periodic orbits are in the Julia set. Components of the Fatou set are called Fatou domains. There is only one

2020 *Mathematics Subject Classification.* Primary 37F20.
Key words and phrases. Complex dynamics, circle dynamics, angle d-tupling, covering map, rotation set, laminations of the unit disk.

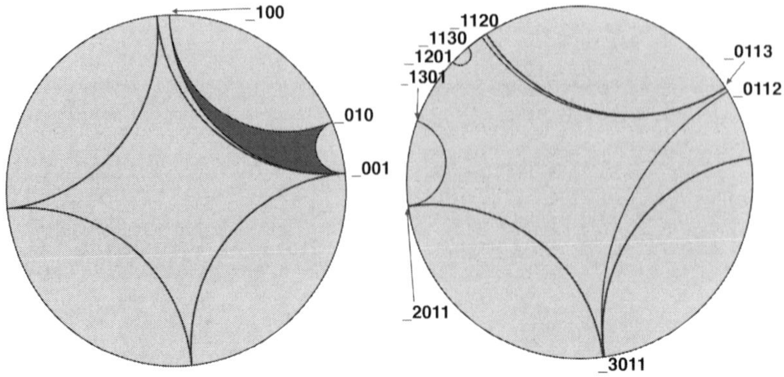

FIGURE 1. Here we have two different examples of d-nary (in this case 4-nary) coordinates. On the left, we only use the digits 0 and 1, but on the right we use all 4 digits: 0, 1, 2, and 3. The underscore that precedes the digits indicates that those digits are repeated. We generate all lamination pictures except Figure 9 using Lamination Builder [**10**].

unbounded Fatou domain whereas there can be zero, one, or countably infinitely many bounded Fatou domains.

An important part of complex dynamics is understanding the structure of parameter spaces. For example, the *unicritical degree d connectedness locus* would be defined as such:

$$\mathcal{M}_d = \{c \in \mathbb{C} \mid J(P_c) \text{ is connected}\}$$

where $P_c = z^d + c$.

The unicritical degree d connectedness locus is a way of studying the unicritical polynomials of degree d with connected Julia sets all at once. There is also a multicritical degree d connectedness locus. Consider the family of monic and centered polynomials of degree d.

$$\mathcal{P}_d := \{(a_0, a_1, \ldots, a_{d-2}) \in \mathbb{C}^{d-1} \mid P(z) = z^d + 0z^{d-1} + a_{d-2}z^{d-2} + \cdots + a_0\}$$

The *degree d connectedness locus* is:

$$\mathcal{C}_d = \{(a_0, a_1, \ldots, a_{d-2}) \in \mathbb{C}^{d-1} \mid P \in \mathcal{P}_d \text{ and } J(P) \text{ is connected}\}$$

We study one of the connections between these two parameter spaces.

1.2. Laminations. Laminations are a combinatorial and topological way to study Julia sets. Unicritical laminations (Definition 2.1) appear, for example, on the boundary of the main cuboid for cubic polynomials. A rotational polygon in a lamination corresponds to a fixed point in the Julia set, and an identity return polygon corresponds to a branch point in the Julia set that returns with no rotation.

Most of the definitions and facts in this section come from [**4**], [**6**], [**9**], and [**14**].

DEFINITION 1.1 (σ_d map). We use "d-nary" coordinates on the circle $\mathbb{S}$. The map $\sigma_d : \mathbb{S} \to \mathbb{S}$ is defined to be $\sigma_d(t) = dt \pmod 1$. For example, σ_2 has binary coordinates, σ_3 has ternary, σ_4 has quarternary, and so on.

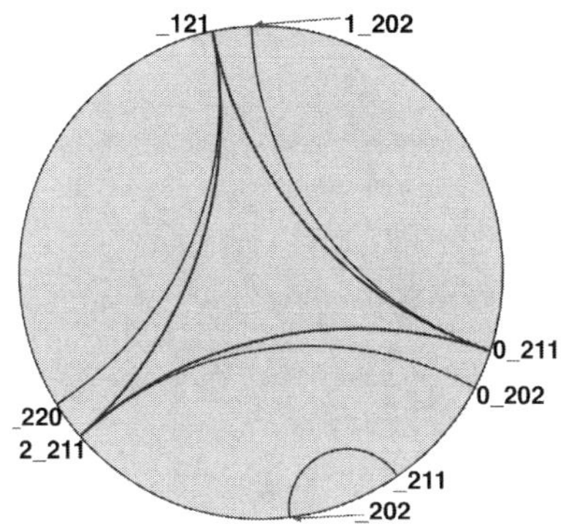

FIGURE 2. Here we have a full sibling collection under σ_3. All of the long "major" leaves map to one short "minor" leaf. The central triangle is made of three critical chords creating an all critical triangle.

DEFINITION 1.2 (Lamination). A *lamination*, $\mathcal{L}$, is a collection of chords of the closed unit disk, $\overline{\mathbb{D}}$, which we call *leaves*, such that:

(1) Any two leaves of $\mathcal{L}$ meet, if at all, in a point of $\mathbb{S}$.
(2) $\mathcal{L}^* := \mathbb{S} \cup \{\cup \mathcal{L}\}$ is a closed subset of $\overline{\mathbb{D}}$.

In the case that condition (1) only is met, we call our collection of chords a *pre-lamination.*

DEFINITION 1.3 (Leaf mapping). We denote mapping a leaf, $\ell = \overline{ab}$, with endpoints a and b, under the map σ_d by

$$\sigma_d(\ell) = \overline{\sigma_d(a)\sigma_d(b)}$$

DEFINITION 1.4 (Critical chords). A chord, $\ell = \overline{ab}$, is called *critical* when both its endpoints map to a single point $\sigma_d(a) = \sigma_d(b) \in \mathbb{S}$.

DEFINITION 1.5 (Sibling leaves). Let $\ell_1 \in \mathcal{L}$ be a leaf and suppose $\sigma_d(\ell_1) = \ell'$, for some nondegenerate leaf $\ell' \in \mathcal{L}$. A leaf $\ell_2 \in \mathcal{L}$, disjoint from ℓ_1, is called a *sibling* of ℓ_1 provided $\sigma_d(\ell_2) = \ell' = \sigma_d(\ell_1)$. A collection $\mathcal{S} = \{\ell_1, \ell_2, \ldots, \ell_d\} \subset \mathcal{L}$ is called a *full sibling collection* provided that for each i, $\sigma_d(\ell_i) = \ell'$ and for all $i \neq j$, $\ell_i \cap \ell_j = \emptyset$.

DEFINITION 1.6 (Sibling invariant lamination). A lamination $\mathcal{L}$ is said to be *sibling d-invariant* (or simply *invariant* if no confusion will result) provided that the following three statements hold:

(1) (Forward invariant) For every $\ell \in \mathcal{L}$, $\sigma_d(\ell) \in \mathcal{L}$.
(2) (Backward invariant) For every nondegenerate $\ell' \in \mathcal{L}$, there is a leaf $\ell \in \mathcal{L}$ such that $\sigma_d(\ell) = \ell'$.
(3) (Sibling invariant) For every $\ell_1 \in \mathcal{L}$ with $\sigma_d(\ell_1) = \ell'$, a nondegenerate leaf, there is a full sibling collection $\{\ell_1, \ell_2, \ldots, \ell_d\} \subset \mathcal{L}$ such that $\sigma_d(\ell_i) = \ell'$.

REMARK 1.7. A sibling d-invariant lamination induces an equivalence relation. Two points on $\mathbb{S}$ are equivalent if they are joined by a finite concatenation of leaves. We consider laminations for which this results in a closed equivalence relation. Thus, the sibling invariant laminations we will be considering have a fourth condition from [**6**] not listed in the definition:

(4) $\mathcal{L}$ has finite equivalence classes, and all leaves are boundary chords of the convex hulls of equivalence classes.

DEFINITION 1.8 (Gap). A *gap* in a lamination, $\mathcal{L}$,is the closure of a component of $\overline{\mathbb{D}} \setminus \mathcal{L}^*$. A gap is *critical* iff two points in its boundary map to the same point. A gap with finitely many leaves in its boundary is usually called a *polygon*. The leaves bounding a finite gap are called the *sides* of the polygon.

DEFINITION 1.9 (Fatou gap). A *Fatou gap* in a lamination is a gap whose boundary intersected with $\mathbb{S}$ contains a Cantor set.

We can see Fatou gaps in a lamination in Figure 3. The white spaces in the lamination are Fatou gaps since as we pull back our polygons farther, we get a gap whose boundary meets $\mathbb{S}$ in a Cantor set.

REMARK 1.10 (Degree of a Fatou gap). The degree of a Fatou gap is the amount of criticality associated with the gap. For example, one critical chord in the gap gives a degree two Fatou gap which means the gap will map forward two-to-one.

A Fatou gap in a lamination (that corresponds to a Julia set) maps to the closure of a bounded Fatou domain in dynamical space. Comparing the Rabbit lamination to the Rabbit Julia set in Figure 3, one can see the correspondence.

DEFINITION 1.11 (All critical polygon). A polygon whose endpoints all map to a single point is called an *all critical polygon*. See Figure 2.

PROPOSITION 1.12 (Order preserving [**6**]). *On both finite and infinite gaps of a d-invariant lamination, the map σ_d preserves circular order.*

PROPOSITION 1.13 (How Fatou gaps meet). *Fatou gaps in a lamination can meet in two ways. Either they meet at a leaf shared in both of their boundaries, or a leaf in one gap meets another leaf in the other gap at a single point.*

PROOF. Suppose we have two gaps G_1 and G_2. The boundary of a gap is the convex hull of where it meets the circle. Note the convex hull of a gap cannot contain, or cross, another convex hull. This means that two gaps can share a piece of their convex hull such as a leaf, or a point joined by two leaves. Note that two gaps cannot meet at a point that is not an endpoint of a leaf for then the two convex hulls would intersect in their interior. □

REMARK 1.14. Finite gaps do not meet since each gap is its own equivalence class.

1.3. Pullback laminations.

DEFINITION 1.15 (Critical portrait). A maximal collection of critical chords for σ_d is called a *critical portrait* (maximal here meaning at least $d-1$ noncrossing critical chords). When a collection of critical chords meet at endpoints, the maximum may exceed $d-1$ by forming all critical polygons.

DEFINITION 1.16 (Critical sectors). A critical sector is a region C in the closed unit disk that is bounded by critical chords and arcs of the circle such that the boundary of C maps onto the circle with degree 1.

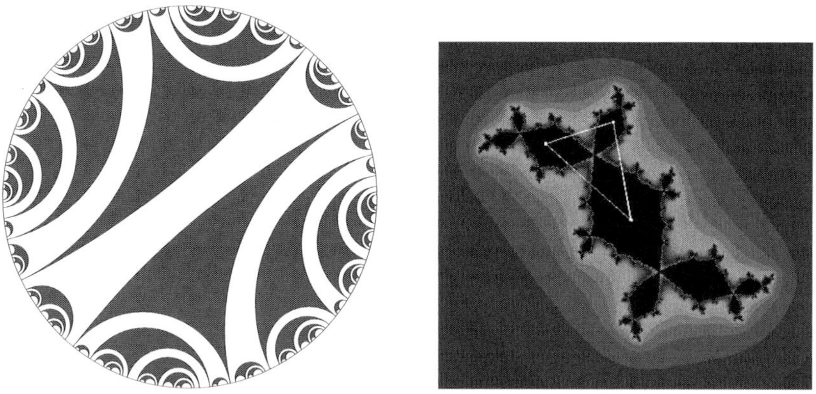

FIGURE 3. Here are corresponding quadratic lamination (left) and Julia set (right) named the "Douady Rabbit". There is a quotient map from the lamination to the Julia set. Under this quotient map the white gaps correlate with the black bounded domains of the Fatou set. The triangles map to "pinch points" between the bounded Fatou domains.

DEFINITION 1.17 (Forward invariant set). A *forward invariant set* consists of periodic leaves or polygons that map forward preserving circular order without intersecting.

DEFINITION 1.18 (Compatible). A critical portrait, C, is *compatible* with a forward invariant set, F, when F meets $\cup C$ only at endpoints of leaves or vertices of polygons.

DEFINITION 1.19 (Branches of the inverse). Let C be a critical portrait. Every critical sector S in $\overline{\mathbb{D}}$ defined by C will have a function $\tau : \partial\mathbb{D} \to \partial S$ that is one to one, and $\sigma_d \circ \tau$ is the identity on $\partial\mathbb{D}$.

Combining the previous definitions we can now define a pullback scheme.

DEFINITION 1.20 (Pullback scheme). Let C be a critical portrait and F a compatible forward invariant set. The corresponding collection $PB(F, C)$ of the branches of the inverse determined by C as in Definition 1.19 gives us a *pullback scheme* for F.

Using Lamination Builder [**10**], we can visualize multiple steps of the pullback scheme.

REMARK 1.21. There can be multiple critical portraits compatible with F that define the pullback scheme differently. There are some critical portraits that are compatible with F, but the related pullback lamination does not satisfy condition (4) of Definition 1.6. The following theorem is well known.

THEOREM 1.22. *Let F be a periodic forward invariant set under σ_d, C a compatible critical portrait, and the pullback scheme $PB(F, C) = \{\tau_1, \tau_2, \ldots, \tau_d\}$. Let $F_0 = F$, and $F_1 = F_0 \cup \tau_1(F_0) \cup \tau_2(F_0) \cup \cdots \cup \tau_d(F_0)$. In general, for any given stage n of the pull back $F_n = F_{n-1} \cup \tau_1(F_{n-1}) \cup \cdots \cup \tau_d(F_{n-1})$. Let $F_\infty = \bigcup_{n=0}^{\infty} F_n$, and let $\mathcal{L} = \overline{F_\infty}$. Then, $\mathcal{L}$ is a sibling d-invariant lamination.*

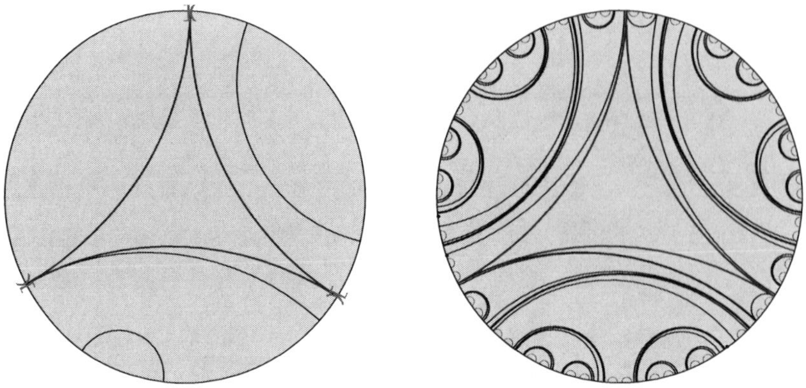

FIGURE 4. (Left) A pullback scheme with an all critical triangle and included branches of the inverse marked by closed ends of half open intervals on the circle. (Right) The resulting pullback with the indicated pullback scheme.

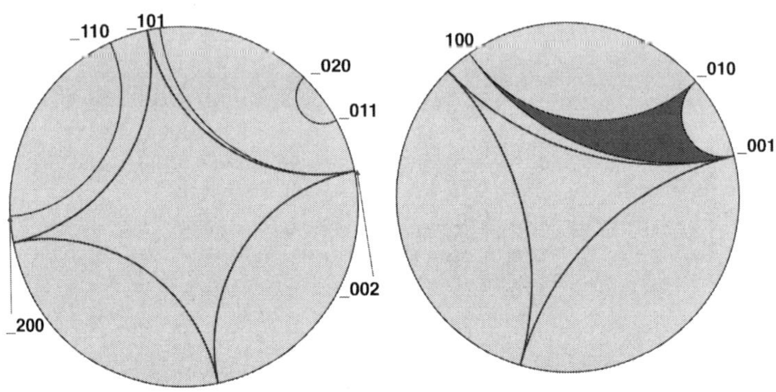

FIGURE 5. Identity return leaf in σ_4 (left), rotational polygon in σ_3 (right).

PROOF SKETCH. Note, F_0 is forward invariant, and the τ_i are homeomorphisms away from critical values. Hence, each F_i is forward invariant, and up to F_{i-1} is backward and sibling invariant. It follows that F_∞ is sibling d-invariant but is not closed. Taking the closure $\overline{F_\infty}$ picks up points in the circle and limit leaves. The only potential problem here is crossing limit leaves. But, if limit leaves were to cross when we take the closure, that means there were some crossings in F_∞ prior to the closure. To ensure $\overline{F_\infty}$ has finite equivalence classes, one must choose the appropriate pullback scheme. □

DEFINITION 1.23 (Grand orbit). Let P be an arbitrary subset of a lamination (for example a point, leaf, polygon, gap, etc.) in a d-invariant lamination. The *grand orbit* $\mathcal{GO}(P)$ is defined to be

$$\mathcal{GO}(P) = \{\sigma_d^{-n}(\sigma_d^k(P)) | n, k \in \mathbb{Z}^+ \cup 0\}.$$

1.4. Periodic polygons. Periodic polygons (and leaves) in a lamination can be rotational, rotation return, or identity return. We define each type below.

1.4.1. *Rotational Polygons.*

DEFINITION 1.24 (Rotational set). Consider $\sigma_n \colon \mathbb{S} \to \mathbb{S}$ for a particular $n \geq 2$. Let $P = \{x_i | 0 < x_1 < x_2 < x_3 < \cdots < x_k < 1\}$ be a finite set in consecutive order in $\mathbb{S}$. We say that P is a *rotational set* (for σ_n) iff

(1) $\sigma_n(P) = P$, and
(2) For $1 \leq j \leq k$, if $\sigma_n(x_j) = x_i$, set $i(j) = i$. Then for all j, $j - i(j) (\text{mod } k)$ is the same.

If (2) holds (but possibly not (1)), we say σ_n is circular order-preserving on P. We call a rotational set which is a single periodic orbit, a *rotational orbit.*

DEFINITION 1.25 (Rotation number). To each rotational set we can assign a *rotation number*, a rational number $0 \leq \dfrac{p}{q} < 1$ in lowest terms.

Let $\mathcal{O} = \{x_1 < x_2 < x_3 \cdots < x_q\}$ be a rotational periodic orbit. Suppose that $\sigma_n(x_1) = x_j$. Set $p = j - 1$. The rotation number of $\mathcal{O}$ is $\dfrac{p}{q}$. Our notation is $\rho(\mathcal{O}) = \rho(x_1) = \rho(x_i) = \dfrac{p}{q}$.

DEFINITION 1.26 (Rotational polygon). A polygon (or a leaf) $P = P_0$ is said to be *rotational* iff $\sigma_d|_{P_0}$ maps P_0 back to itself preserving circular order and $\forall i \neq j$, $\sigma_d(p_i) = p_j$ according to some nonzero rotation number.

DEFINITION 1.27 (Rotation return polygon). A polygon is said to be *rotation return* iff it is a polygon that maps off of itself and when it returns it has a nonzero rotation number.

1.4.2. *Identity return polygons.*

DEFINITION 1.28 (Identity return). A polygon (or a leaf) $P = P_0$ is said to be *identity return* iff its *orbit*

$$\{P_0, P_1 = \sigma_d(P_0), P_2 = \sigma_d(P_1), P_3, \ldots, P_n = P_0\}$$

is periodic (of least period n) and has the properties

(1) the polygons in the orbit are pairwise disjoint,
(2) $\sigma_d^n|_{P_0}$ is the identity (rotation number $= 0$), and
(3) P_i maps to $P_{i+1 \;(\text{mod } n)}$ preserving circular order.

Each vertex (and each side of P) is in a different orbit of period n since the polygons in the orbit are disjoint. Restating Kiwi's theorem [**11**] for polygons in a lamination, we have:

If a periodic polygon P in a d-invariant lamination has exactly k distinct critical values associated with it, then the number of orbits of sides of P is at most $k + 1$. Moreover, if the number of orbits of sides is $k + 1$, then P is identity return.

Thus, d is the maximal number of sides for an identity return polygon in a d-invariant lamination.

In the special case that the polygon has period $n = 1$, we give the following definition:

DEFINITION 1.29 (Fixed polygon). A polygon (or a leaf) P is said to be *fixed* iff $\sigma_d(P) = P$ and $\sigma_d(x_i) = x_i$ where $x_1, x_2, \ldots, x_i$ are the vertices of P (endpoints of leaves of P).

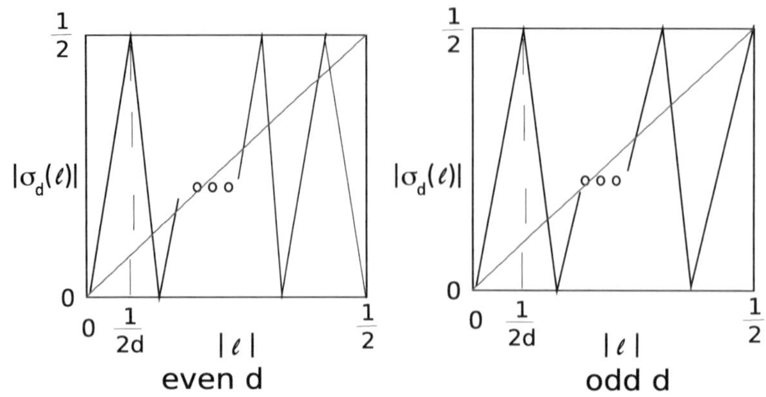

FIGURE 6. Here is the leaf length graph for even and odd d under σ_d from Cosper et al. [**9**].

However, we only discuss correspondences with identity return and rotational polygons here.

1.5. Leaf length. The following definitions are adapted from Cosper et al. [**9**].

DEFINITION 1.30 (Parameterizing the circle). The positive order on the circle $\mathbb{S}$ will be fixed in the counterclockwise direction. Let $|(a,b)|$ be the length in the parameterization of the arc (a,b) in $\mathbb{S}$ from a to b counterclockwise. Given a chord $\overline{ab}$ there are two arcs of $\mathbb{S}$ it subtends. Define the length $|\overline{ab}|$ of $\overline{ab}$ to be the shorter of the two arcs subtended. Note, the maximum length for a leaf is $\frac{1}{2}$.

DEFINITION 1.31 (Leaf length function).

$$|\sigma_d(\overline{ab})| = \begin{cases} d|\overline{ab}| \pmod 1, \text{ if } d|\overline{ab}| \pmod 1 \le \frac{1}{2} \\ 1 - d|\overline{ab}| \pmod 1, \text{if} d|\overline{ab}| \pmod 1 \ge \frac{1}{2}. \end{cases}$$

PROPOSITION 1.32 (Growing leaves). *A leaf, ℓ, of length $|\ell| < \frac{1}{d+1}$ will increase in length under σ_d until $|\sigma_d^k(\ell)| \ge \frac{1}{d+1}$.*

REMARK 1.33. Note that the first positive fixed point is $\frac{1}{d+1}$ which can be seen in Figure 6 where the diagonal intersects the graph for the first time.

DEFINITION 1.34 (Distance between leaves). Take two leaves $\ell_1 = \overline{ab}$ and $\ell_2 = \overline{cd}$ with the order of the points being $a < b < c < d$ in a lamination $\mathcal{L}$. Define the distance between ℓ_1 and ℓ_2 to be

$$dist(\ell_1, \ell_2) = |(b,c)| + |(d,a)|$$

2. Unicritical laminations

DEFINITION 2.1 (Unicritical lamination). A d-invariant lamination which is compatible with an all critical d-gon is called a *unicritical lamination.*

DEFINITION 2.2 (Major and minor leaves). For the orbit of a periodic leaf in a unicritical lamination, the leaf, M, closest to critical length is called the *major*, and the image $\sigma_d(M)$ of the major is called the *minor* usually denoted m.

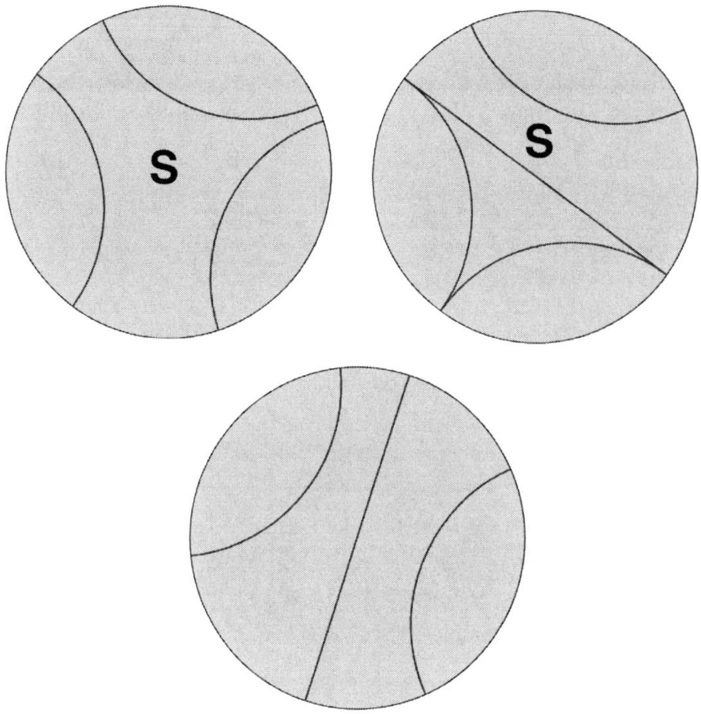

FIGURE 7. We can see on the left and middle figure that S is a maximal critical sector as it has all critical leaves on its boundary and a side of the all critical polygon (if there is one). On the right, none of the four critical sectors will be maximal as no sector has all critical chords on its boundary.

2.1. Major all critical (MAC).

DEFINITION 2.3 (Major all critical (MAC)). The major leaf in a periodic leaf orbit for σ_d which is compatible with an all critical d-gon is called a *major all critical (MAC)* leaf. This MAC leaf may be a single leaf or a side of a polygon.

DEFINITION 2.4 (Canonical MAC lamination). Let M be a MAC leaf. Attach a guiding all critical d-gon to one of the endpoints of M. Pullback M with respect to the guiding all critical d-gon as described in Section 1.3. We call this MAC lamination, $\mathcal{L}(M)$, or just $\mathcal{L}$ when M is understood.

REMARK 2.5 (First return of MAC leaf). If a MAC leaf is just a leaf and not a side of a polygon, the first time the leaf returns to itself is by the identity except when the rotation number is $\frac{1}{2}$.

2.2. Single critical moment (SCM).

DEFINITION 2.6 (Maximal critical sector). Let C be a critical portrait. If a critical sector of C has in its boundary all critical leaves and a side of each all critical polygon, then we call it a *maximal critical sector*. See Figure 7.

DEFINITION 2.7 (Single critical moment (SCM)). A periodic return polygon, P, for σ_d is *single critical moment (SCM)* if it has the following properties:

(1) At one moment in its orbit P must be inside a maximal critical sector.
(2) No side of P makes a closer approach to criticality in a different sector.

DEFINITION 2.8 (Canonical SCM lamination). Let P be an SCM polygon and $\mathcal{C}$ be a collection of $d-1$ guiding critical chords each touching one endpoint of each of the $d-1$ longest sides of P. These longest sides of P are required to be adjacent to each other. Pullback P with respect to $\mathcal{C}$ as described in Section 1.3. We call this the *canonical SCM lamination* for P.

REMARK 2.9. In Definition 2.8, the existence of the collection $\mathcal{C}$ of guiding critical chords follows immediately from Definition 2.7.

DEFINITION 2.10 (Maximally critical SCM polygon). A *maximally critical SCM polygon* is where criticality is broken up as much as possible: all Fatou gaps will be degree two. See Remark 1.10.

We now consider two cases: polygons that return by the identity and polygons that are rotational. We are interested in finding the correspondence between polygons with SCM and MAC orbits. Examples of MAC and SCM laminations can be seen in Figures 13 and 14.

In Barry's PhD thesis [**2**], identity return triangles with a single critical moment were shown to stand in one to one correspondence with MAC leaves of a unicritical cubic lamination. Properties of these leaves and triangles were deduced. Our aim is to generalize this correspondence to certain d-invariant laminations for $d > 3$. In the identity return and rotational polygon cases, we will identify the SCM polygons that correspond one-to-one to MAC laminations.

There are two ways to define the period of an object: the first return of the vertices of the object vs. the first return of the object. If the polygon is identity return, then these are the same since the object and vertices return back to themselves at the same iterate. However, in the rotational and rotation return cases the period of the object and the period of the vertices differ. In fact, the period of the object is a factor of the period of the vertices. We will denote the period of the vertices as k, and the period of the object as r. In the rotational case, r is always 1.

The following is the major theorem of this paper.

THEOREM 2.11 (MAC $\iff$ SCM). *There exists a one to one correspondence between canonical MAC laminations and canonical SCM laminations for σ_d as follows:*

(1) *MAC identity return leaves correspond to SCM identity return d-gons,*
(2) *MAC rotational polygons correspond to SCM rotational $k(d-1)$-gons,*

all with $d-1$ consecutive major leaves.

In this section, we will develop an understanding of the structure of both SCM and MAC laminations as well as any similarities they might have between them. We establish one direction of Theorem 2.11 in this section, and the other direction in the following section.

2.3. Endcaps. We now want to define what we mean by *endcaps* or *endcaps intervals*. Assume we have a forward invariant set, F, and a compatible critical set, C. Take the major leaf from F. (In our cases, our major leaf is unique) Following the process from Section 1.3, generate the first pullback to get the siblings of our initial leaf. One endpoint of each of the leaf and siblings will be touching a critical chord. An **endcap interval** is the interval from the endpoint of the leaf or sibling not touching a critical chord to the associated critical chord. In the special case of the symmetric sibling portrait, all siblings are the same length and each endcap interval is also the same length. Each endcap, where we start with a major leaf, maps one-to-one to the minor since end points of the interval are the end points of the major or one of its siblings.

2.4. Central strips.

DEFINITION 2.12 (Central strip). Consider the *sibling portrait* of a full collection, $\mathcal{S}$, of sibling leaves where no leaf maps to a diameter and at least one leaf is longer than $\frac{1}{d+1}$. The *central strip* is the union of components of $\overline{\mathbb{D}} \setminus \cup\mathcal{S}$ such that these components contain at least two arcs of the circle of length less than $\frac{1}{2d}$.

THEOREM 2.13 (Central strip lemma [9]). *Let C be a central strip of a leaf, ℓ, and its siblings where the length, η, of any arc of C is less than $\frac{1}{d(d+1)}$ (This is a* narrow *central strip). Then, the following hold:*

(1) *The first image $\ell_1 = \sigma_d(\ell)$ cannot re-enter C.*
(2) *The second image $\ell_2 = \sigma_d^2(\ell)$ cannot re-enter C with both endpoints in a single component of $C \cap \mathbb{S}$.*
(3) *If an iterate $\ell_j = \sigma_d^j(\ell)$ of ℓ re-enters C, for least $j > 2$, and has endpoints lying in one component of $C \cap \mathbb{S}$, then iterate ℓ_k, for some $k \leq j-1$, gets at least as close in the endpoint metric as $\dfrac{\eta}{d^{j-k}}$ to a critical chord in $\overline{\mathbb{D}} \backslash C$.*

The proof of the following corollary, first stated in [9], is left to the reader.

COROLLARY 2.14 (Unicritical central strip lemma). *Let C be the central strip of leaf, ℓ, and its siblings for the map σ_d in a unicritical lamination. Then, no image of ℓ can re-enter C with both endpoints in a single component of $C \cap \mathbb{S}$.*

COROLLARY 2.15 (Maximal central strip lemma). *Let P be a maximally critical SCM polygon. Then, the orbit of any major cannot re-enter C with both endpoints in a single component of $C \cap \mathbb{S}$.*

PROOF. Due to the central strip lemma (Theorem 2.13), it is sufficient to show that a major leaf must be within $\dfrac{1}{d(d+1)}$ of critical length. Since we have an SCM polygon, the only critical length is $\dfrac{1}{d}$.

First, let us consider the identity return case. Our SCM polygon consists of $d-1$ majors of length greater than $\dfrac{1}{d}$. There is only room for one other leaf in a disjoint orbit from the $d-1$ majors. Call this leaf, M. By our definition of an SCM polygon (Definition 2.7), $|M| > \dfrac{1}{d+1}$.

By way of contradiction, assume there exists a leaf, ℓ, such that ℓ is not within $\dfrac{1}{d(d+1)}$ of critical length. This implies $|\ell| > \dfrac{1}{d} + \dfrac{1}{d(d+1)} = \dfrac{d+2}{d(d+1)}$. The length of the other majors is $|\sum_{i=1}^{d-2}(\ell_i)| > \dfrac{d-2}{d}$. Thus, the length of our circle used up

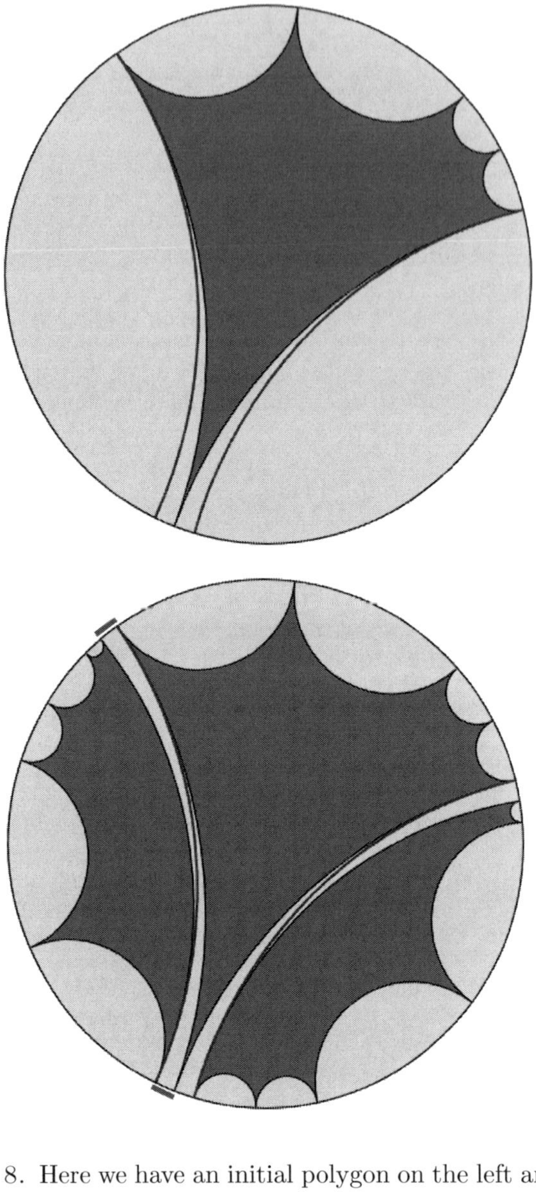

FIGURE 8. Here we have an initial polygon on the left and its first pullback on the right. On the right, the two endcaps associated with the left sibling are highlighted.

by the majors is $|\ell| + |\sum_{i=1}^{d-2}(\ell_i)| > \dfrac{d+2}{d(d+1)} + \dfrac{d-2}{d} = \dfrac{d}{d+1}$. This leaves less than $\dfrac{1}{d+1}$ for M, a contradiction.

Now, let us consider the rotational case. Our SCM polygon consists of $d-1$ majors of length greater than $\dfrac{1}{d}$ and all of the pre-images of those majors.

Again, assume by way of contradiction that there exists a leaf, ℓ, such that ℓ is not within $\dfrac{1}{d(d+1)}$ of critical length. As before, the majors take up more than $\dfrac{d}{d+1}$ of the circle. This leaves $\dfrac{1}{d+1}$ for all the pre-image leaves. If we consider the immediate pre-image of ℓ, we can simply divide by d to find its length since we are one-to-one except at the major. Thus, we have that $|\sigma_d^{-1}(\ell)| = \dfrac{|\ell|}{d} > \dfrac{d+2}{d^2(d+1)}$. Similarly, we can find the length of the immediate pre-images of the other majors: $|\sum_{i=1}^{d-2} \sigma_d^{-1}(\ell_i)| = \dfrac{|\sum_{i=1}^{d-2}(\ell_i)|}{d} > \dfrac{d-2}{d^2}$. Thus, the length of the pre-images of the majors is greater than $\dfrac{d+2}{d^2(d+1)} + \dfrac{d-2}{d^2} = \dfrac{1}{d+1}$. However, we only have less than $\dfrac{1}{d+1}$ left from the majors, a contradiction. □

2.5. Topological Julia set.

DEFINITION 2.16 (First return map). Starting with a d-invariant lamination and a Fatou gap, G, the *first return map*, R, of G is the first iterate of σ_d where G returns to itself.

REMARK 2.17. The idea that G returns to itself after some σ_d^k makes sense, because we know G is periodic.

DEFINITION 2.18 (Topological Julia set). Starting with a d-invariant lamination, $\mathcal{L}$, let $\sim_\mathcal{L}$ be the induced equivalence relation. Let $J(\mathcal{L})$ be the quotient space of $\sim_\mathcal{L}$ and let $h : \mathbb{S} \to J(\mathcal{L})$ be the natural projection. We call $J(\mathcal{L})$, a *topological Julia set.*

REMARK 2.19. We can think of the quotient map as going from the lamination to $J(\mathcal{L})$ shrinking leaves and polygons to points. It follows from the definition of d-invariant lamination that equivalence classes map to equivalence classes, and the pre-image of equivalence classes is a union of equivalence classes.

DEFINITION 2.20 (Induced return map). We call

$$P(x) = h(\sigma(h^{-1}(x))) : J(\mathcal{L}) \to J(\mathcal{L})$$

a *topological polynomial.* If we take $\hat{G} = h(\partial G)$ then R induces a map $\hat{R} : \hat{G} \to \hat{G}$ which we call the *induced return map.*

Where no confusion will result, we will use R for the first return map on both laminations and topological Julia sets.

Note a map $f : \mathbb{S} \to \mathbb{S}$ is *topologically exact* iff for any interval $I \subset \mathbb{S}$, there exists some n such that $f^n(I) = \mathbb{S}$. The following theorem is well known. (See [**3**].)

THEOREM 2.21. *A topologically exact covering map $f : \mathbb{S} \to \mathbb{S}$ is conjugate to σ_d for some $d \geq 2$.*

DEFINITION 2.22 (Hyperbolic lamination). A d-invariant lamination is said to be *hyperbolic* if and only if all compatible critical chords are interior to periodic Fatou gaps.

PROPOSITION 2.23. *Canonical MAC laminations and canonical SCM laminations are hyperbolic.*

2.6. Structure of canonical unicritical laminations.

LEMMA 2.24 (Fatou gaps of unicritical laminations [**13**]). *Let $\mathcal{L}$ be a degree d unicritical lamination with a periodic major, M. The following hold:*

(1) *There is a periodic Fatou gap, G, compatible with an all critical d-gon. Hence, $\mathcal{L}$ is hyperbolic.*
(2) *All boundary leaves of G are pre-images of the orbit of M, and the convex hull of $G \cap \mathbb{S}$ is G.*
(3) *The induced return map $\hat{R} : \hat{G} \to \hat{G}$ is conjugate to σ_d.*

PROPOSITION 2.25 (MAC orbit location). *Let $\mathcal{L}$ be a unicritical lamination with a MAC leaf, M. Then, all of the leaves in the MAC orbit lie outside the central gap bounded by the major and its siblings.*

PROOF. We have a unicritical central strip with small arcs of the d-gon between the major's siblings and the all critical d-gon by Definition 2.3. By Corollary 2.14, this central strip cannot be entered by any part of the forward orbit of the major and a leaf can not have both endpoints in one end cap. If we connected two different endcaps with a leaf, then such a leaf would be closer to critical length than our initial major which contradicts our choice of major leaf. □

THEOREM 2.26. *In a MAC lamination with a rotational polygon, there are no limit leaves and the pullback leaves limit to points on the circle.*

PROOF. Let M be our major leaf in a MAC lamination. In a MAC lamination, we have a symmetric sibling portrait so every critical sector is of length $\dfrac{1}{d}$. A single critical sector maps one-to-one onto the circle. Thus, as we do each step of the pullback lamination (Definition 1.3), there is a pre-image in every critical sector. As we pullback again, the critical sector is broken into d equal intervals each again with a pre-image in it. As we continue the pullback process, our intervals will continually get smaller and cause our pullback leaves to limit to points on the circle. □

Definition 2.27 and Theorem 2.29 are adapted from [**13**].

DEFINITION 2.27 (Co-root). A *co-root* is a point, other than an endpoint of the major, in the boundary of the central gap of the MAC lamination that is fixed under the first return map.

REMARK 2.28. Each co-root is in a different endcap of the siblings of the MAC leaf. See Theorem 2.29.

THEOREM 2.29 (Co-root theorem). *Given a degree d MAC lamination, there will be $d-2$ co-roots in the end caps of the central strip that are not adjacent to the major. The distance between co-roots is greater than $1/d$.*

PROOF. Let G be the degree d Fatou gap of a MAC lamination with major, M, and return map, R. Applying the quotient map h to ∂G takes the leaves bounding our Fatou gap to points as pictured in Figure 9. Thus, h takes ∂G to a topological circle $\hat{G}$ with a first return map $\hat{R}$ conjugate to σ_d (Lemma 2.24). Since $\hat{R}$ is of degree d, it will have $d-1$ fixed points; the first of which we can identify as $h(M)$. The rest of the fixed points under σ_d must be equally spaced around the circle with spacing $\frac{1}{d-1}$ and are between the siblings of M. We claim that these points

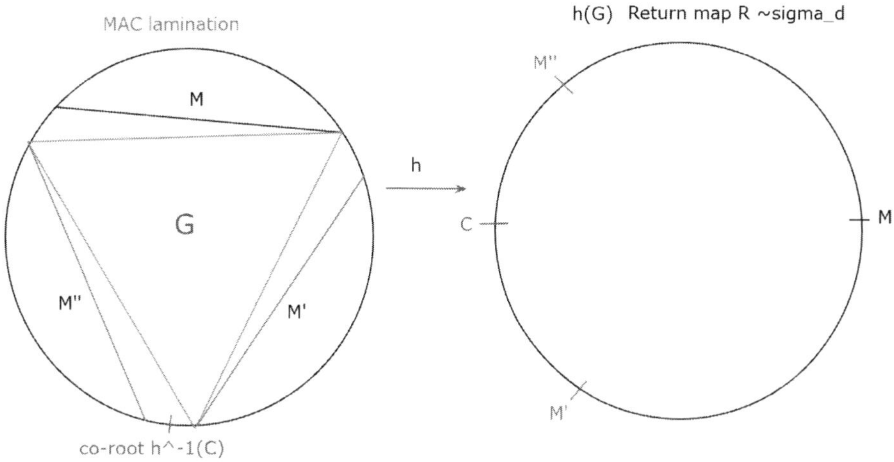

FIGURE 9. Here is a MAC leaf in a lamination bounding a Fatou gap G (Left). The quotient map, h, applied to G gives $h(G)$ (Right). In $h(G)$, the point labeled M is technically $h(M)$. The return map on $h(G)$ is conjugate to σ_d.

under h^{-1} are points in ∂G and not leaves. All of the points in $\hat{G}$ corresponding to leaves in ∂G eventually map to the point $h(M)$ because all of the leaves in ∂G eventually map to M (Lemma 2.24). Points in $\hat{G}$ corresponding to leaves cannot be fixed under $\hat{R}$, otherwise they would not map to $h(M)$. Now, we apply h^{-1} to each of these points to get points in ∂G between the siblings of M which are our co-roots. □

The following algorithm gives rise to the existence of co-roots for all unicritical laminations.

REMARK 2.30 (The generalized Lavaur's algorithm, [**1**, Section 6]). For all unicritical laminations of any degree, we can uniquely find the corresponding minor leaf that generates the lamination. In this process of finding these minors, the algorithm "skips" over certain points of every period. These points that are skipped over are the desired co-roots. However, the Lavaur's algorithm identifies the minor leaf rather than the major. Thus, the "co-roots" it identifies are actually the images of the points described in the co-root definition. These points are in the same orbit so we will use the term interchangeably when no confusion will arise. We will identity a co-root with the leaf it is associated with in its orbit.

LEMMA 2.31. *The forward images of the Fatou gap, G, in a MAC lamination are pairwise disjoint except possibly at the endpoint of M or an image of M.*

PROOF. Given a MAC lamination with major, M, and Fatou Gap, G, along with their respective forward images, want to show the forward images of G are pairwise disjoint except possibly at endpoints of M or its images.

First, we need to show at no point in the forward orbit of G that a gap has two leaves of the MAC orbit in its boundary. By way of contradiction, assume at some iterate k in the forward orbit of G there were two leaves of the MAC orbit in the boundary of G_k. The forward image, G_{k+1}, of G_k will also have two leaves of the MAC orbit in its boundary since the gap will map forward one to one until it returns to G. When we return to G, it will have two leaves in the MAC orbit in its boundary, one of which being M. The other leaf, ℓ, maps into the central strip formed by M and its siblings. The leaf, ℓ, cannot enter the central strip long because it will be closer to critical than M which contradicts the fact that M was our major leaf. Thus, ℓ must enter the central strip short which implies that both endpoints of the leaf are in the same component of the central strip. However, this contradicts Corollary 2.14. Therefore, the forward orbit of G only has one leaf in the MAC orbit in its boundary which is M or an image of M depending on the iterate of G.

Second, we need to show that gaps in the forward orbit of G cannot meet at a leaf. By way of contradiction, assume that G_i and G_k with $i < k < n$ meet a leaf where n is the period of M. All of the leaves in the boundary of G and its images eventually map into the MAC orbit (Lemma 2.24). The leaf shared between the two iterates, G_i and G_k, will eventually become M. Then, G_k will map to G in $n-k$ steps, and $\sigma^{n-k}(G_i)$ will only be attached to G by M. Then, $\sigma^{n-k}(G_i)$ will map to G in k steps before $G = \sigma^{n-k}(G_k)$ will return to G since $k < n$. Since M returns to G in fewer than n steps, a forward image of M must be in G. However, this contradicts the results of the previous paragraph.

In the rotational case, a forward image of G will meet G at an endpoint of M since the rotational MAC leaf will touch two of its forward images. □

Corollary 2.32. *In the identity return case, the forward images of G are always pairwise disjoint.*

Proof. We will verify that no two gaps in the forward orbit of G share a point. Note that by Proposition 1.13, we know two gaps cannot meet at a point that is not an endpoint of leaves. Suppose by way of contradiction, there is a boundary leaf in G_i that shares an endpoint with a boundary leaf in G_k. Then, we have two leaves that will eventually map into the MAC orbit attached at a point. The forward images of these two leaves must also be attached. However, an identity return MAC leaf and its images must be disjoint, thus a contradiction. □

We currently have enough information to prove one direction (MAC to SCM) of this one-to-one correspondence. We want to show, given any degree d MAC lamination, there is a unique (canonical) SCM lamination to which it corresponds.

Lemma 2.33 (MAC to SCM). *Starting with a MAC leaf, M, there is a unique corresponding SCM polygon, $P(M)$, with adjacent majors. In the identity return case, we will have a d-gon and a $k(d-1)$-gon in the rotational case.*

Proof. Let $\mathcal{L}$ be MAC lamination with minor leaf, m, and major leaf, M. By Theorem 2.29 we have the following statements. There are $d-2$ co-roots associated with M. The co-roots of the major always lie on the long arc of the circle subtended by M. The siblings of M are symmetric about the circle lying in different critical sectors. These siblings have endpoints on the long arc of the circle subtended by M. The siblings are between either an endpoint of the major and a co-root or between two co-roots.

Let us first discuss the identity return case. We can construct a d-gon, $P(M)$, using M and its co-roots. We will remove the current critical chords and justify

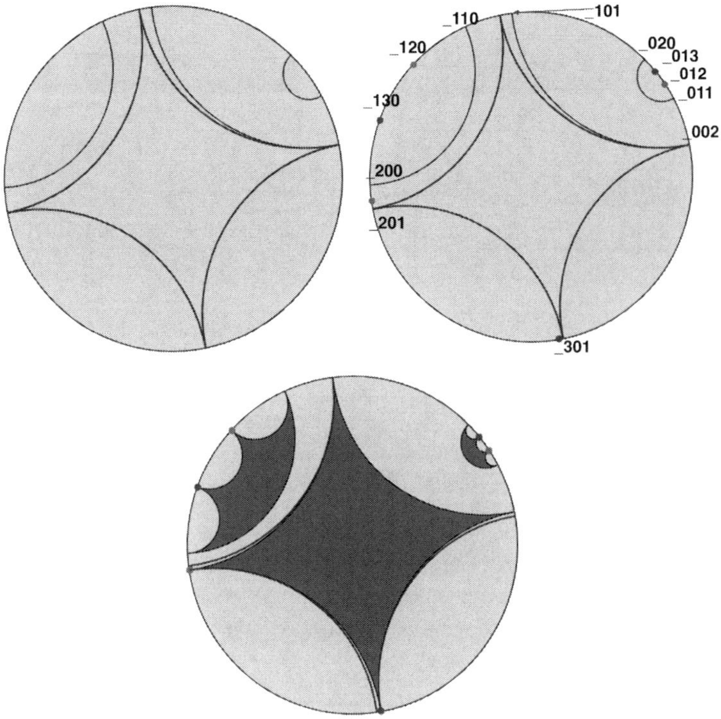

FIGURE 10. On the left, we have a degree 4 MAC identity return leaf and its full forward orbit. In the middle, we have the co-roots and their images added. On the right, we have the resulting SCM polygon.

the existence of a new complete set of compatible $d-1$ critical chords. We consider the short arc subtended by M. Along that arc, one endpoint is counterclockwise of the other. We chose that as our starting point (see Figure 12). Now, take this point and connect it to the nearest co-root in the counterclockwise direction with a leaf. Doing so gives us two connected leaves, or two sides of our d-gon, $P(M)$, we want to construct. There are now $d-3$ co-roots remaining and the other endpoint of the major to connect together. Proceeding in the counterclockwise direction, we connect the rest of the points in similar fashion until we arrive at the other endpoint of M giving us an additional $d-2$ sides. This gives a total of d sides making a d-gon with M as a side and the co-roots as vertices.

The new leaves connecting the endpoints and co-roots are longer than the siblings of the major; this follows from the fact that they're closer to critical because they're in the end cap. By Theorem 2.29 we know that the distance between co-roots is greater than $\frac{1}{d}$. Since these new sides are all longer than critical length, we know that there is room for a critical chord under $d-1$ of the sides of our new d-gon, $P(M)$, putting it into a maximal critical sector.

We will now show that $P(M)$ is SCM with the same period as M. M and its co-roots have the same period since they are fixed under the first return map of the gap, G, for which M is the major (Lemma 2.24). So, all of the vertices of

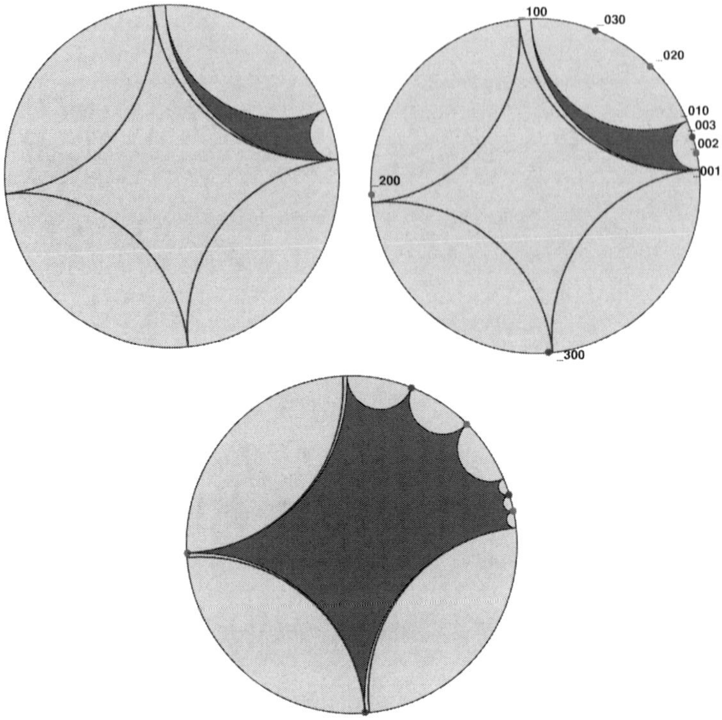

FIGURE 11. On the left, we have a degree 4 MAC rotational polygon. In the middle, we have the co-roots and their images added. On the right, we have the resulting SCM polygon.

$P(M)$ have the same period. Since $P(M)$ lies entirely in G, and G maps forward preserving circular order, the vertices, and thus, the sides of $P(M)$ map forward in order.

Since M is identity return and a side of $P(M)$, then $P(M)$ must also be identity return. To verify that $P(M)$ is the closest to critical in its orbit we look at its side lengths compared to the side lengths of its forward images. Since $P(M)$ is contained in G it will map with the forward orbit of G. This forward orbit follows the forward orbit of M so that G maps immediately under the image of M and stays on the short arc of the images of M until returning to M.

Note that the shortest side of $P(M)$ is M because the other leaves in $P(M)$ are closer to critical that M.

Thus, when mapping forward under σ_d, M will be the longest side of the forward images of $P(M)$ until returning to $P(M)$. The sides of the images of $P(M)$ are bounded under the corresponding images of M. The sides of $P(M)$ will not be able to approach criticality until M does, and this is the iterate with M as a side. Therefore $P(M)$ is the closest to critical in its orbit. Now we have $P(M)$ is identity return, in a maximal critical sector, and is the closest to critical in its forward orbit. Thus, $P(M)$ is an SCM d-gon by Definition 2.7. The construction guarantees its uniqueness and completes the identity return case.

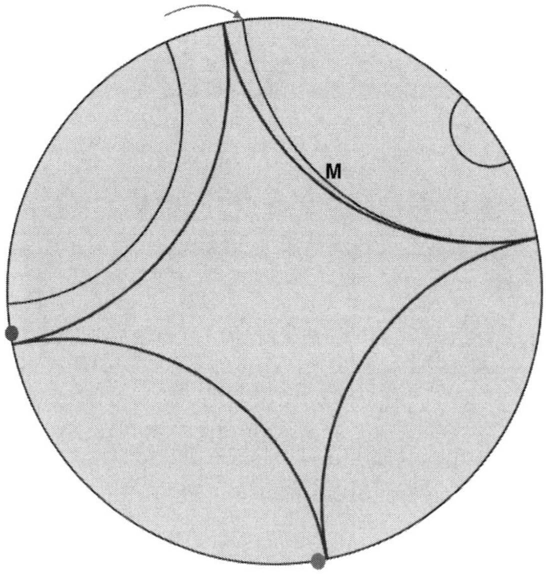

FIGURE 12. The endpoint of M at the clockwise end of the short arc subtended by M in the identity return case with the two respective co-root points.

Now, let us consider the rotational case. Again, we will remove the current critical chords and put in compatible ones with the new polygon. We construct a $k(d-1)$-gon using M and its co-roots where k is the period of the endpoints of the rotation return polygon. We choose the endpoint of M as before. We connect it and the nearest co-root in the counterclockwise direction. This is one leaf of our $k(d-1)$-gon. There are now $d-3$ co-roots remaining and the other endpoint of the major to connect together. Proceeding in the clockwise direction we connect the rest of the points in similar fashion until we arrive at the other endpoint of M. This creates $d-1$ new leaves that subtend M. We maintain the endpoints of M, but the original leaf has been replaced by the leaves generated by the co-roots and the endpoints of M. As we repeat this process for the other k sides of the rotational polygon and the corresponding co-root images, we have a total of $k(d-1)$ sides of our new polygon. The original k sides of rotational polygon have each been replaced by $d-1$ new sides.

Similarly to the identity return case, we have adjacent majors and our $k(d-1)$-gon is in a maximal critical sector so we have room for $d-1$ compatible critical chords. Our polygon is the same period as M and rotational with the same rotation number as M. Thus, $P(M)$ is a rotational SCM $k(d-1)$-gon by Definition 2.7. The construction again guarantees its uniqueness and completes the rotational case.

□

THEOREM 2.34 (MAC lamination to SCM lamination). *Let $\mathcal{L}(M)$ be a MAC lamination with the MAC leaf M. There is a canonical SCM lamination $\mathcal{S}(M)$ that contains an SCM polygon $P(M)$ as follows:*

(1) *Identity Return: SCM d-gon with M as one of its sides.*

FIGURE 13. Here are the finite pullbacks for Figure 10. On the left, we have pullbacks of our identity return unicritical leaf, and on the right, we have pullbacks for our SCM identity return polygon.

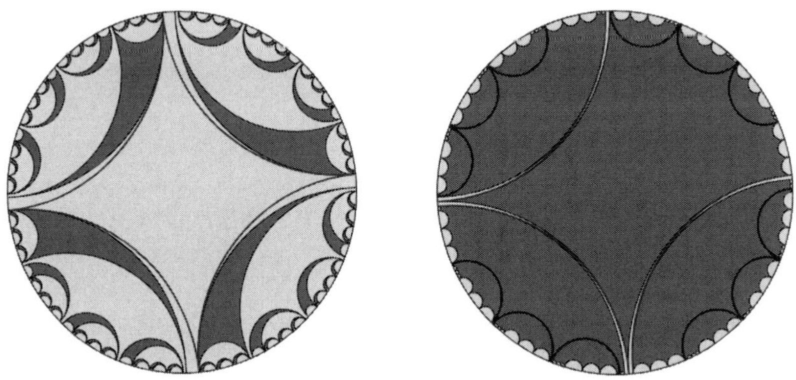

FIGURE 14. Here are the finite pullbacks for Figure 11. On the left, we have pullbacks of our rotational unicritical polygon, and on the right, we have pullbacks for our SCM rotational polygon.

(2) *Rotational: SCM $k(d-1)$-gon where M is a chord interior to the polygon.*

PROOF. Let $\mathcal{L}(M)$ be a MAC lamination with the MAC leaf M. We apply Lemma 2.33 to obtain the respective SCM polygon $P(M)$, and in the identity return case, we have M as one of its sides. In the pullback lamination of $P(M)$ with guiding critical chords described in Definition 2.8, the canonical lamination $\mathcal{S}(M)$ is uniquely determined. □

THEOREM 2.35 (Identity return MAC gaps). *In the identity return case, the only gap attached to the major of a MAC orbit is the infinite gap containing the all critical d-gon. In other words the major and its siblings are single sided limit leaves.*

PROOF. Start with a MAC lamination, $\mathcal{L}$, with major, M, and central gap, G. Since $\mathcal{L}$ is a MAC lamination, all the criticality is used up by the all critical d-gon in G. The major along with its siblings are the longest leaves in the lamination. We can determine a few things about the side of M opposite G. There is no room outside G for a chord, or leaf, of critical length. This means there is no room for a collapsing polygon anywhere since it requires a critical chord in its interior. Also, infinite gaps such as Siegel gaps and Fatou gaps (degree k-covering gaps) all require a critical length in their interior or on their boundary. These gaps cannot exist in $\mathcal{L}$ since all the criticality is used up by G.

On the other side of M, there could either be a finite gap or an infinite gap. We show each is impossible.

The only infinite gap case left to consider is a pre-image of G. Suppose there was such a gap G_{-k}, a kth pre-image of G. Then, there is a pre-image of M on the boundary of G_{-k} that is not a leaf in the MAC orbit or else G_{-k} would be in the forward orbit of G. The gap G_{-k} cannot reside in the forward orbit of G or else G would share a side with one of its forward images which is prohibited by Lemma 2.31. Then let us consider when G_{-k} maps into the forward orbit of G. As G_{-k} maps around, it maps one-to-one until the iterate where the side in the MAC orbit maps back to M. When G_{-k} maps onto G the points in the boundary of G_{-k} map forward preserving circular order, so M must flip. This contradicts that M returns by the identity.

There remain three finite gap cases to consider: eventually collapsing polygons, eventually periodic polygons, and periodic polygons. First, we cannot have an eventually collapsing polygon because the only gap containing criticality is G which does not collapse. Second, we consider a pre-periodic polygon, but since this polygon would be attached to the already periodic leaf, M, the polygon must have the same period. Therefore, the leaf could not be pre-periodic, but it could be periodic.

Third, consider a finite gap with periodic endpoints attached to M. Now apply the construction from Lemma 2.33 and Theorem 2.34 to M to generate an SCM lamination $\mathcal{S}(M)$ with an SCM d-gon $P(M)$. This would cause M to now have an SCM d-gon on one side and a polygon on the other. All of the points in question are of the same period by assumption and should return by the identity. But, we see M is interior to a larger polygon with at least $d+1$ sides that returns to itself by the identity contradicting Kiwi's theorem [**11**].

We are left with the fact that the other side of M contains no gap which means there are only leaves limiting to M on that side. Hence M is a one-sided limit leaf in its MAC lamination.

Since we have the symmetric sibling portrait, the siblings of M are also one-sided limit leaves by a similar argument. □

COROLLARY 2.36. *Let $\mathcal{L}$ be a MAC lamination with major, M, which is identity return. The following hold:*

(1) *The leaf, M, and its siblings are single sided limit leaves in $\mathcal{L}$.*
(2) *There are no isolated leaves in $\mathcal{L}$.*

PROOF. By Section 1.3, every leaf in $\mathcal{L}$ is either a pullback of the forward orbit of M or a limit of pullbacks of the forward orbit of M. By Theorem 2.35 M is a single sided limit leaf, and pulling M back we have its siblings and their pre-images are also single sided limit leaves giving us (1). Then, we take the closure of the pre-lamination and thus every leaf added is itself a limit leaf giving us (2). □

3. SCM to unicritical laminations

3.1. Structure of canonical MAC laminations. Recall in Definition 2.8 we state an SCM polygon, P, is in a maximal critical sector surrounded by $d-1$ critical chords. This will be our starting place to expand our understanding of these polygons in order to find a one-to-one correspondence to MAC laminations. We already have one direction of the correspondence from MAC to SCM. Now, we want to go in the other direction from a canonical SCM lamination to a MAC lamination.

THEOREM 3.1 (Properties of canonical SCM laminations). *Let $\mathcal{S}$ be a canonical SCM lamination with either a SCM d-gon or $k(d-1)$-gon, P. Then $\mathcal{S}$ and P have the following properties.*

(1) *P must have exactly $d-1$ sides of length between $\frac{1}{d}$ and $\frac{1}{d-1}$ and at least one side of length less than $\frac{1}{d}$.*
(2) *For each of the $d-1$ longest sides, $\ell_1, \dots, \ell_{d-1}$, there is a degree 2 Fatou gap bounded by the ℓ_i and its sibling ℓ_i'.*
(3) *In the identity return case, the maximal critical sector has exactly one arc of the circle has two vertices of P lying on it, namely the endpoints of M.*

PROOF. Definition 2.8 gives us an SCM polygon P and a collection of $d-1$ guiding critical chords each touching one endpoint of each of the $d-1$ longest sides of P. For item (1), each of the sides of P are at their closest approach to criticality (2.7), meaning each side of P is within $\frac{1}{d(d+1)}$ of critical length $\frac{1}{d}$ (2.15). Thus, the longest these sides could be is

$$\frac{1}{d} + \frac{1}{d(d+1)} = \frac{d+2}{d(d+1)}.$$

Through a simple algebraic argument, we have that

$$\frac{d+2}{d(d+1)} < \frac{1}{d-1}.$$

Thus, the $d-1$ longest sides of P are less than $\dfrac{1}{d-1}$. All of the majors of P are longer than $\frac{1}{d}$ since each subtends a critical chord. There is at least one other side of P that does not have a critical chord that it subtends and is therefore less than $\dfrac{1}{d}$. The sibling S' emanates from the other endpoint of the critical chord. Together S and S' bound a degree two Fatou gap. Item (2) follows immediately from Definition 2.8. Item (3) follows from the pigeon hole principle. □

THEOREM 3.2 (MAC and SCM laminations compared). *Let $\mathcal{L}(M)$ be a MAC lamination and $\mathcal{S}(M)$ be the corresponding canonical SCM lamination as in Theorem 2.34. In the identity return case, the only difference between $\mathcal{L}$ and $\mathcal{S}$ is the grand orbits of the sides added to M to make $P(M)$. In the rotational case, the grand orbit of the leaves of $\mathcal{L}(M)$ are internal to polygons of $P(M)$.*

PROOF. In the identity return case, we first want to verify that $\overline{\mathcal{GO}(\partial G)} = \overline{\mathcal{GO}(M)}$ where G is the central gap of $\mathcal{L}(M)$. The boundary leaves of G are made entirely of pre-images of M by Lemma 2.24, thus we know that pulling back ∂G is the same as pulling back M. Therefore $\overline{\mathcal{GO}(\partial G)} = \overline{\mathcal{GO}(M)}$, which means the two grand orbits give us the same lamination $\mathcal{L}$. By a similar argument $\mathcal{S} = \overline{\mathcal{GO}(P(M))}$.

Next, we need to verify that $\mathcal{S}$ and $\mathcal{L}$ have the same limit leaves. We begin by noting that $\mathcal{GO}(M) \subset \mathcal{GO}(P(M))$ which means $\mathcal{L} \subset \mathcal{S}$. Now we must show that any limit leaf in $\mathcal{S}$ is also in $\mathcal{L}$. The SCM d-gon $P(M)$ is contained in the central gap G of $\mathcal{L}$. Then, $\mathcal{GO}(P(M)) \subset \mathcal{GO}(G)$ because $P(M)$ subdivides G. This means there are no limit leaves of $P(M)$ lying outside the $\mathcal{GO}(G)$ that are not already in $\mathcal{L}$. That only leaves the possibility of new limit leaves appearing interior to $\mathcal{GO}(G)$ in $\mathcal{S}$. Consider a limit leaf ℓ interior to ∂G. The leaf ℓ must lie in a degree two Fatou gap of $\mathcal{S}$, or else it is in $\mathcal{GO}(P(M))$ by Theorem 3.1. If ℓ connects end caps then S and S' do not bound the degree 2 Fatou gap, contradicting $\mathcal{S}$ being canonical by part (2) of Theorem 3.1. Therefore, ℓ must be in one end cap of the Fatou gap limited to by pre-images of $P(M)$. The forward images of ℓ are limits of the forward images of these pre-images of $P(M)$ including $P(M)$ itself. This reduces to the case where ℓ subdivides a Fatou gap. Hence all of the limit leaves in $\mathcal{S}$ are contained in $\mathcal{L}$.

The rotational case follows from Lemma 2.33. □

THEOREM 3.3 (Canonical SCM to MAC laminations). *Let $\mathcal{S}$ be a canonical SCM lamination containing an SCM polygon, P, that is either identity return or rotational.*

Then, there exists a unique corresponding MAC lamination, $\mathcal{L}$, with a leaf M as its MAC leaf.

If it is identity return, our SCM polygon has M as its shortest side. If it is rotational, M is internal to the SCM polygon.

The following is an outline of the proof of the canonical SCM lamination to MAC lamination theorem which will be carried out through a sequence of lemmas.

- First, in the rotational case, we recover M and add it and all of its pull-backs back into the lamination. In the identity return case, M is already a side of P.
- We will remove the grand orbit of $P \setminus M$, meaning we will get rid of all of the pullbacks and forward images that are not in the grand orbit of M.
- Next, we need to show that this 'reduced' version of our subset of the SCM lamination is still a lamination.
- Lastly, after ensuring what we have left is still a lamination we want to show it is a MAC lamination where our side M is the major leaf.

LEMMA 3.4 (Recovering M). *Let $\mathcal{S}(M)$ be a canonical SCM lamination and $P(M)$ the corresponding polygon (identity return or rotational). We can recover our MAC major, M, defining a d-invariant lamination $\widehat{\mathcal{S}}$. Furthermore, $\widehat{\mathcal{S}} \supset \mathcal{S}$.*

PROOF. In both cases, our goal it to connect our chain of major leaves to get our desired leaf, M. Recall that an SCM polygon has $d-1$ adjacent major leaves (Definition 2.8).

In the identity return case, our SCM polygon only has one other leaf on its boundary which already connects our chain of major leaves. This is our leaf, M.

In the rotational case, our SCM polygon will have, in addition to the majors, the full forward orbit of each of those major leaves as its sides. As before, we will connect our chain of major leaves by leaf. In this case, this leaf will be internal to our SCM polygon. We add this leaf and its full forward and backward orbit to our lamination. We will denote this new lamination as $\widehat{\mathcal{S}}$. □

REMARK 3.5. Note in the rotational case, our polygon in $\widehat{\mathcal{S}}$ has interior leaves and will be resolved in subsequent lemmas. This does not change equivalence classes from $\mathcal{S}$.

The goal of Lemma 3.6 is to remove the co-root leaves of P.

LEMMA 3.6. *Let $\widehat{\mathcal{S}}$ be as described in Lemma 3.4. Then $\widehat{\mathcal{L}} = \overline{\widehat{\mathcal{S}} \setminus (\mathcal{GO}(P \setminus M))}$ has a central gap G bounded by M and its siblings. Moreover, $\widehat{\mathcal{L}}$ is compatible with an all critical d-gon.*

PROOF. Start with $\widehat{\mathcal{S}}$ as in Lemma 3.4. Since P was in a maximal critical sector, the only compatible sibling portrait is the symmetric one for $\widehat{\mathcal{S}}$. Now, if we remove $\mathcal{GO}(P \setminus M)$, then the central strip of the disk bounded by M and its siblings has no long leaves in it. We claim the central strip contains a Fatou gap, G, bounded by M and its siblings. Because $\widehat{\mathcal{S}}(M)$ was canonical, we know that there are no long leaves subdividing the central strip except for the leaves of P that were removed. In the central strip, there is room for an all critical d-gon with each vertex attached to an appropriate endpoint of M and its siblings. This follows from the fact that our majors we removed only left less than $\frac{1}{d}$ in our circle for M, and all forward images of M are underneath it. Thus, $\widehat{\mathcal{L}}$ has a central gap G bounded by M and its siblings, and G is compatible with an all critical d-gon. □

LEMMA 3.7. *Let $\widehat{\mathcal{L}}$ be obtained from $\widehat{\mathcal{S}}$ by Lemma 3.6. Then, $\widehat{\mathcal{L}}$ is a d-invariant lamination.*

PROOF. To show that $\widehat{\mathcal{L}}$ from Lemma 3.6 is a d-invariant lamination we need to show the following properties:

(1) $\widehat{\mathcal{L}}$ is a lamination (it does not contain crossing leaves).
(2) $\widehat{\mathcal{L}}$ has only finite equivalence classes.
(3) $\widehat{\mathcal{L}}$ is closed.
(4) $\widehat{\mathcal{L}}$ is forward, backward, and sibling invariant.

Our definition of $\widehat{\mathcal{L}}$ in Lemma 3.6 removed leaves from the d-invariant lamination $\widehat{\mathcal{S}}$. Since removing leaves does not cause any intersection of leaves we have (1). Similarly, for (2) since $\widehat{\mathcal{S}}$ was d-invariant with finite equivalence classes and removing leaves from a lamination can only decrease the size of equivalence classes, $\widehat{\mathcal{L}}$ only has finite equivalence classes. (3) follows by definition of $\widehat{\mathcal{L}}$. In the construction of $\widehat{\mathcal{L}}$, we removed the entire grand orbit of leaves which leaves $\widehat{\mathcal{L}}$ to be forward, backward, and sibling invariant. □

LEMMA 3.8. *Let $\widehat{\mathcal{L}}$ be the d-invariant lamination obtained from $\widehat{\mathcal{S}}$ by Lemma 3.7. The leaf, M, in $\widehat{\mathcal{L}}$ is a MAC leaf, and $\widehat{\mathcal{L}}$ is the MAC lamination $\mathcal{L}(M)$.*

PROOF. Let $\widehat{\mathcal{L}}$ be the d-invariant lamination obtained from $\widehat{\mathcal{S}}$ by Lemma 3.7. Let us first consider when M is identity return. We know the leaf, M, from the construction of $\widehat{\mathcal{L}}$ is identity return since the d-gon, P, in $\widehat{\mathcal{S}}$ was identity return and M and its siblings bound a gap G containing an all critical d-gon by Lemma 3.6. Rotating the all critical d-gon to the vertices of M and its siblings shows that we have an identity return major leaf that approaches an all critical d-gon, and by Definition 2.3 M is a MAC leaf.

Now that we have M is a MAC leaf, we want to show $\mathcal{L}(M)$ is equivalent to $\widehat{\mathcal{L}}$ giving us that $\widehat{\mathcal{L}}$ is a MAC lamination. Recall $\widehat{\mathcal{L}} = \overline{\mathcal{GO}(M)}$, and $\mathcal{L}(M)$ is the pullback lamination of M. These are two equivalent sets of leaves by Section 1.3.

The rotational case follows with a similar argument. □

PROOF OF THEOREM 3.3. Lemmas 3.6–3.8 give us a uniquely defined construction starting with an arbitrary canonical SCM lamination $\mathcal{S}$ then leading to a MAC lamination $\widehat{\mathcal{L}}$. Since there were no choices at any step of the construction, it is uniquely determined. Thus, the construction proves the theorem. □

4. Future work: Noncanonical SCM to unicritical laminations

Now that we have a one to one correspondence between canonical MAC laminations and canonical SCM laminations in the identity return and rotational case, we would like to extend this correspondence to the rotation return case as well. In addition, we want to understand how non-canonical SCM and canonical SCM laminations are related. This is future work on *tuning* a lamination by another lamination "inserted in its Fatou gaps." In our construction from SCM to MAC, we showed that no new limit leaves were added. However, we would like to show that the limit leaves in both laminations are exactly the same.

In addition, we would like to extend our results to "locally unicritical" laminations. In [**5**], the idea of "flower-like sets" suggests that our correspondence would apply in laminations containing such sets. This idea follows from the fact that in a portion of the circle, we are unicritical and invariant which is what we call "locally unicritical."

References

[1] Sourav Bhattacharya, Alexander Blokh, and Dierk Schleicher, *Unicritical laminations*, Fund. Math. **258** (2022), no. 1, 25–63, DOI 10.4064/fm18-2-2022. MR4422233

[2] Brandon L. Barry, *On the simplest lamination of a given identity return triangle*, ProQuest LLC, Ann Arbor, MI, 2015. Thesis (Ph.D.)–The University of Alabama at Birmingham. MR3419326

[3] Alexander Blokh, James M. Malaugh, John C. Mayer, Lex G. Oversteegen, and Daniel K. Parris, *Rotational subsets of the circle under z^d*, Topology Appl. **153** (2006), no. 10, 1540–1570, DOI 10.1016/j.topol.2005.04.010. MR2216120

[4] Alexander M. Blokh, Clinton P. Curry, and Lex G. Oversteegen, *Locally connected models for Julia sets*, Adv. Math. **226** (2011), no. 2, 1621–1661, DOI 10.1016/j.aim.2010.08.011. MR2737795

[5] A. Blokh, L. Oversteegen, V. Timorin, and Y. Wang, *A model of the cubic connectedness locus*, Preprint, (2022).

[6] Alexander M. Blokh, Debra Mimbs, Lex G. Oversteegen, and Kirsten I. S. Valkenburg, *Laminations in the language of leaves*, Trans. Amer. Math. Soc. **365** (2013), no. 10, 5367–5391, DOI 10.1090/S0002-9947-2013-05809-6. MR3074377

[7] Alexander Blokh, Lex Oversteegen, Ross Ptacek, and Vladlen Timorin, *Laminations from the main cubioid*, Discrete Contin. Dyn. Syst. **36** (2016), no. 9, 4665–4702, DOI 10.3934/dcds.2016003. MR3541500

[8] Douglas K. Childers, *Wandering polygons and recurrent critical leaves*, Ergodic Theory Dynam. Systems **27** (2007), no. 1, 87–107, DOI 10.1017/S0143385706000526. MR2297088

[9] David J. Cosper, Jeffrey K. Houghton, John C. Mayer, Luka Mernik, and Joseph W. Olson, *Central strips of sibling leaves in laminations of the unit disk*, Topology Proc. **48** (2016), 69–100. MR3340275

[10] C. Falcione. Lamination Builder. `https://csfalcione.github.io/lamination-builder/`
[11] Jan Kiwi, *Wandering orbit portraits*, Trans. Amer. Math. Soc. **354** (2002), no. 4, 1473–1485, DOI 10.1090/S0002-9947-01-02896-3. MR1873015
[12] John Milnor, *Dynamics in one complex variable*, 3rd ed., Annals of Mathematics Studies, vol. 160, Princeton University Press, Princeton, NJ, 2006. MR2193309
[13] Dierk Schleicher, *On fibers and local connectivity of Mandelbrot and Multibrot sets*, Fractal geometry and applications: a jubilee of Benoît Mandelbrot. Part 1, Proc. Sympos. Pure Math., vol. 72, Amer. Math. Soc., Providence, RI, 2004, pp. 477–517, DOI 10.1090/pspum/072.1/2112117. MR2112117
[14] Dierk Schleicher (ed.), *Complex dynamics*, A K Peters, Ltd., Wellesley, MA, 2009. Families and friends, DOI 10.1201/b10617. MR1500163

Department of Mathematics, University of Alabama at Birmingham, Birmingham, Alabama 35924-0001
Email address: beburd@uab.edu

Department of Mathematics, University of Alabama at Birmingham, Birmingham, Alabama 35924-0001
Email address: caleb.falcione@gmail.com

Department of Mathematics, University of Alabama at Birmingham, Birmingham, Alabama 35924-0001
Email address: CameronGHale@outlook.com

Department of Mathematics, University of Alabama at Birmingham, Birmingham, Alabama 35924-0001
Email address: jcmayer@uab.edu

Contemporary Mathematics
Volume **797**, 2024
https://doi.org/10.1090/conm/797/15953

On fractal dimension of the graph of nonstationary fractal interpolation function

Subhash Chandra and Syed Abbas

ABSTRACT. This article aims to study fractal dimension of the nonstationary fractal interpolation function (FIF) on function spaces. We estimate the Hausdorff dimension and the box dimension of nonstationary FIF on convex-Lipschitz space. We also compute the upper box dimension of the nonstationary FIF on oscillation spaces. Furthermore, we study the fractal dimension of the nonstationary FIF and the Katugampola fractional integral on Hölder space and bounded variational space.

1. Introduction and preliminaries

Fractal interpolation is a recent technique that generalizes the traditional interpolation methods. A fractal interpolation function (FIF) is a continuous function that interpolates a given set of data and whose graph is a fractal set. The theory of FIF was introduced by Barnsley [**Bar86**, **Bar88**] using the concept of iterated function system (IFS). FIF has more advantages than the classical interpolants for approximating and fitting the naturally occurring functions having self-similarity. Therefore, a large number of articles on this topic have been published so far. Recently, David and others [**DLM**, **LDV**] introduced a more general class of sequences of operators and studied their convergence properties using the idea of a nonstationary subdivision algorithm. The authors in [**DLM**] have used the sequence of IFS systems in the field of nonstationary subdivision schemes. Recently, Massopust [**M**] studied the nonstationary FIF.

The fractal dimension is an important aspect of fractal geometry, and there are many recent developments on it that can be seen in the literature. There are many notions of fractal dimensions, such as Hausdorff dimension, box dimension, packing dimension, quantization dimension, and Assouad dimension. The study of dimensions of graphs began with Weierstrass type functions [**Hun98**, **MW86**]. In [**MW86**], Mauldin and Williams studied the fractal dimension of Weierstrass type functions which satisfy the convex-Lipschitz condition. In this order, we introduced a new space, namely convex-Lipschitz space and we studied the fractal dimension of the graph of fractal functions in [**CA1**]. In [**JV**], Jha and Verma studied the

2020 *Mathematics Subject Classification.* Primary 28A80, 41A30, 33C47.

Key words and phrases. Nonstationary fractal interpolation functions, convex-Lipschitz space, Hölder space, oscillation spaces, Hausdorff dimension, box dimension.

The work of the first author was financially supported by the CSIR, India with fellowship number 09/1058(0012)/2018-EMR-I..

fractal dimension of α-fractal function on various function spaces such as Hölder space, oscillation space, and the space of bounded variation. These works are the motivation to conduct the current study. In this work, we consider nonstationary FIF which is the generalized version of FIF, and we study the fractal dimension of nonstationary FIF and the Katugampola fractional integral using various function spaces such as convex-Lipschitz space, oscillation spaces, Hölder space and bounded variational space.

Let us recall some basic definitions and terminologies which act as preludes to the current article. Here, we consider a sequence of transformations $\{T_i\}_{i\in\mathbb{N}}$, $T_i : X \to X$ to introduce the concept of trajectories.

DEFINITION 1.1 (Invariant set). A subset $\mathcal{P}$ of X is said to be an invariant set of $\{T_i\}_{i\in\mathbb{N}}$ if $\forall\ i \in \mathbb{N}$ and $\forall\ x \in \mathcal{P}$, $T_i(x) \in \mathcal{P}$.

LEMMA 1.2 ([**LDV**, Lemma 3.7]). *Consider a sequence of transformations* $\{T_i\}_{i\in\mathbb{N}}$. *If there exists* q *in* X *such that for every* $x \in X$

$$d(T_i(x), q) \leq \mu d(x, q) + M,\ \ 0 \leq \mu < 1, M \geq 0,$$

then the ball of radius $\frac{M}{1-\mu}$ *centered at* q, $B(q, \frac{M}{1-\mu})$, *is an invariant domain of* $\{T_i\}_{i\in\mathbb{N}}$.

Let (X, d) be a complete metric space, we consider a sequence of set valued maps $W_k : H(X) \to H(X)$ associated with systems of function systems (SFS) as

$$W_k(A) = \bigcup_{i=1}^{n_k} w_{i,k}(A), A \in H(X), \tag{1.1}$$

where $W_k = \{w_{1,k}, w_{2,k}, \dots, w_{n_k,k}\}$ is a family of contractions constituting an IFS on (X, d). We denote the collection of all nonempty compact subsets of X by $H(X)$.

DEFINITION 1.3 (Forward and backward trajectories). Let X be a metric space and $\{T_k\}_{k\in\mathbb{N}}$ be a sequence of Lipschitz maps on X. We define forward and backward procedures

$$\Phi_k := T_k \circ T_{k-1} \circ \dots \circ T_1 \text{ and } \Psi_k := T_1 \circ T_2 \circ \dots \circ T_k.$$

PROPOSITION 1.4 ([**NV**]). *Let* $\{T_k\}_{k\in\mathbb{N}}$ *be a sequence of Lipschitz maps with Lipschitz constant* c_i *on a complete metric space* X. *If there exists* $x \in X$ *such that the sequence* $\{d(x, T_k(x))\}$ *is bounded, and* $\sum_{k=1}^{\infty}\prod_{i=1}^{k} c_i < \infty$ *then the sequence* $\{\Psi_k(x)\}$ *converges for all* $x \in X$ *to a unique limit* $\overline{x}$.

Here, we give a brief introduction to nonstationary FIF. We recommend the reader to visit [**LDV, M**], for more details.

1.1. Nonstationary fractal interpolation function. Let the set of interpolation points is given by $\{(x_n, y_n) : n = 1, 2, \dots, N\}$. We set $I = [x_1, x_N]$, $J = \{1, 2, \dots, N-1\}$, and $I_j = [x_j, x_{j+1}]$ for $j \in J$. Let $S_j : I \to I_j, j \in J$ be affine map of the form $S_j(x) = c_j x + d_j$ with

$$S_j(x_1) = x_j, S_j(x_N) = x_{j+1}, j \in J.$$

Let $K = I \times \mathbb{R}$. Now, we define $F_{j,k} : K \to \mathbb{R}$ as

$$F_{j,k}(x, y) = \alpha_{j,k} y + q_{j,k}(x),$$

with

$$F_{j,k}(x_1, y_1) = y_j, F_{j,k}(x_N, y_N) = y_{j+1}, j \in J$$

and $\alpha_{j,k} \in (-1,1)$ for all $j \in J, k \in \mathbb{N}$, $q_{j,k} \in \mathcal{C}(I)$. Define $W_{j,k} : K \to I_j \times \mathbb{R}$ by

$$W_{j,k}(x,y) = (S_j(x), F_{j,k}(x,y)),$$

which forms a sequence of IFSs

$$\mathcal{I}_k = \{K; W_{j,k} : j \in J\}.$$

Now, we define a sequence of Read–Bajraktarević (RB) operators $T^{\alpha_k} : C(I) \to C(I)$ as

$$(T^{\alpha_k} f)(x) = F_{j,k}(S_j^{-1}(x), f(S_j^{-1}(x))), j \in J,$$

where $\alpha_k = \{\alpha_{j,k}\}, j \in J, k \in \mathbb{N}$. Let us denote $\|\alpha_k\|_\infty = \max_{j\in J}\{|\alpha_{j,k}|\} < 1$.

THEOREM 1.5. *Let us consider the sequence of operators T^{α_k} on $\mathcal{C}(I)$. Then for every $f \in \mathcal{C}(I)$, the sequence $\{T^{\alpha_1} \circ T^{\alpha_2} \circ \cdots \circ T^{\alpha_k} f\}$ converges to map f^* of $\mathcal{C}(I)$.*

PROOF. The proof follows from Proposition 1.4. □

Here, f^* refers to nonstationary FIF.

Let F be a nonempty subset of $\mathbb{R}^n$. The diameter of F is given by

$$|F| = \sup\{\|x-y\|_2 : x, y \in F\}.$$

If $\{F_i\}$ is a countable (or finite) collection of sets having diameter at most δ which cover set $E \subseteq \mathbb{R}^n$, then we say that $\{F_i\}$ is a δ-cover of E. For $\delta > 0$ and a non-negative real number s, we define

$$H_\delta^s(E) = \inf\left\{\sum_{i=1}^{\infty}|F_i|^s \colon \{F_i\} \text{ is a } \delta - \text{cover of } E\right\}. \tag{1.2}$$

DEFINITION 1.6. The s-dimensional Hausdorff measure of set E is given by $H^s(E) = \lim_{\delta\to 0} H_\delta^s(E)$.

DEFINITION 1.7 (Hausdorff dimension). Let $s \geq 0$ and $E \subseteq \mathbb{R}^n$. The Hausdorff dimension of E is defined as

$$\dim_H(E) = \inf\{s : H^s(E) = 0\} = \sup\{s : H^s(E) = \infty\}.$$

DEFINITION 1.8 (Box dimension). Let E be a nonempty bounded subset of $\mathbb{R}^n$ and let $N_\delta(E)$ be the smallest number of sets of diameter at most δ that can cover the set E. Then

$$\underline{\dim}_B(E) = \underline{\lim}_{\delta\to 0} \frac{\log N_\delta(E)}{-\log\delta},$$

is said to be lower box dimension of E. And

$$\overline{\dim}_B(E) = \overline{\lim}_{\delta\to 0} \frac{\log N_\delta(E)}{-\log\delta},$$

is known as upper box dimension of E. If both, lower and upper box dimensions are same, then that quantity is known as box dimension of E and defined as

$$\dim_B(E) = \lim_{\delta\to 0} \frac{\log N_\delta(E)}{-\log\delta}.$$

The rest of the work is organized as follows: In Section 2, we deal with the convex-Lipschitz space and estimate the Hausdorff and box dimensions of nonstationary FIF. In Section 3, we find the upper bound of the box dimension of the nonstationary FIF using oscillation spaces. Section 4, is devoted to the Hölder space, in this section, we study the box dimension of the nonstationary FIF and the Katugampola fractional integral associated with nonstationary FIF. In the last section, we estimate the Hausdorff and box dimension of the nonstationary FIF and the Katugampola fractional integral associated with nonstationary FIF with the help of bounded variational space.

2. Convex-Lipschitz space

In this section, first we define convex-Lipschitz space by using the definition of convex-Lipschitz function and then we study fractal dimensions on it.

DEFINITION 2.1 (Convex-Lipschitz function). [**MW86**] Let θ be a function from $\mathbb{R}^+$ to $\mathbb{R}^+$. A function ϕ is called convex Lipschitz of order θ on an interval $[a,b]$ provided there exists a constant M with if $a \leq x < x+y \leq b$ and $0 \leq \beta \leq 1$, then

$$|\Delta(x,y,\beta)| := |\phi(x+\beta y) - (\beta\phi(x+y) + (1-\beta)\phi(x))| \leq M\theta(y).$$

Let us define the convex Lipschitz space as follows

$$V^\theta(I) = \{\phi : I \to \mathbb{R} : \phi \text{ is convex Lipschitz of order } \theta\}.$$

It can be seen that $V^\theta(I)$ is a vector space over field $\mathbb{R}$.

Let $\phi \in V^\theta(I)$, we define $\|\phi\|_{V^\theta} = \|\phi\|_\infty + [\phi]^*$, where

$$[\phi]^* = \sup_{a \leq x < x+y \leq b} \frac{|\phi(x+\beta y) - (\beta\phi(x+y) + (1-\beta)\phi(x))|}{\theta(y)}.$$

Then $\|\cdot\|_{V^\theta}$ forms a norm on $V^\theta(I)$. It is easy to check the following properties of the norm. Let $\phi, \psi \in V^\theta(I)$ and $\lambda \in \mathbb{R}$.

(i)

$$\|\phi\|_{V^\theta} = 0 \iff \|\phi\|_\infty = 0,$$

and

$$\sup_{a \leq x < x+y \leq b} \frac{|\phi(x+\beta y) - (\beta\phi(x+y) + (1-\beta)\phi(x))|}{\theta(y)} = 0$$
$$\iff \phi = 0.$$

(ii)

$$\begin{aligned}\|\lambda\phi\|_{V^\theta} &= |\lambda|\|\phi\|_\infty + \sup_{a \leq x < x+y \leq b} \frac{|\lambda||\phi(x+\beta y) - (\beta\phi(x+y) + (1-\beta)\phi(x))|}{\theta(y)} \\ &= |\lambda|\|\phi\|_{V^\theta}.\end{aligned}$$

(iii)

$$\begin{aligned}\|\phi+\psi\|_{V^\theta} &= \|\phi+\psi\|_\infty \\ &\quad + \sup_{a \leq x < x+y \leq b} \frac{|(\phi+\psi)(x+\beta y) - (\beta(\phi+\psi)(x+y) + (1-\beta)(\phi+\psi)(x))|}{\theta(y)} \\ &\leq \|\phi\|_{V^\theta} + \|\psi\|_{V^\theta}.\end{aligned}$$

THEOREM 2.2. *The space $(V^\theta(I), \|\cdot\|_{V^\theta})$ is a Banach space.*

PROOF. Let $\phi : I \to \mathbb{R}$ and (ϕ_n) be a sequence of continuous functions and converges to ϕ uniformly, then

$$\frac{|\phi_n(x+\beta y)-\phi(x+\beta y)-[\beta\phi_n(x+y)-\beta\phi(x+y)+(1-\beta)\phi_n(x)-(1-\beta)\phi(x)]|}{\theta(y)}$$

$$= \frac{|\phi_n(x+\beta y)-\lim_{k\to\infty}\phi_k(x+\beta y)}{\theta(y)}$$

$$-\frac{[\beta\phi_n(x+y)-\beta\lim_{k\to\infty}\phi_k(x+y)+(1-\beta)\phi_n(x)-(1-\beta)\lim_{k\to\infty}\phi_k(x)]|}{\theta(y)}$$

$$= \lim_{k\to\infty}\frac{|\phi_n(x+\beta y)-\phi_k(x+\beta y)-[\beta\phi_n(x+y)-\beta\phi_k(x+y)+(1-\beta)\phi_n(x)}{\theta(y)}$$

$$-\frac{(1-\beta)\phi_k(x)]|}{\theta(y)}$$

$$\leq \liminf_{k\to\infty}\sup_{a\leq x<x+y\leq b}\frac{|\phi_k(x+\beta y)-\phi(x+\beta y)-[\beta\phi_k(x+y)-\beta\phi(x+y)}{\theta(y)}$$

$$+\frac{(1-\beta)\phi_k(x)-(1-\beta)\phi(x)]|}{\theta(y)}.$$

Now, let the sequence (ϕ_n) is Cauchy in $V^\theta(I)$ with respect to $\|\cdot\|_{V^\theta}$. Equivalently, for any $\epsilon > 0$, and $n_0 \in \mathbb{N}$, we can write

$$\|\phi_n - \phi_k\|_{V^\theta} < \epsilon \;\forall\; n, k \geq n_0.$$

By definition of $\|\cdot\|_{V^\theta}$, one gets $\|\phi_n - \phi_k\|_\infty < \epsilon \;\forall\; n, k \geq n_0$. Because the space $(C(I), \|\cdot\|_\infty)$ is Banach , there is a continuous function ϕ satisfying $\|\phi_n - \phi\|_\infty \to 0$ as $n \to \infty$. We claim that $\phi \in V^\theta(I)$ and $\|\phi_n - \phi\|_{V^\theta} \to 0$ as $n \to \infty$. Let $n \geq n_0$. By using the above calculation, we have

$$\|\phi_n-\phi\|_\infty + \frac{|\phi_n(x+\beta y)-(\beta\phi_n(x+y)+(1-\beta)\phi_n(x))-[\phi(x+\beta y)-(\beta\phi(x+y)}{\theta(y)}$$

$$+\frac{(1-\beta)\phi(x))]|}{\theta(y)}$$

$$= \|\phi_n-\phi\|_\infty + \frac{|\phi_n(x+\beta y)-\phi(x+\beta y)}{\theta(y)}$$

$$-\frac{[\beta\phi_n(x+y)-\beta\phi(x+y)+(1-\beta)\phi_n(x)-(1-\beta)\phi(x)]|}{\theta(y)}$$

$$= \lim_{k\to\infty}\left(\|\phi_n-\phi_k\|_\infty + \frac{|\phi_n(x+\beta y)-\phi_k(x+\beta y)}{\theta(y)} - \frac{[\beta\phi_n(x+y)-\beta\phi_k(x+y)+(1-\beta)\phi_n(x)-(1-\beta)\phi_k(x)]|}{\theta(y)}\right)$$

$$\leq \lim_{k\to\infty}\left(\|\phi_n-\phi_k\|_\infty + \sup_{a\leq x<x+y\leq b}\frac{|\phi_n(x+\beta y)-\phi_k(x+\beta y)-[\beta\phi_n(x+y)}{\theta(y)} - \frac{\beta\phi_k(x+y)+(1-\beta)\phi_n(x)-(1-\beta)\phi_k(x)]|}{\theta(y)}\right)$$

$$\leq \sup_{k\geq n_0}\left(\|\phi_n-\phi_k\|_\infty+\sup_{a\leq x<x+y\leq b}\frac{|\phi_n(x+\beta y)-\phi_k(x+\beta y)-[\beta\phi_n(x+y)}{\theta(y)}\right.$$
$$\left.-\frac{\beta\phi_k(x+y)+(1-\beta)\phi_n(x)-(1-\beta)\phi_k(x)]|}{\theta(y)}\right)$$
$$=\sup_{k\geq n_0}\|\phi_n-\phi_k\|_{V^\theta}$$
$$\leq \epsilon.$$

Hence, we obtain $\phi-\phi_{n_0}\in V^\theta(I)$. So, we have $\phi=\phi-\phi_{n_0}+\phi_{n_0}\in V^\theta(I)$ and $\|\phi_n-\phi\|_{V^\theta}\leq\epsilon\ \forall\ n\geq n_0$, completing the proof. □

DEFINITION 2.3 ([**Fal03**, Section 2.5]). Let $E\subset\mathbb{R}^n$ and dimension function $\theta:\mathbb{R}^+\to\mathbb{R}^+$ that is increasing and continuous. Analogously to (1.2), we define

$$H^\theta_\delta(E)=\inf\left\{\sum\theta(|F_i|):\{F_i\}\text{ is a }\delta-\text{cover of }E\right\}.$$

This leads to a measure, taking $H^\theta(E)=\lim_{\beta\to 0}H^\theta_\delta(E)$. If $\theta(t)=t^s$, this is the usual definition of s-dimensional Hausdorff measure.

THEOREM 2.4 ([**MW86**]). *Let $\theta:\mathbb{R}^+\to\mathbb{R}^+$ be a continuous map such that*
(i) *if $t>0$, $\theta(t)>0$,*
(ii) $\overline{\lim}_{t\to 0}\frac{t}{\theta(t)}<\infty$, *and*
(iii) $\exists\ \gamma\geq 0$ *such that* $\lim_{t\to 0}\frac{\theta(ct)}{\theta(t)}=c^\gamma$ *for all $c>0$. If ϕ is a continuous map on $[0,1]$ which is convex Lipschitz of order θ, then ϕ has σ-finite h measure, where $h(y)=\frac{y^2}{\theta(y)}$.*

Following Theorem 2.4, let us note the next result.

THEOREM 2.5. *Under the hypotheses of Theorem 2.4, we have*
- *If $\theta(y)=y^\alpha$, then $\dim_H(Gr(\phi))\leq\overline{\dim}_B(Gr(\phi))\leq 2-\alpha$.*
- *If $\theta(y)=y\ln(1/y)$, then $\dim_H(Gr(\phi))=\dim_B(Gr(\phi))=1$.*

PROOF. By definitions of the Hausdorff measure and Hausdorff dimension, the proof follows. □

THEOREM 2.6. *Let $q_{j,k}\in V^\theta(I)$ and $\alpha j,k=\alpha 1,k\in(-1,1)$. Then, the associated fractal interpolation function $f^*\in V^\theta(I)$ provided that*

$$\max_k\left\{\|\alpha_k\|_\infty,(N-1)\max_j\sup_{y\in[0,b-a]}\|\alpha_k\|_\infty\frac{\theta(y/c_j)}{\theta(y)}\right\}<1.$$

PROOF. We first define $V^\theta_\phi(I):=\{\phi\in V^\theta:\phi(x_1)=y_1,\ \phi(x_N)=y_N\}$. It is patent that $V^\theta_\phi(I)$ is a closed subset of $V^\theta(I)$. It follows that $V^\theta_\phi(I)$ is a complete metric space with respect to the metric induced by the norm $\|\cdot\|_{V^\theta}$. Let us define a sequence of mappings $T^k:V^\theta_\phi(I)\to V^\theta_\phi(I)$ by

$$(T^k\phi)(x)=\alpha_{j,k}\phi(S_j^{-1}(x))+q_{j,k}(S_j^{-1}(x)),x\in I_j,j\in J.$$

By definition of T^k, it follows that the mapping T^k is well defined. Now, for $\phi,\ \psi\in V^\theta_\phi(I)$, we have

$$\|T^k\phi-T^k\psi\|_{V^\theta}$$

$$= \|T^k\phi - T^k\psi\|_\infty + [T^k\phi - T^k\psi]^* \le |\alpha_{j,k}| \|\phi(S_j^{-1}(x)) - \psi(S_j^{-1}(x))\|_\infty$$
$$+ (N-1)\max_j \sup_{a\le x<x+y\le b} \frac{|T^k\phi(x+\beta y) - (\beta T^k\phi(x+y) + (1-\beta)T^k\phi(x))}{\theta(y)}$$
$$- \frac{T^k\psi(x+\beta y) + (\beta T^k\psi(x+y) + (1-\beta)T^k\psi(x))|}{\theta(y)}$$
$$= |\alpha_{j,k}| \|\phi(\frac{1}{c_j}x - \frac{d_j}{c_j}) - \psi(\frac{1}{c_j}x - \frac{d_j}{c_j})\|_\infty + (N-1)\max_j \sup_{a\le x<x+y\le b}$$
$$|\alpha_{j,k}| \Bigg(\frac{|\phi(S_j^{-1}(x+\beta y)) - (\beta\phi(S_j^{-1}(x+y)) + (1-\beta)\phi(S_j^{-1}(x)))}{\theta(y)}$$
$$- \frac{\psi(S_j^{-1}(x+\beta y)) + (\beta\psi(S_j^{-1}(x+y)) + (1-\beta)\psi(S_j^{-1}(x)))|}{\theta(y)} \Bigg)$$
$$= |\alpha_{j,k}| \|\phi(\frac{1}{c_j}x - \frac{d_j}{c_j}) - \psi(\frac{1}{c_j}x - \frac{d_j}{c_j})\|_\infty + (N-1)\max_j \sup_{a\le x<x+y\le b}$$
$$\frac{|\alpha_{j,k}||(\phi-\psi)(S_j^{-1}(x+\beta y)) - (\beta(\phi-\psi)(S_j^{-1}(x+y)) + (1-\beta)(\phi-\psi)(S_j^{-1}(x)))|}{\theta(y)}$$
$$= |\alpha_{j,k}| \|\phi(\frac{1}{c_j}x - \frac{d_j}{c_j}) - \psi(\frac{1}{c_j}x - \frac{d_j}{c_j})\| + (N-1)\max_j$$
$$\sup_{a\le x<x+y\le b} \frac{|\alpha_{j,k}||(\phi-\psi)(\frac{1}{c_j}(x+\beta y) - \frac{d_j}{c_j}) - (\beta(\phi-\psi)(\frac{1}{c_j}(x+y) - \frac{d_j}{c_j})}{\theta(y)}$$
$$+ \frac{(1-\beta)(\phi-\psi)(\frac{1}{c_j}x - \frac{d_j}{c_j}))|}{\theta(y)}$$

Set $X = \frac{1}{c_j}x - \frac{d_j}{c_j}$ and $Y = \frac{1}{c_j}y$. Then

$$\begin{aligned}
&\|T^k\phi - T^k\psi\|_{V^\theta} \\
&\quad = |\alpha_{j,k}| \|\phi(X) - \psi(X)\| + (N-1)\max_j \sup_{a\le c_jX+d_j<c_j(X+Y)+d_j\le b} \\
&\quad \frac{|\alpha_{j,k}||(\phi-\psi)(X+\beta Y) - (\beta(\phi-\psi)(X+Y) + (1-\beta)(\phi-\psi)(X)|}{\theta(c_jY)} \times \frac{\theta(Y)}{\theta(Y)} \\
&\quad \le |\alpha_{j,k}| \|\phi-\psi\|_\infty + (N-1)\max_j |\alpha_{j,k}| \frac{\theta(Y)}{\theta(c_jY)} [\phi-\psi]^* \\
&\quad \le |\alpha_{j,k}| \|\phi-\psi\|_\infty + (N-1)\max_j \sup_{y\in[0,b-a]} |\alpha_{j,k}| \frac{\theta(y/c_j)}{\theta(y)} [\phi-\psi]^* \\
&\quad \le \max_k \{\|\alpha_k\|_\infty, (N-1)\max_j \sup_{y\in[0,b-a]} \|\alpha_k\|_\infty \frac{\theta(y/c_j)}{\theta(y)}\} \|\phi-\psi\|_{V^\theta}.
\end{aligned}$$

Since $\max_k \left\{ \|\alpha_k\|_\infty, (N-1)\max_j \sup_{y\in[0,b-a]} \|\alpha_k\|_\infty \frac{\theta(y/c_j)}{\theta(y)} \right\} < 1$, the mapping T^k is contraction on $V_\phi^\theta(I)$. One can see that the sequence $\{\|T^k\phi - \phi\|_{V^\theta}\}$ is bounded. In view of Proposition 1.4, the backward trajectories $\Psi_k := T^1 \circ T^2 \circ \ldots \circ T^k$ of

(T^k) converges for every $\phi \in V_\phi^\theta(I)$ to a unique attractor $f^* \in V_\phi^\theta(I)$. This proves the result. □

THEOREM 2.7. *Let $q_{j,k} \in V^\theta(I)$ and $\alpha j, k = \alpha 1, k \in (-1, 1)$ such that*

$$\max_k \left\{ \|\alpha_k\|_\infty, (N-1) \max_j \sup_{y \in [0,b-a]} \|\alpha_k\|_\infty \frac{\theta(y/c_j)}{\theta(y)} \right\} < 1.$$

Then we have the following bounds for the fractal dimension of the graph of the associated fractal interpolation function f^*:

- *If $\theta(y) = y^\alpha$ then $\dim_H(Gr(f^*)) \leq \overline{\dim}_B(Gr(f^*)) \leq 2 - \alpha$.*
- *If $\theta(y) = y \ln(1/y)$ then $\dim_H(Gr(f^*)) = \dim_B(Gr(f^*)) = 1$.*

PROOF. The proof is immediate from the above theorem and Theorem 2.5, hence omitted. □

REMARK 2.8. This work partially generalizes Mauldin and Williams's [**MW86**] work. In this work, we consider nonstationary FIF which is the generalized version of FIF and use sequence of mappings T^k. If we take $T^k = T$, then the above result coincides with some results of [**CA1**]. By using the result of [**MW86**], one can find the fractal dimension of some Weierstrass type functions but with the help of the above result, we can find the fractal dimension of any finite composition of Weierstrass type functions.

3. Oscillation spaces

Here, we introduce oscillation spaces briefly, reader may refer to [**C, DJ**] for details. Let $Q \subset [0,1]$ p-adic subinterval such that $Q = \left[jp^{-m}, (j+1)p^{-m}\right]$ and $0 \leq j < p^{-m}$, for some integers $m \geq 0$. The oscillation $\phi : [0,1] \to \mathbb{R}$ over Q is given by

$$\begin{aligned} R_\phi(Q) &= \sup_{x,y \in Q} |\phi(x) - \phi(y)| \\ &= \sup_{x \in Q} \phi(x) - \inf_{y \in Q} \phi(y), \end{aligned}$$

and the total oscillation having order m, is given by

$$Osc(m, \phi) = \sum_{|Q| = p^{-m}} R_\phi(Q),$$

and above sum lies over all $Q \subset [0,1]$ p-adic intervals having length $|Q|= p^{-m}$.

Let us define oscillation space $V^s(I)$, $s \in \mathbb{R}$ by

$$V^s(I) = \left\{ \phi \in \mathcal{C}(I) : \sup_{m \in \mathbb{N}} \frac{Osc(m, \phi)}{p^{m(1-s)}} < \infty \right\}.$$

Also one can define

$$V^{s-}(I) = \{\phi \in \mathcal{C}(I) : \phi \in V^{s-\epsilon}(I) \ \forall \, \epsilon > 0\},$$

and

$$V^{s+}(I) = \{\phi \in \mathcal{C}(I) : \phi \notin V^{s+\epsilon}(I) \ \forall \, \epsilon > 0\}.$$

THEOREM 3.1 ([**C**, Theorem 4.1]). *Let ϕ be a real-valued continuous function defined on I, we have*

$$\overline{\dim}_B(Gr(g)) \leq 2-s \iff \phi \in V^{s-}(I) \text{ if } 0 < s \leq 1$$

and

$$\overline{\dim}_B(Gr(g)) \geq 2-s \iff g \in V^{s+}(I) \text{ if } 0 \leq s < 1.$$

THEOREM 3.2. *Let $q_{j,k} \in V^s(I)$, $\alpha_{j,k} \in (-1,1)$ and*

$$\max_k \left\{ \|\alpha_k\|_\infty, \sum_{j\in J} |\alpha_{j,k}| \right\} < 1.$$

Then, the associated fractal interpolation function $f^* \in V^s(I)$. Moreover,

$$\overline{\dim}_B(Gr(f^*)) \leq 2-s.$$

PROOF. Let $V_\phi^s(I) = \{\phi \in V^s(I) : \phi(x_1) = y_1,\ \phi(x_N) = y_N\}$. Here, the space V_ϕ^s is a closed subset of $V^s(I)$. This yields that $V_\phi^s(I)$ is complete metric space with respect to the metric induced by the norm $\|\cdot\|_{V^s}$. Now, define a sequence of mappings $T^k : V_\phi^s(I) \to V_\phi^s(I)$ by

$$(T^k\phi)(x) = \alpha_{j,k}\phi(S_j^{-1}(x)) + q_{j,k}(S_j^{-1}(x)), x \in I_j, j \in J.$$

From the definition of T^k, it is is well defined. Now, for $\phi, \psi \in V_\phi^s(I)$, we have

$$\begin{aligned}
\|T^k\phi - T^k\psi\|_{V^s} =& \|T^k\phi - T^k\psi\|_\infty + \sup_{m\in\mathbb{N}} \frac{Osc(m, T^k\phi - T^k\psi)}{p^{m(1-s)}} \\
\leq& |\alpha_{j,k}| \|\phi(\frac{1}{c_j}x - \frac{d_j}{c_j}) - \psi(\frac{1}{c_j}x - \frac{d_j}{c_j})\|_\infty \\
&+ \sum_{j\in J} |\alpha_{j,k}| \sup_{m\in\mathbb{N}} \frac{Osc(m, \phi(\frac{1}{c_j}x - \frac{d_j}{c_j}) - \psi(\frac{1}{c_j}x - \frac{d_j}{c_j}))}{p^{m(1-s)}}.
\end{aligned}$$

Set $X = \frac{1}{c_j}x - \frac{d_j}{c_j}$. Then

$$\begin{aligned}
\|T^k\phi - T^k\psi\|_{V^s} \leq& |\alpha_{j,k}| \|\phi(X) - \psi(X)\|_\infty + \sum_{j\in J} |\alpha_{j,k}| \sup_{m\in\mathbb{N}} \frac{Osc(m, f(X) - g(X))}{p^{m(1-\beta)}} \\
\leq& \max_k \left\{ \|\alpha_k\|_\infty, \sum_{j\in J} |\alpha_{j,k}| \right\} \|\phi - \psi\|_{V^s}.
\end{aligned}$$

Since

$$\max_k \left\{ \|\alpha_k\|_\infty, \sum_{j\in J} |\alpha_{j,k}| \right\} < 1,$$

then T^k is contraction on $V_\phi^s(I)$. One can see that the sequence $\{\|T^k\phi - \phi\|_{V^s}\}$ is bounded. In view of Proposition 1.4, the backward trajectories $\Psi_k := T^1 \circ T^2 \circ \ldots \circ T^k$ of (T^k) converges for every $\phi \in V_\phi^s(I)$ to a unique attractor $f^* \in V_\phi^s(I)$, completing the proof. $\square$

4. Hölder space

First, we define the Hölder space as follows.

$$\mathcal{H}^\sigma(I) = \{\phi : I \to \mathbb{R} : \phi \text{ is Hölder continuous with exponent } \sigma\}.$$

If we equip the space $\mathcal{H}^\sigma(I)$ with norm

$$\|\phi\|_{\mathcal{H}} = \|\phi\|_\infty + [\phi]_\sigma,$$

where

$$[\phi]_\sigma = \sup_{x\neq y} \frac{|\phi(x) - \phi(y)|}{|x-y|^\sigma},$$

then it forms a Banach space.

THEOREM 4.1. *Let*

$$q_{j,k} \in \mathcal{H}^\sigma(I), \alpha_{j,k} \in (-1,1) \text{ and } (N-1)\max_k\{\|\alpha_k\|_\infty, \frac{\|\alpha_k\|_\infty}{c^\sigma}\} < 1.$$

Then, the associated fractal interpolation function $f^* \in \mathcal{H}^\sigma(I)$. Moreover,

$$\overline{\dim}_B(Gr(f^*)) \leq 2 - \sigma.$$

PROOF. Let $\mathcal{H}^\sigma_\phi(I) = \{\phi \in \mathcal{H}^\sigma(I) : \phi(x_1) = y_1, \phi(x_N) = y_N\}$. It can be seen that $\mathcal{H}^\sigma_\phi(I)$ is closed subset of $\mathcal{H}^\sigma(I)$. Since $\mathcal{H}^\sigma(I)$ is Banach space, it yields $\mathcal{H}^\sigma_\phi(I)$ is complete metric space with respect to the metric induced by $\|\cdot\|_{\mathcal{H}}$. Let us define a sequence of mappings $T^k : \mathcal{H}^\sigma_\phi(I) \to \mathcal{H}^\sigma_\phi(I)$ by

$$(T^k\phi)(x) = \alpha_{j,k}\phi(S_j^{-1}(x)) + q_{j,k}(S_j^{-1}(x)), x \in I_j, j \in J.$$

First we shall show that T^k is well defined. Let us note that

$$\begin{aligned}
[T^k\phi]_\sigma &= (N-1)\max_{j\in J} \sup_{x\neq y, x,y\in I_j} \frac{|T^k\phi(x) - T^k\phi(y)|}{|x-y|^\sigma} \\
&\leq (N-1)\max_{j\in J}[\sup_{x\neq y, x,y\in I_j} \frac{|\alpha_{j,k}||\phi(S_j^{-1}(x)) - \phi(S_j^{-1}(y))|}{|x-y|^\sigma} \\
&\quad + \sup_{x\neq y, x,y\in I_j} \frac{|q_{j,k}(S_j^{-1}(x)) - q_{j,k}(S_j^{-1}(y))|}{|x-y|^\sigma}] \\
&= (N-1)\max_{j\in J}[\sup_{x\neq y, x,y\in I_j} \frac{|\alpha_{j,k}||\phi(S_j^{-1}(x)) - \phi(S_j^{-1}(y))|}{c_j^\sigma|S_j^{-1}(x)) - q_{j,k}(S_j^{-1}(y)|^\sigma} \\
&\quad + \sup_{x\neq y, x,y\in I_j} \frac{|q_{j,k}(S_j^{-1}(x)) - q_{j,k}(S_j^{-1}(y))|}{c_j^\sigma|S_j^{-1}(x)) - q_{j,k}(S_j^{-1}(y)|^\sigma}] \\
&\leq (N-1)\|\alpha_k\|_\infty \frac{[\phi]_\sigma}{c^\sigma} + (N-1)\frac{[q_{j,k}]}{c^\sigma},
\end{aligned}$$

where $c = \max_{j\in J}\{c_j\}$. Now, for $\phi, \psi \in \mathcal{H}^\sigma_\phi(I)$, we have

$$\begin{aligned}
\|T^k\phi - T^k\psi\|_{\mathcal{H}} &= \|T^k\phi - T^k\psi\|_\infty + [T^k\phi - T^k\psi]_\sigma \\
&\leq (N-1)\|\alpha_k\|_\infty\|\phi - \psi\|_\infty + (N-1)\frac{\|\alpha_k\|_\infty}{c^\sigma}[\phi - \psi]_\sigma \\
&\leq (N-1)\max_k\{\|\alpha_k\|_\infty, \frac{\|\alpha_k\|_\infty}{c^\sigma}\}\|\phi - \psi\|_{\mathcal{H}}.
\end{aligned}$$

Since $(N-1)\max_k\{\|\alpha_k\|_\infty, \frac{\|\alpha_k\|_\infty}{c^\sigma}\} < 1$, the map T^k is contraction on $\mathcal{H}^\sigma_\phi(I)$. One can see that the sequence $\{\|T^k\phi - \phi\|_{\mathcal{H}}\}$ is bounded. In view of Proposition 1.4, the backward trajectories $\Psi_k := T^1 \circ T^2 \circ \ldots \circ T^k$ of (T^k) converges for every $\phi \in \mathcal{H}^\sigma_\phi(I)$ to a unique attractor $f^* \in \mathcal{H}^\sigma_\phi(I)$. Since $f^* \in \mathcal{H}^\sigma_\phi(I)$, from [**Fal03**, Corollary 11.2], we get $\overline{\dim}_B(Gr(f^*)) \leq 2 - \sigma$. □

4.1. Katugampola integral of nonstationary FIF. First, let us introduce the Katugampola fractional integral. The Katugampola fractional integral is the unification of the Riemann–Liouville fractional integral and the Hadamard fractional integral and defined as follows:

DEFINITION 4.2. Let f be a an integrable function on $[0,1]$. The Katugampola fractional integral of f is defined by

$$({}^p_0\mathcal{I}^\mu f)(x) = \frac{(p+1)^{1-\mu}}{\Gamma(\mu)} \int_0^x (x^{p+1} - r^{p+1})^{\mu-1} r^p f(r)dr,$$

where $\mu > 0$ and $p \neq -1$.

Liang and Su [**LS07**] initiated the theory of the fractal dimension of fractional integrals. Liang gave interesting results on fractal dimensions of functions of bounded variation in [**Lia10**]. Also, he proved that the fractal dimension of the Riemann–Liouville fractional integral of a function of bounded variation is one. In this order, Verma and Viswanathan [**VV**] proved that fractal dimension of the Katugampola fractional integral of a function which is of bounded variation, is one. We studied the box dimension of the mixed Katugampola fractional integral associated with two-dimensional continuous functions in [**CA**]. In the upcoming results, we study the fractal dimension of the Katugampola fractional integral of nonstationary FIF on function spaces.

THEOREM 4.3. *Let $q_{j,k} \in \mathcal{H}^\sigma(I)$, $\alpha_{j,k} \in (-1,1)$ and*

$$(N-1)\max_k\{\|\alpha_k\|_\infty, \frac{\|\alpha_k\|_\infty}{c^\sigma}\} < 1.$$

Then the upper box dimension of the Katugampola fractional integral of f^* is at most $2-\sigma$.

PROOF. Let $0 \leq x < x+h \leq 1$, $-1 < p < 0$ and $0 < \mu < 1$. We have the following

$$({}^p_0\mathcal{I}^\mu f^*)(x+h) - ({}^p_0\mathcal{I}^\mu f^*)(x) = \frac{(p+1)^{1-\mu}}{\Gamma(\mu)} \int_0^{x+h} ((x+h)^{p+1} - r^{p+1})^{\mu-1} r^p f^*(r)dr$$
$$- \frac{(p+1)^{1-\mu}}{\Gamma(\mu)} \int_0^x (x^{p+1} - r^{p+1})^{\mu-1} r^p f^*(r)dr.$$

By applying change of variable in the second integral, we get

$$\begin{aligned}({}^p_0\mathcal{I}^\mu f^*)(x+h) - ({}^p_0\mathcal{I}^\mu f^*)(x) &= \frac{(p+1)^{1-\mu}}{\Gamma(\mu)} \int_0^{x+h} ((x+h)^{p+1} - r^{p+1})^{\mu-1} r^p f^*(r)dr \\ &\quad - \frac{(p+1)^{1-\mu}}{\Gamma(\mu)} \int_{((x+h)^{p+1}-x^{p+1})^{\frac{1}{p+1}}}^{x+h} (x^{p+1} - r^{p+1})^{\mu-1} r^p \\ &\quad f^*((r^{p+1} + x^{p+1} - (x+h)^{p+1})^{\frac{1}{p+1}})dr.\end{aligned}$$

So, we obtain

$$({}_0^p\mathcal{I}^\mu f^*)(x+h) - ({}_0^p\mathcal{I}^\mu f^*)(x) = \frac{(p+1)^{1-\mu}}{\Gamma(\mu)} \int_{((x+h)^{p+1}-x^{p+1})^{\frac{1}{p+1}}}^{x+h} (x^{p+1} - r^{p+1})^{\mu-1} r^p$$
$$\times (f^*(r) - f^*((r^{p+1} + x^{p+1} - (x+h)^{p+1})^{\frac{1}{p+1}}))dr$$
$$+ \frac{(p+1)^{1-\mu}}{\Gamma(\mu)} \int_0^{((x+h)^{p+1}-x^{p+1})^{\frac{1}{p+1}}} ((x+h)^{p+1} - r^{p+1})^{\mu-1} r^p f^*(r)dr$$
$$= I_1 + I_2.$$

Now, we estimate I_1 and I_2 as follows:

$$|I_1| = \Bigg| \frac{(p+1)^{1-\mu}}{\Gamma(\mu)} \int_{((x+h)^{p+1}-x^{p+1})^{\frac{1}{p+1}}}^{x+h} (x^{p+1} - r^{p+1})^{\mu-1} r^p \times (f^*(r) - f^*((r^{p+1} + x^{p+1} - (x+h)^{p+1})^{\frac{1}{p+1}}))dr \Bigg|.$$

Since $f \in \mathcal{H}^\sigma(I)$, for some positive constant C, we have

$$|I_1| \leq \frac{C(p+1)^{1-\mu}}{\Gamma(\mu)} \int_{((x+h)^{p+1}-x^{p+1})^{\frac{1}{p+1}}}^{x+h} (x^{p+1} - r^{p+1})^{\mu-1} r^p |r - (r^{p+1} + x^{p+1} - (x+h)^{p+1})^{\frac{1}{p+1}})|dr.$$

Because of the mean value theorem and Bernoulli's inequality, $(1+x)^t \leq 1 + tx$, for $0 \leq t \leq 1,\ x \geq -1$, we get

$$|I_1| \leq \frac{C(p+1)^{1-\mu}}{\Gamma(\mu)} h^\sigma.$$

In similar steps, we can get

$$|I_2| \leq \frac{C(p+1)^{1-\mu}}{\Gamma(\mu)} h^\sigma.$$

Consequently, we have

$$|({}_0^p\mathcal{I}^\mu f^*)(x+h) - ({}_0^p\mathcal{I}^\mu f^*)(x)| \leq |I_1| + |I_2| \leq \frac{2C(p+1)^{1-\mu}}{\Gamma(\mu)} h^\sigma,$$

for sufficiently small $h > 0$. Now, for sufficiently small $0 < \delta < 1$, let $N_\delta(Gr({}_0^p\mathcal{I}^\mu f^*))$ be the number of boxes having diameter δ that cover the graph of ${}_0^p\mathcal{I}^\mu f^*$, we have

$$\begin{aligned} N_\delta(Gr({}_0^p\mathcal{I}^\mu f^*)) &\leq \sum_{i=1}^{\lceil \frac{1}{\delta} \rceil} \left(1 + \left\lceil \frac{\mathcal{R}_{{}_0^p\mathcal{I}^\mu f^*}[(i-1)\delta, i\delta]}{\delta} \right\rceil \right) \\ &\leq \sum_{i=1}^{\lceil \frac{1}{\delta} \rceil} \left(2 + \left\lceil \frac{\mathcal{R}_{{}_0^p\mathcal{I}^\mu f^*}[(i-1)\delta, i\delta]}{\delta} \right\rceil \right) \\ &= 2\left\lceil \frac{1}{\delta} \right\rceil + \sum_{i=1}^{\lceil \frac{1}{\delta} \rceil} \frac{\mathcal{R}_{{}_0^p\mathcal{I}^\mu f^*}[(i-1)\delta, i\delta]}{\delta} \\ &\leq 2\left\lceil \frac{1}{\delta} \right\rceil + \sum_{i=1}^{\lceil \frac{1}{\delta} \rceil} \frac{2C(p+1)^{1-\mu}}{\Gamma(\mu)} \delta^{\sigma-1}. \end{aligned}$$

From the definition of the box dimension, we have

$$\overline{\dim}_B Gr({}_0^p\mathcal{I}^\mu f^*, [0,1]) = \overline{\lim}_{\delta\to 0}\frac{\log N_\delta(Gr({}_0^p\mathcal{I}^\mu f^*))}{-\log\delta} \leq 2-\sigma,$$

completing the proof. □

5. Bounded variational space

DEFINITION 5.1. Let $\phi : I \to \mathbb{R}$ be a function. For each partition $P : t_0 < t_1 < t_2 \cdots < t_n$ of the interval I, we define

$$V(\phi, I) = \sup_P \sum_{i=1}^{n} |\phi(t_i) - \phi(t_{i-1})|,$$

where the supremum is taken over all partitions P of the interval I. If $V(\phi, I) < \infty$, we say that ϕ is of bounded variation.

Let us define

$$\mathcal{BV}(I) := \{\phi : I \to \mathbb{R} : \phi \text{ is of bounded variation on } I\}.$$

Now, we define a norm $\|\phi\|_{\mathcal{BV}} := |\phi(t_0)| + V(\phi, I)$ on $\mathcal{BV}(I)$. The space $(\mathcal{BV}(I), \|\cdot\|_{\mathcal{BV}})$ is a Banach space (see, for instance [**JV**]).

LEMMA 5.2 ([**Lia10**]). *Let $f \in \mathcal{C}(I) \cap \mathcal{BV}(I)$, then*

$$\dim_H(Gr(f)) = \dim_B(Gr(f)) = 1.$$

THEOREM 5.3. *Let $q_{j,k} \in \mathcal{BV}(I)$, $\alpha_{j,k} \in (-1,1)$ and $|\alpha_{j,k}| < \frac{1}{(N-1)}$. Suppose that $P' : x_1 < x_2 < \cdots < x_N$ is a partition of I. Then, the associated fractal interpolation function $f^* \in \mathcal{BV}(I)$. Moreover, $\dim_H(Gr(f^*)) = \dim_B(Gr(f^*)) = 1$.*

PROOF. Let $\mathcal{BV}_\phi(I) = \{\phi \in \mathcal{BV}(I) : \phi(x_1) = y_1, \phi(x_N) = y_N\}$. Here, the space $\mathcal{BV}_\phi(I)$ is a closed subset of $\mathcal{BV}(I)$. This yields that $\mathcal{BV}_\phi(I)$ is complete metric space with respect to the metric induced by the norm $\|\cdot\|_{\mathcal{BV}}$. Now, define a sequence of mappings $T^k : \mathcal{BV}_\phi(I) \to \mathcal{BV}_\phi(I)$ by

$$(T^k\phi)(x) = \alpha_{j,k}\phi(S_j^{-1}(x)) + q_{j,k}(S_j^{-1}(x)), x \in I_j, j \in J.$$

From the definition of T^k, it is is well defined.

Let $P'' : t_0 < t_1 < t_2 \cdots < t_m$, $m \in \mathbb{N}$ be a partition of I_j. Then, for $\phi, \psi \in \mathcal{BV}_\phi(I)$, we have

$$\begin{aligned}
&|(T^k\phi - T^k\psi)(t_i) - (T^k\phi - T^k\psi)(t_{i-1})| \\
&= |\alpha_{j,k}(\phi(S_j^{-1}(t_i)) - \psi(S_j^{-1}(t_i))) - \alpha_{j,k}(\phi(S_j^{-1}(t_{i-1})) - \psi(S_j^{-1}(t_{i-1})))| \\
&\leq |\alpha_{j,k}||(\phi-\psi)(S_j^{-1}(t_i)) - (\phi-\psi)(S_j^{-1}(t_{i-1}))|.
\end{aligned}$$

Summing over $i = 1$ to m, we get

$$\begin{aligned}
&\sum_{i=1}^{m} |(T^k\phi - T^k\psi)(t_i) - (T^k\phi - T^k\psi)(t_{i-1})| \\
&\leq |\alpha_{j,k}| \sum_{i=1}^{m} |(\phi-\psi)(S_j^{-1}(t_i)) - (\phi-\psi)(S_j^{-1}(t_{i-1}))| \\
&\leq |\alpha_{j,k}| \|\phi - \psi\|_{\mathcal{BV}}.
\end{aligned}$$

The above inequality holds for any partition of I_j. Hence, we obtain

$$\|T^k\phi - T^k\psi\|_{\mathcal{BV}} \leq (N-1)|\alpha_{j,k}|\|\phi - \psi\|_{\mathcal{BV}}.$$

Since $|\alpha_{j,k}| < \frac{1}{(N-1)}$, the map T^k is contraction on $\mathcal{BV}_\phi(I)$. One can see that the sequence $\{\|T^k\phi - \phi\|_{\mathcal{BV}}\}$ is bounded. In view of Proposition 1.4, the backward trajectories $\Psi_k := T^1 \circ T^2 \circ \ldots \circ T^k$ of (T^k) converges for every $\phi \in \mathcal{BV}_\phi(I)$ to a unique attractor $f^* \in \mathcal{BV}_\phi(I)$. Since $f^* \in \mathcal{BV}_\phi(I)$, from Lemma 5.2, we have $\dim_H(Gr(f^*)) = \dim_B(Gr(f^*)) = 1$. □

In Theorem 5.4, we show that the Hausdorff dimension and the box dimension of the Katugampola fractional integral is also one.

THEOREM 5.4. *Let $q_{j,k} \in \mathcal{BV}(I)$, $\alpha_{j,k} \in (-1,1)$ and $|\alpha_{j,k}| < \frac{1}{(N-1)}$. Suppose that $P' : x_1 < x_2 < \cdots < x_N$ is a partition of I. Then, ${}_0^p\mathcal{I}^\mu f^* \in \mathcal{BV}(I)$. Moreover, $\dim_H(Gr({}_0^p\mathcal{I}^\mu f^*)) = \dim_B({}_0^p\mathcal{I}^\mu f^*)) = 1$.*

PROOF. From the above theorem, we have $f^* \in \mathcal{BV}(I)$. This result with [**VV**, Theorem 3.5] yields ${}_0^p\mathcal{I}^\mu f^* \in \mathcal{BV}(I)$. Hence, using Lemma 5.2, we get the desired result. □

Acknowledgments

We are thankful to the anonymous reviewers and editor for their constructive comments and suggestions, which helped us to improve the manuscript considerably.

References

[Bar86] M. F. Barnsley, *Fractal functions and interpolation*, Constr. Approx. **2** (1986), no. 4, 303–329, DOI 10.1007/BF01893434. MR892158

[Bar88] M. Barnsley, *Fractals everywhere*, Academic Press, Inc., Boston, MA, 1988. MR977274

[C] A. Carvalho, *Box dimension, oscillation and smoothness in function spaces*, J. Funct. Spaces Appl. **3** (2005), no. 3, 287–320, DOI 10.1155/2005/405979. MR2163628

[CA] S. Chandra and S. Abbas, *Box dimension of mixed Katugampola fractional integral of two-dimensional continuous functions*, Fract. Calc. Appl. Anal. **25** (2022), no. 3, 1022–1036, DOI 10.1007/s13540-022-00050-2. MR4447116

[CA1] S. Chandra and S. Abbas, *On fractal dimensions of fractal functions using function spaces*, Bull. Aust. Math. Soc. **106** (2022), no. 3, 470–480, DOI 10.1017/S0004972722000685. MR4510139

[DJ] A. Deliu and B. Jawerth, *Geometrical dimension versus smoothness*, Constr. Approx. **8** (1992), no. 2, 211–222, DOI 10.1007/BF01238270. MR1152878

[DLM] N. Dyn, D. Levin, and P. Massopust, *Attractors of trees of maps and of sequences of maps between spaces with applications to subdivision*, J. Fixed Point Theory Appl. **22** (2020), no. 1, Paper No. 14, 24, DOI 10.1007/s11784-019-0750-7. MR4036525

[Fal03] K. Falconer, *Fractal geometry*, 2nd ed., John Wiley & Sons, Inc., Hoboken, NJ, 2003. Mathematical foundations and applications, DOI 10.1002/0470013850. MR2118797

[Hun98] B. R. Hunt, *The Hausdorff dimension of graphs of Weierstrass functions*, Proc. Amer. Math. Soc. **126** (1998), no. 3, 791–800, DOI 10.1090/S0002-9939-98-04387-1. MR1452806

[JV] S. Jha and S. Verma, *Dimensional analysis of α-fractal functions*, Results Math. **76** (2021), no. 4, Paper No. 186, 24, DOI 10.1007/s00025-021-01495-2. MR4303614

[NV] M. A. Navascués and S. Verma, *Nonstationary α-fractal surfaces*, Mediterr. J. Math. **20** (2023), no. 1, Paper No. 48, 18, DOI 10.1007/s00009-022-02242-9. MR4527888

[LS07] Y. S. Liang and W. Y. Su, *The relationship between the fractal dimensions of a type of fractal functions and the order of their fractional calculus*, Chaos Solitons Fractals **34** (2007), no. 3, 682–692, DOI 10.1016/j.chaos.2006.01.124. MR2327602

[Lia10] Y. S. Liang, *Box dimensions of Riemann-Liouville fractional integrals of continuous functions of bounded variation*, Nonlinear Anal. **72** (2010), no. 11, 4304–4306, DOI 10.1016/j.na.2010.02.007. MR2606788

[LDV] D. Levin, N. Dyn, and P. V. Viswanathan, *Nonstationary versions of fixed-point theory, with applications to fractals and subdivision*, J. Fixed Point Theory Appl. **21** (2019), no. 1, Paper No. 26, 25, DOI 10.1007/s11784-019-0659-1. MR3903845

[MW86] R. D. Mauldin and S. C. Williams, *On the Hausdorff dimension of some graphs*, Trans. Amer. Math. Soc. **298** (1986), no. 2, 793–803, DOI 10.2307/2000650. MR860394

[M] P. R. Massopust, *Nonstationary fractal interpolation*, Mathematics 7(8) (2019), 666.

[VV] S. Verma and P. Viswanathan, *A note on Katugampola fractional calculus and fractal dimensions*, Appl. Math. Comput. **339** (2018), 220–230, DOI 10.1016/j.amc.2018.07.035. MR3852116

School of Mathematical and Statistical Sciences, Indian Institute of Technology Mandi, Kamand, Himachal Pradesh 175005, India
Email address: sahusubhash77@gmail.com

School of Mathematical and Statistical Sciences, Indian Institute of Technology Mandi, Kamand (H.P.) - 175005, India
Email address: sabbas.iitk@gmail.com

Contemporary Mathematics
Volume **797**, 2024
https://doi.org/10.1090/conm/797/15954

Recurrence properties of superadditive processes and universally good weights

Doğan Çömez

Abstract. In this article recurrence properties of a class of strongly bounded superadditive processes are studied at some length. In particular, four generalizations of Fürstenberg's multiple recurrence theorem are extended to the setting of such superadditive processes. Having these recurrence properties in hand, it is observed that such superadditive processes define universally good weights for various type of ergodic averages.

1. Introduction and motivation

One of the most fundamental properties of a measure preserving transformation T on a probability space (X, Σ, μ, T) is expressed by Poincare in the celebrated Recurrence Theorem (PRT):

For every $A \in \Sigma$ *with* $\mu(A) > 0$, $\mu(A \cap \bigcap_{n=0}^{\infty} \bigcup_{k=n}^{\infty} T^{-k}A) = \mu(A)$.

This is equivalent to saying that for any $A \in \Sigma$ with $\mu(A) > 0$, we have $\mu(A \cap T^{-n}A) > 0$ for some $n \geq 1$. More explicitely, almost all trajectories in a measure preserving dynamical system eventually come arbitrarily close to the original position. A quantitative version of this result is due to Khintchine (KRT):

For every $A \in \Sigma$ *and for every* $\epsilon > 0$, *there exists a relatively dense subset* D *of* $\mathbb{N}$ *such that* $\mu(A \cap T^{-n}A) \geq \mu(A)^2 - \epsilon$ *for every* $n \in D$.

One can interpret Khintchine's recurrence theorem as saying that the return of a point to its original neighborhood, guaranteed by Poincare's recurrence theorem, takes place in a fairly regular fashion. Hence, the sequence of the return times has bounded gaps, i.e., a syndetic set.

In order to give an alternative proof of Szemeredi's theorem, Fürstenberg proved a far reaching generalization of Poincare's recurrence theorem known as the Fürstenberg's multiple recurrence theorem (FMRT):

For every $A \in \Sigma$ *with* $\mu(A) > 0$ *and for any* $j \geq 1$,

$$\liminf_{n} \frac{1}{n+1} \sum_{k=0}^{n} \mu(A \cap T^{-k}A \cap T^{-2k}A \cap \ldots \cap T^{-jk}A) > 0, \quad \text{and}$$

$$\mu(A \cap T^{-n}A \cap T^{-2n}A \cap \ldots \cap T^{-jn}A) > 0 \quad \text{for some } n \geq 1.$$

Furthermore, this result admits a more general version:

2020 *Mathematics Subject Classification.* Primary 28D05, 37A99; Secondary 47A35, 60A10.
Key words and phrases. Recurrence, superadditive processes, universally good sequences.

THEOREM A ([**7**], 1977). *If T is a weakly mixing invertible measure preserving transformation (impt) on a probability space (X, Σ, μ, T), then for every integer $k \geq 1$ and for all $\{f_j\}_{j=1}^k \subset L_\infty(X)$, the multiple averages*

$$\frac{1}{N}\sum_{n=1}^{N} f_1(T^n x) f_2(T^{2n} x) \dots f_k(T^{kn} x)$$

converge in L_2 to $\Pi_{j=1}^k \int f_j$ as $N \to \infty$.

Over the years various generalizations of the PRT and FMRT have been obtained. Some of these generalizations are as follows:

THEOREM B ([**11**], 2005). *If T is an impt on a probability space (X, Σ, μ, T), then for every integer $k \geq 1$, for any integer polynomials $p_1, p_2, \dots, p_k$ with $p_i(0) = 0$, and for all $\{f_j\}_{j=1}^k \subset L_\infty(X)$,*

$$\frac{1}{N}\sum_{n=1}^{N} f_1(T^{p_1(n)} x) f_2(T^{p_2(n)} x) \dots f_k(T^{p_k(n)} x) \quad \textit{converges in } L_2 \textit{ as } N \to \infty. <$$

When $p_m(n) = mn$, $1 \leq m \leq k$, Theorem B is proved in ([**4**], 2005). Also, they proved that if T is totally ergodic and p_i's are rationally independent, then the L_2-limit is $\int f_1 \cdots \int f_k$.

THEOREM C ([**13**], 2008). *If $T_1, T_2, \dots, T_k$ are commuting impts on a probability space (X, Σ, μ, T), then for all $\{f_j\}_{j=1}^k \subset L_\infty(X)$,*

$$\frac{1}{N}\sum_{n=1}^{N} f_1(T_1^n x) f_2(T_2^n x) \dots f_k(T_k^n x) \quad \textit{converges in } L_2 \textit{ as } N \to \infty.$$

In the same vein, KRT also has been generalized:

THEOREM D ([**8**], 2006). *If T is an impt on a probability space (X, Σ, μ, T), then for every integer $k \geq 1$, for any rationally independent integer polynomials $p_1, p_2, \dots, p_k$ with $p_i(0) = 0$, and for all $A \in \Sigma$ with $\mu(A) > 0$, the set*

$$\{n \in \mathbb{Z} : \mu(A \cap T^{p_1(n)} A \cap T^{p_2(n)} A \cap \cdots \cap T^{p_k(n)} A) \geq \mu(A)^{k+1} - \epsilon\} \textit{ is syndetic.}$$

The condition that p_i are rationally independent is essential; for otherwise, the result is not valid for $k \geq 4$ ([**4**], 2005), ([**8**], 2006). However, in the case that p_i are not necessarily rationally independent the following is known:

THEOREM E ([**3**],2000). *Let T be an impt on a probability space (X, Σ, μ, T). If $f \in L_\infty(X)$, $f \geq 0$, then*

$$\lim_{N\to\infty} \frac{1}{N^2} \sum_{m,n=1}^{N} \int f(x) f(T^m x) f(T^n x) f(T^{n+m} x) d\mu \geq (\int f d\mu)^4.$$

Clearly this implies that for any $\epsilon > 0$, and for all $A \in \Sigma$ with $\mu(A) > 0$, the set

$$\{(m, n) \in \mathbb{Z}^2 : \mu(A \cap T^m A \cap T^n A \cap T^{m+n} A) \geq \mu(A)^4 - \epsilon\} \text{ is syndetic;}$$

which can be viewed as a two parameter extension of KRT.

Let $\{A_n\}$ be a sequence of measurable subsets of $\mathbb{R}$ with $A_n \uparrow A$ and also satisfies $\sum m(A_n \backslash A_{n-1}) < \infty$, where m is the Lebesgue measure. Then for any ergodic measure preserving transformation T on $\mathbb{R}$, when a "return-time sequence" $\{n_k\}$ is defined by

$$n_k = \inf\{l \in \mathbb{Z} : T^l(x) \in A_k, l \geq n_{k-1}\},$$

then it was shown in [**5**] that this sequence is good a.e. in L_1. Prompted with this result, naturally, one asks whether statements analogous to PRT, KRT, FMRT and to those in the theorems stated above can hold for such a family $\{A_n\}$. Indeed we have the following observation:

THEOREM 1.1. *Let $\{A_n\}_n$ be a sequence of measurable subsets of a compact interval $X \subset \mathbb{R}$ with $A_n \uparrow A$, and T is a measure preserving transformation on X. Then*

(a) $m(A \cap \cap_{n=0}^{\infty} \cup_{k=n}^{\infty} T^{-k} A_k) = m(A)$.
(b) *For any $\epsilon > 0$, there is a relatively dense subset $L \subset \mathbb{N}$ and $K > 0$ such that whenever $k \geq K$, $m(A \cap T^{-n} A_k) = m(A)^2 - \epsilon$ for all $n \in L$.*

PROOF. Letting $E_k = A \cap T^{-k} A_k$, we have $\limsup E_k = A \cap \cap_{n=0}^{\infty} \cup_{k=n}^{\infty} T^{-k} A_k$. Since $m(E_k) < \infty$, we also have $\limsup_n m(E_n) \leq m(\limsup_n E_n) \leq m(A)$. Now,

$$\begin{aligned} m(E_k) &= m(T^{-k} A_k \cap T^{-k} A \cap A) + m(T^{-k} A_k \cap (A \backslash T^{-k} A)) \\ &\geq m(T^{-k} A_k \cap T^{-k} A) \\ &= m(T^{-k}(A_k \cap A)) \\ &= m(T^{-k} A_k) = m(A_k). \end{aligned}$$

Hence $\lim m(A_k) \leq \limsup_k m(E_k) \leq m(\limsup_k E_k) \leq m(A)$. Since $m(A_k) \uparrow m(A)$, (a) follows.

For (b), given $\epsilon > 0$, by KRT, pick $L \subset \mathbb{N}$ and $K > 0$ such that whenever $k \geq K$,

$$m(A \cap T^{-k} A) = m(A)^2 - \frac{\epsilon}{2} \quad \text{and} \quad m(A_k \backslash A) < \frac{\epsilon}{2}.$$

Then,

$$\begin{aligned} m(A)^2 - \frac{\epsilon}{2} < m(A \cap T^{-n} A) &= m(A \cap T^{-n}(A_k \cup (A \backslash A_k))) \\ &= m[(A \cap T^{-n} A_k) \cup (A \cap T^{-n}(A \backslash A_k)] \\ &\leq m(A \cap T^{-n} A_k) + m(T^{-n}(A \backslash A_k) \\ &\leq m(A \cap T^{-n} A_k) + m(A \backslash A_k) \leq m(A \cap T^{-n} A_k) + \frac{\epsilon}{2}. \end{aligned}$$

Hence, (b) follows. □

Theorem 1.1(a) implies that $m(A \cap T^{-k} A_k) > 0$ for some $k \geq 1$. Also, observe that, if $f_k := T^k \mathbf{1}_{A_k}$, then the family $\{f_k\}$ defines a certain type of superadditive process, called as T-admissible process (see below for definitions). Hence (a) and (b) can be rewritten in terms of such processes. Indeed, it turns out that all the results mentioned above can be extended to the setting of a class of superadditive processes relative to measure preserving transformations that include admissible processes.

2. Recurrence theorems for superadditive processes

In this section, we will state and prove the generalizations of the recurrence theorems stated above to the setting of a class of superadditive processes.

Let (X, Σ, μ) be a probability space and T be an impt on X. By a process $F = \{F_n\}_n$ we will mean a family of functions $\{f_i\}_{i\geq 0}$ with partial sums $F_n = \sum_{i=0}^{n-1} f_i$. For convenience, we will use notations $F = \{F_n\}$ and $F = \{f_i\}$ interchangeably. A process $F = \{F_n\} \subset L_p(X)$ is called a T-*superadditive process* if $F_{m+n} \geq F_m + F_n \circ T^m$ for all $m, n \geq 0$. Hence, any additive process $\{\sum_{i=0}^{n-1} f \circ T^i\}_{n\geq 1}$ is T-superadditive. F is called *positive* if $F_n \geq 0$, for all $n \geq 1$, *bounded* if $\gamma_F := \sup_n \frac{1}{n}\|F_n\|_1 < \infty$, and *strongly p-bounded* if $\sup_i \|f_i\|_p < \infty$. For simplicity, strongly ∞-bounded processes will be called *strongly bounded*. Clearly, for $1 \leq p \leq \infty$, a strongly p-bounded process is strongly 1-bounded, which, in turn, is bounded. If there exists $g \in L_1$ such that, for all $n > 0$,

$$\frac{1}{n}F_n \leq \frac{1}{n}\sum_{i=0}^{n-1} T^i g,$$

then F is called *dominated*; the function g is called an *exact dominant* if $\int |g| d\mu = \gamma_F$. Any bounded T-superadditive process $F \subset L_1(X)$ admits an exact dominant [**1**]; hence, it follows that a strongly p-bounded superadditive process has an exact dominant.

By Kingman's decomposition [**10**], any bounded T-superadditive process $F = \{F_n\}$ can be expressed as a difference of two processes as $F = G - H$, where $G = \{G_n\}$ is an additive process and $H = \{H_n\}$ is a subadditive process (that is, for all $m, n \geq 0$, $H_{m+n} \leq H_m + H_n \circ T^m$). In fact, H is a *purely subadditive* process, in the sense that, it does not dominate any nonzero T-additive process and $\lim_n \frac{1}{n}H_n = 0$ a.e. (and hence in L_1-norm) [**1**], [**10**]. Furthermore, $G_n = \sum_{k=0}^{n-1} T^k\delta$, where δ is the exact dominant of the process, and the exact dominant of purely subadditive process is 0 a.e. For the purely subadditive part of a superadditive process, for each $n \geq 0$, $H_n = \sum_{i=0}^{n-1} h_i \geq 0$ necessarily; however, this does not imply that $h_i \geq 0$ for every $i \geq 0$. There is a fairly large class of superadditive processes, namely admissible processes, with this property. A process $F = \{f_n\}_{n\geq 0} \subset L_1(X)$ is called T-*admissible* if $f_n \circ T \leq f_{n+1}$ for all $n \geq 0$. Obviously, a T-admissible process is T-superadditive. Furthermore, if $F = \{f_n\}$ be a positive strongly 1-bounded T-admissible process, then

(a) there exists $\delta \in L_1^+$ such that $f_n \leq \delta \circ T^n$ for all $n \geq 0$ and $\|\delta\|_1 = \sup_n \|f_n\|_1$,
(b) if $H = \{h_n\}$ is the purely subadditive part, then $h_n \geq 0$ for all $n \geq 0$, and
(c) $\lim_n h_n = 0$ a.e. (and in 1-norm) [**5**].

In [**5**] extension of various subsequential and weighted ergodic theorems, such as Wiener–Wintner theorem and return times theorem, for admissible processes were obtained. By (b), admissible processes satisfy the condition $h_n \geq 0$ for all $n \geq 0$, for its purely subadditive part, which plays an important role in some of the proofs. On the other hand, as the following example (simple) shows, there are superadditive processes satisfying this condition that are not admissible.

EXAMPLE 2.1. Let X be a one-point space with $T : X \to X$ being the identity map, and let $H = \{h_k\}$ be a process where $h_0 = 1,\ h_1 = h_2 = \frac{1}{2},\ h_3 = 1$, and

$$h_i = \frac{1}{4},\ 4 \leq i \leq 7,\ h_8 = \frac{1}{2},\ h_i = \frac{1}{8},\ 9 \leq i \leq 16,\ h_{17} = \frac{1}{4},\ h_i = \frac{1}{16},\ 18 \leq i \leq 33,\ \ldots$$

Then H is a purely positive T-subadditive process which is not T-admissible.

A subadditive version of Theorem A would be stated as follows: if T is an impt on a probability space X, and $H^{(j)} = \{h_n^{(j)}\} \subset L_\infty,\ 1 \leq j \leq k$, are T-subadditive processes, then $\frac{1}{N}\sum_{n=1}^N h_n^{(1)} h_{2n}^{(2)} \ldots h_{kn}^{(k)} \to 0$ in L_2-norm as $N \to \infty$. However, as the next example shows this is not the case.

EXAMPLE 2.2. Consider a one-point space with T being the identity map, and let $H = \{h_i\}$ be a process with $h_i(x) = (-1)^i$. Then $H_n = 1$ for n odd and 0 otherwise. Therefore, H is a subadditive process which does not dominate any non-zero additive process. It is not a positive process, and is not admissible either. Thus, $\frac{1}{N}\sum_{n=1}^N h_n \to 0$ Observe that, for $k = 3$, since $h_{2n} = 1,\ h_{3n} = (-1)^n$, we have

$$\frac{1}{N}\sum_{n=1}^{N} h_n h_{2n} h_{3n} = \frac{1}{N}\sum_{n=1}^{N}(-1)^n(-1)^n \to 1 \neq 0.$$

Furthermore, there are subadditive processes for which the assertion of Theorem A is not valid.

EXAMPLE 2.3. Again, consider a one-point space with identity map and $H = \{f_n\}$ be as in Example 2.2. Now, construct another process $G = \{g_n\}$ as follows:

$$g_0 = f_0,\ g_2 = -f_1,\ g_4 = f_2,\ g_6 = f_3,\ g_8 = f_4,\ g_{10} = -f_5,$$

and

$$g_{2n} = \begin{cases} f_n, \text{ for } n = 3^i 2 + j \text{ when } 0 \leq j \leq 3^i 2 - 1 \\ -f_n, \text{ for } n = 3^i 4 + j \text{ when } 0 \leq j \leq 3^i 2 - 1, \end{cases}$$

$g_1 = g_3 = -1,\ g_5 = 0,\ g_7 = g_9 = 1$, and $g_{2k+1} = 0$ for $k \geq 2$. Hence, both processes H and G are purely subadditive. Also, it follows that

$$\frac{1}{n}\sum_{k=0}^{n-1} f_k g_{2k} = \begin{cases} 0 \text{ if } n = 3^i 2, \\ \frac{1}{2} \text{ if } n = 3^i 4. \end{cases}$$

Therefore, $\liminf_n \frac{1}{n}\sum_{k=0}^{n-1} f_k g_{2k} = 0$ whereas $\limsup_n \frac{1}{n}\sum_{k=0}^{n-1} f_k g_{2k} = \frac{1}{2}$.

These examples suggest that, if one wants to extend Theorems A, B, C and E to the superadditive setting, additional conditions on its subadditive part is needed. Also, in Example 2.3, the processes have the property that $\sum_{k=0}^n \|h_k h_{k+j}\|_1 = O(n)$ for $0 \leq j \leq n-1$. Hence, having these observations, one remedy would be imposing the condition

$$(\dagger) \qquad \sum_{k=0}^{n-1} |h_k h_{k+j}| = O(n^\alpha),\ 0 \leq j \leq n-1, \text{for some } 0 < \alpha < 1,$$

for the purely subadditive part. As another condition, one may assume that

$$(*) \qquad h_k \geq 0 \text{ a.e. for all } k \geq 1,$$

which include admissible processes. Notice that subadditive processes satisfying $(\dagger)$ need not satisfy $(*)$ and, as Example 1 indicates, they need not be admissible either. On the other hand, subadditive processes satisfying $(*)$ include admissible processes.

The statements below extend Theorems A, B, C and E to the setting of superadditive processes whose purely subadditive part satisfies $(\dagger)$ or $(*)$.

THEOREM 2.4. *Let (X, Σ, μ) be a probability measure space and T be an impt on X. For any $k \geq 1$, if $F^{(j)} = \{f_n^{(j)}\} \subset L_\infty$, $1 \leq j \leq k$, are strongly bounded T-superadditive processes whose subadditive part satisfy $(\dagger)$ or $(*)$, then*

$$\frac{1}{N}\sum_{n=1}^{N} f_n^{(1)} f_{2n}^{(2)} \dots f_{kn}^{(k)} \text{ converges in } L_2\text{-norm as } N \to \infty.$$

Furthermore, if T is weakly mixing, then the L_2-limit is $\int \delta^{(1)} \int \delta^{(2)} \dots \int \delta^{(k)}$, where $\delta^{(j)}$ is an exact dominant of $F^{(j)}$.

PROOF. Since the assertion is valid for additive processes, we can assume that each process $F^{(j)}$ is nonnegative. By the hypotheses on the processes $F^{(j)}$, each of them has an exact dominant $\delta^{(j)}$, respectively. Thus, in order to prove the statement, it is enough to prove that

$$\lim_N \|\frac{1}{N}\sum_{n=1}^{N} f_n^{(1)} f_{2n}^{(2)} \dots f_{kn}^{(k)} - \frac{1}{N}\sum_{n=1}^{N} T^n\delta^{(1)}T^{2n}\delta^{(2)} \dots T^{kn}\delta^{(k)}\|_2 = 0.$$

Now, by the decomposition theorem, for each $j = 1, \dots, k$, we have $f_n^{(j)} = T^n\delta^{(j)} - h_n^{(j)}$, where $\{h_n^{(l)}\}$ is the purely subadditive part of the T^l-superadditive process $F^{(j)} = \{f_n^{(j)}\}$. Consequently,

$$\|\frac{1}{N}\sum_{n=1}^{N} f_n^{(1)} f_{2n}^{(2)} \dots f_{kn}^{(k)} - \frac{1}{N}\sum_{n=1}^{N} T^n\delta^{(1)}T^{2n}\delta^{(2)} \dots T^{kn}\delta^{(k)}\|_2$$

is bounded by $(2^k - 1)$-terms, where each of the these terms are of the form

$$\frac{1}{N}\sum_{n=1}^{N} \dots h_{ln}^{(l)}(x) T^{mn}\delta^{(m)} \dots,$$

for some $1 \leq l, m \leq k$, but at least one of the factors is $h_{ln}^{(l)}$. Hence, it remains to show that all these $(2^k - 1)$-terms converge to zero. First consider the term

$$\frac{1}{N}\sum_{n=1}^{N} h_n^{(1)} h_{2n}^{(2)} T^{3n}\delta^{(3)} \dots T^{kn}\delta^{(k)}.$$

Let $v_n = h_n^{(1)} h_{2n}^{(2)} T^{3n}\delta^{(3)} \dots T^{kn}\delta^{(k)}$. Since F is strongly bounded, we can assume that $\|v_n\|_2 \leq 1$ for all $n \geq 1$. Then, for $1 \leq m, n \leq N$,

$$\begin{aligned}
&| < v_{n+m}, v_n > | \\
&= |\int (h_{m+n}^{(1)} h_{2(m+n)}^{(2)} T^{3(m+n)}\delta^{(3)} \dots T^{k(m+n)}\delta^{(k)})(h_n^{(1)} h_{2n}^{(2)} T^{3n}\delta^{(3)} \dots T^{kn}\delta^{(k)}) d\mu| \\
&\leq C^3 \|\delta\|_\infty^{2(k-2)} \int h_n^{(1)} d\mu,
\end{aligned}$$

where $C = \sup_n \|h_n^{(2)}\|_\infty$. Therefore,

$$|\frac{1}{N}\sum_{n=1}^{N} < v_{m+n}, v_n > | \leq C^2\|\delta\|_\infty^{2(k-2)}\frac{1}{N}\sum_{i=1}^{N}\int |h_n^{(1)}h_{n+m}^{(1)}|d\mu.$$

By the condition (†), for some constant $0 < D < \infty$,

$$\frac{1}{N}\sum_{i=1}^{N}\int |h_n^{(1)}h_n^{(1)}|d\mu \leq DN^{\alpha-1} \to 0 \text{ in } L_1 \text{ as } \ N \to \infty.$$

Thus, by the van der Corput lemma, it follows that

$$\frac{1}{N}\sum_{n=1}^{N} h_n^{(1)}h_{2n}^{(2)}T^{3n}\delta^{(3)}\ldots T^{kn}\delta^{(k)} \to 0 \text{ in } L_2.$$

When the subadditive part satisfies $(*)$, since $\{h_n^{(j)}\} \subset L_\infty$, for all $1 \leq j \leq k$,

$$\begin{aligned}
&< v_{n+m}, v_n > \\
&= \int (h_{m+n}^{(1)}h_{2(m+n)}^{(2)}T^{3(m+n)}\delta^{(3)}\ldots T^{k(m+n)}\delta^{(k)})(h_n^{(1)}h_{2n}^{(2)}T^{3n}\delta^{(3)}\ldots T^{kn}\delta^{(k)})d\mu \\
&\leq C^3\|\delta\|_\infty^{2(k-2)}\int h_n^{(1)}d\mu,
\end{aligned}$$

where $C = \sup_n \|h_n^{(2)}\|_\infty$. Therefore,

$$\frac{1}{N}\sum_{n=1}^{N} < v_{m+n}, v_n > \leq C^3\|\delta\|_\infty^{2(k-2)}\frac{1}{N}\sum_{i=1}^{N}\int h_n^{(1)}d\mu.$$

Since $\frac{1}{N}H_n^{(1)} \to 0$ in L_1, by the van der Corput lemma, it follows that

$$\frac{1}{N}\sum_{n=1}^{N} h_n^{(1)}h_{2n}^{(2)}T^{3n}\delta^{(3)}\ldots T^{kn}\delta^{(k)} \to 0 \text{ in } L_2.$$

All the remaining $(2^k - 1)$-terms of this kind are handled similarly with L_2-limit zero; hence, the assertion follows. □

COROLLARY 2.5. *Let T be a totally ergodic impt on a probability space (X, Σ, μ). Then, for any $k \geq 1$, and strongly bounded T-superadditive processes*

$$F^{(j)} = \{f_n^{(j)}\}_{n=0}^{\infty} \subset L_\infty^+$$

whose subadditive part satisfy (†) or $(*)$ and having exact dominant $\delta^{(j)}$, $j = 1, \ldots, k$,

$$\lim_N \|\frac{1}{N}\sum_{n=1}^{N} f_n^{(1)}f_{2n}^{(2)}f_{3n}^{(3)}\ldots f_{kn}^{(k)} - \prod_{j=1}^{k}\int \delta^{(j)}\|_2 = 0.$$

PROOF. By the hypothesis, we have $f_i^{(j)} = T^i\delta^{(j)} - h_i^{(j)}$. Therefore,

$$\begin{aligned}
&\|\frac{1}{N}\sum_{i=1}^{N} f_i^{(1)}f_{2i}^{(2)}f_{3i}^{(3)}\ldots f_{ki}^{(k)} - \prod_{j=1}^{k}\int \delta^{(j)} \ \|_2 \\
&\leq \|\frac{1}{N}\sum_{i=1}^{N} T^i\delta^{(1)}T^{2i}\delta^{(2)}\ldots T^{ki}\delta^{(k)} - \prod_{j=1}^{k}\int \delta^{(j)} \ \|_2 + \ (2^k - 1)\text{-terms},
\end{aligned}$$

where each of the (2^k-1)-terms are of the form $\|\frac{1}{N}\sum_{i=1}^{N}\dots h_{li}^{(l)}(x)T^{mi}\delta^{(m)}\dots\|_2$, for some $1\le l,m\le k$, but at least one of the factors is of the type h_{li}^{l}. By Theorem A the first term on the right hand side convereges to zero. As in the proof of Theorem 2.4 the remaining (2^k-1)-terms also converge to zero too; hence, the assertion follows. □

THEOREM 2.6. *Let (X,Σ,μ) be a probability measure space and T be an impt on X. Then for any $k\ge 1$, integer polynomials $p_1, p_2, \dots, p_k$ with $p_i(0)=0$ and positive coefficients, strongly bounded T-admissible processes $F^{(j)}=\{f_n^{(j)}\}\subset L_\infty$, $1\le j\le k$,*

$$\frac{1}{N}\sum_{n=1}^{N} f^{(1)}_{p_1(n)}f^{(2)}_{p_2(n)}\dots f^{(k)}_{p_k(n)} \text{ converges in } L_2\text{-norm as } N\to\infty.$$

PROOF. Since the assertion is valid for additive processes, we can assume that each process $F^{(j)}$ is nonnegative. By the hypotheses on the processes $F^{(j)}$, each of them has an exact dominant $\delta^{(j)}$, respectively. Thus, in order to prove the theorem, it is enough to prove that

$$\lim_N \|\frac{1}{N}\sum_{n=1}^{N} f^{(1)}_{p_1(n)}f^{(2)}_{p_2(n)}\dots f^{(k)}_{p_k(n)} - \frac{1}{N}\sum_{n=1}^{N} T^{p_1(n)}\delta^{(1)}T^{p_2(n)}\delta^{)2)}\dots T^{p_k(n)}\delta^{(k)}\|_2=0.$$

Now, by the decomposition theorem, for each $j=1,\dots,k$, we have $f_n^{(j)}=T^n\delta^{(j)}-h_n^{(j)}$, where $\{h_n^{(l)}\}$ is the purely subadditive part of the superadditive process $F^{(l)}=\{f_n^{(l)}\}$. Consequently,

$$\|\frac{1}{N}\sum_{n=1}^{N} f^{(1)}_{p_1(n)}f^{(2)}_{p_2(n)}\dots f^{(k)}_{p_k(n)} - \frac{1}{N}\sum_{n=1}^{N} T^{p_1(n)}\delta^{(1)}T^{p_2(n)}\delta^{(2)}\dots T^{p_k(n)}\delta^{(k)}\|_2$$

is bounded by (2^k-1)-terms, where each of the these terms are of the form

$$\frac{1}{N}\sum_{n=1}^{N}\dots h^{(l)}_{p_l(n)}(x)T^{p_m(n)}\delta^{(m)}\dots,$$

for some $1\le l,m\le k$, but at least one of the factors is $h^{(l)}_{p_l(n)}$. Hence, it remains to show that all these (2^k-1)-terms converge to zero. First consider the term

$$\frac{1}{N}\sum_{i=1}^{N} h^{(1)}_{p_1(i)}h^{(2)}_{p_2(i)}T^{p_3(i)}\delta^{(3)}\dots T^{p_k(i)}\delta^{(k)}.$$

Let $v_i = h^{(1)}_{p_1(i)}h^{(2)}_{p_2(i)}T^{p_3(i)}\delta^{(3)}\dots T^{p_k(i)}\delta^{(k)}$. Since F is strongly bounded, we can assume that $\|v_i\|_2\le 1$. Then, for $0\le i,n\le N$,

$$
\begin{aligned}
&< v_{i+n}, v_i > \\
&= \int (h^{(1)}_{p_1(i+n)}h^{(2)}_{p_2(i+n)}T^{p_3(i+n)}\delta^{(3)}\dots T^{p_k(i+n)}\delta^{(k)})(h^{(1)}_{p_1(i)}h^{(2)}_{p_2(i)}T^{p_3(i)}\delta^{(3)}\dots T^{p_k(i)}\delta^{(k)})d\mu \\
&\le C^3\|\delta\|_\infty^{2(k-2)}\int h^{(1)}_{p_1(i)}d\mu,
\end{aligned}
$$

where $C = \sup_n \|h_n^{(2)}\|_\infty$. Therefore,

$$\frac{1}{N}\sum_{i=1}^{N} < v_{i+n}, v_i > \leq C^3 \|\delta\|_\infty^{2(k-2)} \frac{1}{N}\sum_{i=1}^{N} \int |h^{(1)}_{p_1(i)}| d\mu.$$

Since $\{H_n^{(1)}\}$ is purely subadditive part of the T-admissible process $F^{(1)}$, $h_n \to 0$ a.e. and $\int h_n \to 0$. Consequently, $\frac{1}{N}\sum_{i=1}^{N} \int h^{(1)}_{p_1(i)} d\mu \to 0$. Again, by the van der Corput lemma, it follows that

$$\frac{1}{N}\sum_{i=1}^{N} h^{(1)}_{p_1(i)} h^{(2)}_{p_2(i)} T^{p_3(i)}\delta^{(3)} \dots T^{p_k(i)}\delta^{(k)} \to 0 \text{ in } L_2\text{-norm.}$$

Similarly, the remaining terms also converge to 0. □

REMARK 2.7. It follows from Corollary 2.5 above that if F is an admissible process relative to an invertible weakly mixing measure preserving transformation, then

$$\frac{1}{n}\sum_{k=1}^{n} T^{2k}\delta f_k \to (\int \delta)^2 \quad \text{in} \quad L_2,$$

and hence, $\int \delta f_k > 0$ for some $n \geq 1$, when $\delta = \chi_A$, $f_k = T^k \chi_{A_k} = \chi_{T^{-k}A_k}$, where $\{A_k\}$ and A are as in Theorem 1.1. This can be regarded as a generalization of PRC to the setting of admissible processes.

THEOREM 2.8. *Let T_1, $T_2, \dots,$ T_k be commuting impts on a probability space (X, Σ, μ, T). If $F^{(j)} = \{f_n^{(j)}\} \subset L_\infty(X)$, $1 \leq j \leq k$, are strongly bounded T_j-superadditive processes whose subadditive part satisfy (†) or (∗), then*

$$\frac{1}{N}\sum_{n=1}^{N} f_n^{(1)} f_n^{(2)} \dots f_n^{(k)} \quad \textit{converges in } L_2 \textit{ as } N \to \infty.$$

PROOF. The proof follows essentially the same lines as the proof of Theorem 2.4; namely, it is enough to show that

$$\|\frac{1}{N}\sum_{n=1}^{N} f_n^{(1)} f_n^{(2)} \dots f_n^{(k)} - \frac{1}{N}\sum_{n=1}^{N} T_1^n\delta^{(1)} T_2^n\delta^{(2)} \dots T_k^n\delta^{(k)}\|_2 \to 0.$$

By the Kingman decomposition, this term is bounded by $(2^k - 1)$-terms of the form

$$\|\frac{1}{N}\sum_{n=1}^{N} \dots h_n^{(l)} T_m^n \delta^{(m)} \dots \|_2, \text{ for some } 1 \leq l, m \leq k,$$

but at least one of the factors is of the type $h_n^{(l)}$. Now, exactly the same argument as in the proof of Theorem 2.4 above implies that all of these $(2^k - 1)$-terms converge to zero. □

Let $F = \{f_i\} \subset L_\infty(X)$ be a nonnegative strongly bounded T-superadditive process whose purely subadditive part $H = \{h_i\}$ satisfy (†) or (∗). If δ is an exact dominant, then, by Kingman decomposition, $f_i = T^i\delta - h_i \geq 0$ for $i \geq 0$. Therefore,

$$\begin{aligned}\delta f_m f_n f_{n+m} &= \delta T^m \delta T^n \delta T^{m+n}\delta + f_m f_n (f_{n+m} - T^{m+n}\delta)\\ &\quad + f_m(f_n - T^n\delta)T^{m+n}\delta + (f_m - T^m\delta)T^n\delta T^{m+n}\delta\\ &= \delta T^m \delta T^n \delta T^{m+n}\delta + f_m f_n h_{m+n} + f_m h_n T^{m+n}\delta + h_m T^n \delta T^{m+n}\delta\\ &\geq \delta T^m \delta T^n \delta T^{m+n}\delta.\end{aligned}$$

Consequently, as an immediate consequence of Theorem E we obtain

THEOREM 2.9. *Let T be an impt on a probability space (X, Σ, μ, T) and $F = \{f_i\} \subset L_\infty(X)$ be a non-negative strongly bounded T-superadditive process whose subadditive part satisfy (†) or (∗). If δ is an exact dominant, then*

$$\lim_{N\to\infty} \frac{1}{N^2} \sum_{m,n=1}^{N} \int \delta f_m f_n f_{n+m} d\mu \geq \left(\int \delta d\mu\right)^4.$$

3. Sequences induced by superadditive processes

As a natural consequence of Theorem 2.4 (and Theorem 1.1 in a special case), the type of superadditive processes studied in the previous section define a type of return time sequences. These sequences different than the return time sequences typically studied in the literature. Hence, one questions if the return time sequences defined by such superadditive processes are good a.e. (or in the norm) for dynamical systems. This result has been obtained in [**5**] when the superadditive processes involved are admissible. There, in addition, Wiener–Wintner theorem is extended to the setting of sequences induced by admissible processes. Furthermore, it is shown that sequences induced by admissible processes in L_p, $1 \leq p \leq \infty$, are a.e. and norm good in L_q, where q is the dual index.

Let $(X, \mathcal{M}, \mu)$ be a probability space and $T : X \to X$ be an ergodic mpt, $F = \{f_n\} \subset L_1$ be a T-superadditive process with additive part $G = \{g_n\}$ and purely subadditive part $H = \{h_n\}$ satisfying (†) or (∗). By Wiener–Wintner theorem, there exists a unique set $N_1 \subset X$ of measure zero such that for all $x \in X \setminus N_1$ and for any complex number $|\lambda| = 1$, the averages $\frac{1}{n}\sum_{k=0}^{n-1} \lambda^k g_k(x)$ converge. Consider the averages $\frac{1}{n}\sum_{i=0}^{n-1} \lambda^i h_i(x)$, where $\{h_i\}$ is the purely subadditive part satisfy (†) and $|\lambda| = 1$. Now,

$$\begin{aligned}\|\sum_{i=0}^{n-1} \lambda^i h_i\|_2^2 &= < \frac{1}{n}\sum_{i=0}^{n-1} \lambda^i h_i \ , \frac{1}{n}\sum_{j=0}^{n-1} \lambda^j h_j >\\ &= \frac{1}{n^2}\sum_{i,j=0}^{n-1} \lambda^{i-j} < h_i, h_j >\\ &= \frac{1}{n^2}\int \sum_{i,j=0}^{n-1} \lambda^{i-j} h_i h_j\\ &= \frac{1}{n^2}\int \sum_{j=0}^{n-1} h_j^2 + \frac{1}{n^2}\int \Big[\sum_{k=1}^{n-1}(\lambda^k + \overline{\lambda^k})(\sum_{j=k}^{n-1} h_j h_{j-k})\Big].\end{aligned}$$

Therefore,

$$|<\frac{1}{n}\sum_{i=0}^{n-1}\lambda^i h_i\ ,\frac{1}{n}\sum_{j=0}^{n-1}\lambda^j h_j>|\leq\frac{2}{n^2}\int\Big[\sum_{k=1}^{n-1}\sum_{j=0}^{n-1}|h_j h_{j-k}|\Big],$$

which, by the condition (†), implies that $\lim_n<\frac{1}{n}\sum_{i=0}^{n-1}\lambda^i h_i\ ,\frac{1}{n}\sum_{j=0}^{n-1}\lambda^j h_j>=0$. By the van der Corput lemma, $\frac{1}{n}\sum_{i=0}^{n-1}\lambda^i f_i(x)$ converges for all $|\lambda|=1$. When the subadditive part satisfy (∗), then, using Koopman's lemma, $\frac{1}{N}\sum_{i=1}^{N}\int h_i d\mu\to 0$ a.e., which, in turn, implies that

$$|\frac{1}{n}\sum_{i=0}^{n-1}\lambda^i h_i(x)|\leq\frac{1}{n}\sum_{i=0}^{n-1}|h_i(x)|\to 0\ \text{ for all } |\lambda|=1.\ \text{ Hence, we proved}$$

THEOREM 3.1. *Let $F=\{f_n\}\subset L_1$ be a bounded T-superadditive process whose subadditive part satisfies (†) or (∗). There exists a set $N\in\Sigma$, with $\mu(N)=0$ such that for $x\in X\backslash N$*

$$\frac{1}{n}\sum_{i=0}^{n-1}\lambda^i f_i(x)\quad \textit{converges for all}\ \ \lambda\in\mathbb{K},$$

where $\mathbb{K}$ is the unit circle in the complex plane.

An important consequence of the Theorem 3.1 is that, for a.e. x, the sequence $\{f_i(x)\}$, where $F=\{f_i\}\subset L_p$ is a bounded superadditive process whose subadditive part satisfy (†) or (∗) has Fourier coefficients, i.e., is a Hartman almost periodic sequence [**12**]. Furthermore, by the a.e. convergence of the averages of such processes [**10**], it is easy to see that $\{f_i(x)\}\in W_1$ for a.e. x, where the sequence classes W_p are defined as $W_p=\{\mathbf{a}=\{a_i\}:\limsup_n\frac{1}{n}\sum_{i=0}^{n-1}|a_i|^p<\infty\}$, $1\leq p<\infty$, and W_∞ is the class of bounded sequences. It is known that, even in the additive case, a Hartman sequence need not be a good sequence for a.e. convergence [**12**]. On the other hand, if $F\subset L_\infty$ is a bounded admissible process, by the observations above and [**2**, Corollary 4.3], for a.e. x, $\{f_i(x)\}$ is a sequence good a.e. for any dynamical system (Ω,Σ',ν,S), where S has Lebesgue spectrum. When the superadditive processes inducing the sequences are T-admissible, the fact that they are good in the norm and a.e convergence is studied in [**5**]. Thus, it is natural to ask whether the sequence of weights induced by the superadditive processes whose subadditive part satisfy (†) or (∗) are good a.e. or in the mean in L_p.

Now we make some observations when the purely subadditive part of the process satisfy (∗). By the celebrated return times theorem of Bourgain, there exists a unique set $N_1\subset X$ of measure zero such that for all $x\in X\setminus N_1$ the sequence $\{g_n(x)\}$ is good a.e. Thus, given a T-superadditive process F, in order to show that it defines a good sequence for a.e. converge all we need to show is that, except on a set of μ-measure zero, the purely subadditive part H also defines a good sequence for a.e. convergence. Since $\frac{1}{n}\sum_{k=0}^{n-1}h_k(x)\to 0$ a.e., let $N_2\subset X$ be the set on which this convergence fails. Hence, $\mu(N_2)=0$. Now, if $(Y,\mathcal{B},\nu,S)$ is a dynamical system and $u\in L_\infty(Y)$, then, since $h_n(x)\geq 0$ a.e.,

$$|\frac{1}{n}\sum_{k=0}^{n-1}h_k(x)u(Sy)|\leq\|u\|_\infty\Big|\frac{1}{n}\sum_{k=0}^{n-1}h_k(x)\Big|\to 0\ \text{a.e.}.$$

Therefore, for all $x \in X \setminus N_2$, $\frac{1}{n}\sum_{k=0}^{n-1} h_k(x)u(Sy) \to 0$ a.e. y. This convergence also holds in norm. Hence, if $F = \{f_k\}$ is a bounded superadditive process with nonnegative subadditive part, then the sequence $\{f_k(x)\}$ is good a.e and in norm for additive processes. It follows from Kingman decomposition, again, that the sequence $\{f_k(x)\}$ is good a.e and in norm for any bounded superadditive process; namely, if $(Y, \mathcal{B}, \nu, S)$ is a mpt and for any bounded $U = \{u_k\}$ is a bounded S-superadditive process, then for any $x \in X \setminus N$,

$$\lim_n \frac{1}{n}\sum_{k=0}^{n-1} f_k(x)u_k(y) \text{ exists a.e. } y \in Y \text{ (in norm).}$$

Extending further, since $\frac{1}{n}\sum_{i=0}^{n-1} h_i \to 0$ a.e., such processes induce linear sequences as defined in [**6**]; hence, they are good weights for multiple recurrence as well.

In the case of processes satisfying (†), however, it appears that their study require different techniques, which is currently an open problem.

References

[1] M. A. Akcoglu and L. Sucheston, *A ratio ergodic theorem for superadditive processes*, Z. Wahrsch. Verw. Gebiete **44** (1978), no. 4, 269–278, DOI 10.1007/BF01013192. MR509202

[2] A. Bellow and V. Losert, *The weighted pointwise ergodic theorem and the individual ergodic theorem along subsequences*, Trans. Amer. Math. Soc. **288** (1985), no. 1, 307–345, DOI 10.2307/2000442. MR773063

[3] Vitaly Bergelson, *The multifarious Poincaré recurrence theorem*, Descriptive set theory and dynamical systems (Marseille-Luminy, 1996), London Math. Soc. Lecture Note Ser., vol. 277, Cambridge Univ. Press, Cambridge, 2000, pp. 31–57. MR1774423

[4] Vitaly Bergelson, Bernard Host, and Bryna Kra, *Multiple recurrence and nilsequences*, Invent. Math. **160** (2005), no. 2, 261–303, DOI 10.1007/s00222-004-0428-6. With an appendix by Imre Ruzsa. MR2138068

[5] Doğan Çömez and Semyon N. Litvinov, *Ergodic averages with generalized weights*, Studia Math. **173** (2006), no. 2, 103–128, DOI 10.4064/sm173-2-1. MR2204785

[6] Tanja Eisner, *Linear sequences and weighted ergodic theorems*, Abstr. Appl. Anal., posted on 2013, Art. ID 815726, 5, DOI 10.1155/2013/815726. MR3055959

[7] Harry Furstenberg, *Ergodic behavior of diagonal measures and a theorem of Szemerédi on arithmetic progressions*, J. Analyse Math. **31** (1977), 204–256, DOI 10.1007/BF02813304. MR498471

[8] Nikos Frantzikinakis and Bryna Kra, *Ergodic averages for independent polynomials and applications*, J. London Math. Soc. (2) **74** (2006), no. 1, 131–142, DOI 10.1112/S0024610706023374. MR2254556

[9] Bernard Host and Bryna Kra, *Nonconventional ergodic averages and nilmanifolds*, Ann. of Math. (2) **161** (2005), no. 1, 397–488, DOI 10.4007/annals.2005.161.397. MR2150389

[10] J. F. C. Kingman, *The ergodic theory of subadditive stochastic processes*, J. Roy. Statist. Soc. Ser. B **30** (1968), 499–510. MR254907

[11] A. Leibman, *Pointwise convergence of ergodic averages for polynomial sequences of translations on a nilmanifold*, Ergodic Theory Dynam. Systems **25** (2005), no. 1, 201–213, DOI 10.1017/S0143385704000215. MR2122919

[12] Michael Lin, James Olsen, and Arkady Tempelman, *On modulated ergodic theorems for Dunford-Schwartz operators*, Proceedings of the Conference on Probability, Ergodic Theory, and Analysis (Evanston, IL, 1997), Illinois J. Math. **43** (1999), no. 3, 542–567. MR1700609

[13] Terence Tao, *Norm convergence of multiple ergodic averages for commuting transformations*, Ergodic Theory Dynam. Systems **28** (2008), no. 2, 657–688, DOI 10.1017/S0143385708000011. MR2408398

[14] Terence Tao, *Norm convergence of multiple ergodic averages for commuting transformations*, Ergodic Theory Dynam. Systems **28** (2008), no. 2, 657–688, DOI 10.1017/S0143385708000011. MR2408398

Department of Mathematics, North Dakota State University, Fargo, North Dakota 58108-6050, USA
Email address: Dogan.Comez@ndsu.edu

Contemporary Mathematics
Volume **797**, 2024
https://doi.org/10.1090/conm/797/15955

Toral band Fatou components for the Weierstrass $\wp$ function

Jane Hawkins and Lorelei Koss

ABSTRACT. We study the existence of unbounded Fatou components for the iterated Weierstrass $\wp$ function and discuss its dependence on lattice shapes. We show that there are open regions in shape space for lattices in $\mathbb{C}$ for which scalings of the lattice (real and complex) yield unbounded toral band Fatou components.

1. Preliminary discussion

An elliptic function is a doubly periodic meromorphic function in $\mathbb{C}$ with a period lattice Λ. The dynamics of iterated elliptic functions fit into the broader landscape of complex dynamics of meromorphic functions, worked on by many, e.g., [**1, 2, 4, 6, 7, 21, 23**]. These studies have led to an interesting array of fractal Julia sets for elliptic functions (see, for example, [**9**]–[**15**] and [**17**]–[**20**]). In this paper we study one of the two basic building blocks of elliptic functions, the Weierstrass $\wp$ function, denoted $\wp$. The other basic function is its derivative $\wp'$, as every elliptic function with period lattice Λ can be written as a rational expression of $\wp_\Lambda$ and $\wp'_\Lambda$. The function $\wp_\Lambda$ is an even elliptic function of order 2.

There are two main points discussed in this paper; neither has been addressed in the literature before. It is well known from the references above that the Julia and Fatou sets depend on the lattice, and not only the lattice shape (such as a square or a rectangle), but also the side length. Here we look at the shape more closely; we give results to show that the more horizontal or vertical looking lattices allow for a wider variety of Julia sets and Fatou components, while the more regular shapes like square and rhombus shapes whose short diagonals are longer than a side of the rhombus, have less variety.

Our discussion adds another variable that seems to affect the dynamics, namely how the lattice is situated relative to the real and imaginary axes. For example, horizontally situated lattices seem to have a wider variety of vertical Fatou components than vertically placed ones (defined below). While lattices give rise to many different fundamental regions or period parallelograms, (see Figure 1), some definitely appear to be more horizontal, others more vertical, and others look more blockish. For elliptic functions, the Julia sets are doubly periodic, so while we do not expect unbounded Fatou components for $\wp_\Lambda$ and they cannot occur for square

2020 *Mathematics Subject Classification.* Primary 37F10, 37F12, 30D05.

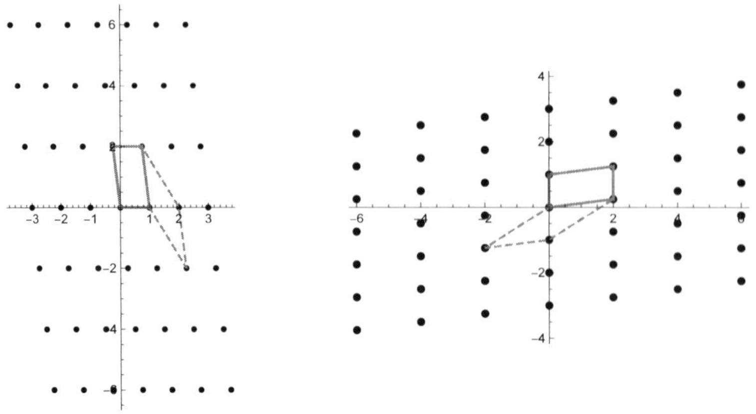

FIGURE 1. The lattice Λ on the left appears to be more vertical than the one on the right, which is $i\Lambda$. This can be seen by drawing a few period parallelograms.

and triangular lattices, they occur more frequently than first thought when changing both the lattice shape and its orientation. These together yield many types of Fatou and Julia sets even when just looking at the single function $\wp$. Moreover, the occurrence of toral bands, once thought to be rare in parameter space, is now established to be very common, occurring in open sets of shapes.

1.1. Background notation and results. Lattices determine double periods for elliptic functions. By $\Lambda = [\lambda_1, \lambda_2]$ we denote the group

$$\Lambda = \{m\lambda_1 + n\lambda_2 \colon m, n \in \mathbb{Z}\} \subset \mathbb{C}. \tag{1.1}$$

Assuming $\lambda_1, \lambda_2 \in \mathbb{C}$ are non-zero and linearly independent over $\mathbb{R}$, we call Λ a *lattice*. A lattice Λ acts on $\mathbb{C}$ by translation, each $\omega \in \Lambda$ inducing the transformation:

$$T_\omega : z \mapsto z + \omega.$$

We denote the coset of $\mathbb{C}/\Lambda$ containing z by $[z]$.

A closed, connected parallelogram Q is a *fundamental parallelogram* for Λ if

(1) for each $z \in \mathbb{C}$, Q contains at least one point in the same Λ-orbit as z;
(2) no two points in the interior of Q are in the same Λ-orbit.

In this paper we focus on parallelograms, but other regions are possible for Q that satisfy (1) and (2). A *real lattice* Λ is one such that $\overline{\Lambda} = \Lambda$, where $\overline{z}$ denotes complex conjugation and $\overline{\Lambda} = \{\overline{\lambda} : \lambda \in \Lambda\}$.

For a general elliptic function over Λ, denoted f_Λ (or f), z_0 is a critical point if $f'_\Lambda(z_0) = 0$ and $f_\Lambda(z_0)$ is a critical value. The *Fatou set* $F(f)$ is the set of points $z \in \hat{\mathbb{C}}$, the Riemann sphere, such that $\{f^k \colon k \in \mathbb{N}\}$ is defined and normal in some neighborhood of z. The *Julia set* $J(f)$ is the complement of the Fatou set in $\hat{\mathbb{C}}$. For each lattice Λ, every elliptic function with period lattice Λ is of Class S [**4**] and does not have wandering domains [**2**] or Baker domains [**23**]. We summarize the types of Fatou components that can occur.

THEOREM 1.1. *If f_Λ is an elliptic function with period lattice Λ, then every component of $F(f_\Lambda)$ is preperiodic, and each forward invariant Fatou component contains one of the following:*

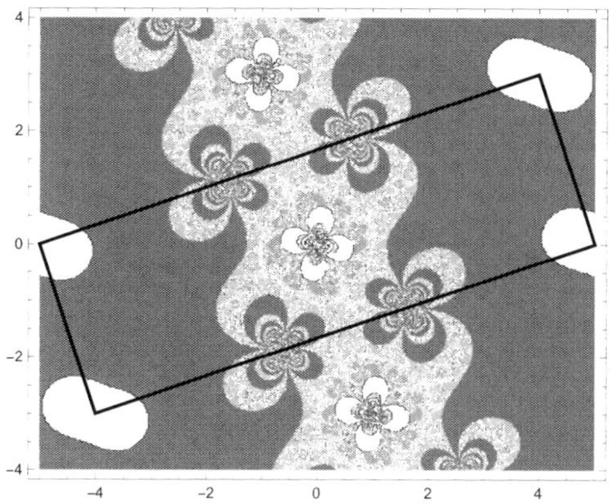

FIGURE 2. This shows a toral band for $\wp_\Xi$ with a rectangular lattice $\Xi = k[1, 3i]$, and $k = 1 + 3i$. $J(\wp_\Xi)$ lies in the darker points in the lacy region.

(1) *a linearizing neighborhood of an attracting periodic point;*
(2) *a Böttcher neighborhood of a superattracting periodic point;*
(3) *an attracting Leau petal for a periodic parabolic point. The periodic point is in* $J(f_\Lambda)$*;*
(4) *a periodic Siegel disk containing an irrationally neutral periodic point.*
(5) *a Herman ring.*

Order 2 elliptic functions do not have Herman rings, so Theorem 1.1(5) does not occur for $\wp_\Lambda$ [**10, 22**]. We define the types of unbounded Fatou components that are possible in this setting.

DEFINITION 1.2. Assume that $f_\Lambda = f$ is an elliptic function over the lattice $\Lambda = [\lambda_1, \lambda_2]$. Then

(1) A Fatou component A_0 of f is a *toral band* if A_0 contains an open subset U which is simply connected in $\mathbb{C}$, but U projects to a topological band around the torus $\mathbb{C}/\Lambda$ containing a homotopically nontrivial curve.
(2) A toral band A_0, is a *double toral band* if it projects to a set in the torus $\mathbb{C}/\Lambda$ that contains closed paths that generate the fundamental group $\pi_1(\mathbb{C}/\Lambda)$.
(3) We say A_0 is a *vertical toral band* if there is a line $L \subset A_0$ where L is parallel to the imaginary axis; and A_0 is a *horizontal toral band* if there is a line $L \subset A_0$ where L is parallel to the real axis.
(4) If A_0 is a toral band that is not double, it is called a *single toral band.*
(5) A toral band is *nonperiodic* if it maps onto a cycle of Fatou components but is not part of the cycle.

REMARK 1.3. Not every toral band for $\wp_\Lambda$ needs to be either horizontal or vertical; a toral band is often parallel to one of the edges of a fundamental parallelogram region. For example, a rectangular lattice scaled by a complex number is still rectangular, but its sides are not parallel to any axis, as shown in Figure 2.

The *Weierstrass elliptic function* is defined for each $z \in \mathbb{C}$, by

$$\wp_\Lambda(z) = \frac{1}{z^2} + \sum_{\lambda \in \Lambda \setminus \{0\}} \left(\frac{1}{(z-\lambda)^2} - \frac{1}{\lambda^2} \right). \tag{1.2}$$

The map $\wp_\Lambda$ is an even elliptic function with poles of order 2 at lattice points. Its derivative $\wp'$ is an odd elliptic function of order 3, also periodic with respect to Λ, and obtained by term by term differentiation of (1.2).

Using the notation from (1.1), we define $c_1 = \lambda_1/2$, $c_2 = \lambda_2/2$, and $c_3 = \lambda_1/2+\lambda_2/2$ to denote the half lattice points in a fundamental parallelogram Q. This is sometimes ambiguous since generators can change; however the critical points of $\wp_\Lambda$ are $[c_j]$ for $j = 1, 2, 3$. The critical values of $\wp_\Lambda$ are denoted by $e_j = \wp_\Lambda(c_j)$ for $j = 1, 2, 3$. They are distinct for each lattice, and they satisfy the relationship

$$e_1 + e_2 + e_3 = 0, \quad e_1e_3 + e_2e_3 + e_1e_2 = \frac{-g_2}{4}, \quad e_1e_2e_3 = \frac{g_3}{4}. \tag{1.3}$$

We define g_2 and g_3 as follows. Given a lattice Λ, define

$$S_n = \sum_{\lambda \in \Lambda, \lambda \neq 0} \lambda^{-n}, \tag{1.4}$$

which converges absolutely for all $n \geq 3$. We then set

$$g_2 = 60\, S_4, \text{ and } g_3 = 140\, S_6,$$

so that $\wp'(z)^2 = 4\wp_\Lambda(z)^3 - g_2\wp_\Lambda(z) - g_3$. Let $\triangle = g_2^3 - 27g_3^2$ denote the discriminant of the cubic polynomial $p(x) = 4x^3 - g_2z - g_3$, which satisfies $\triangle \neq 0$ since the e_js are distinct.

2. Toral bands for $\wp_\Lambda$ real and nonreal lattices

A goal of this paper is to show that there are many toral bands occurring in $F(\wp_\Lambda)$ for an open set of lattices in shape space (see Figure 7). Earlier papers ruled out toral bands for the square and triangular shapes, [**5**, **11**], each of which is represented by a single point in shape space. We also argue, with some supporting results, that because of the shape of the primary region for τ space, if we use constants k that are purely imaginary, then we see more vertical toral bands. The presence of toral bands impacts the geometry of the complementary Julia set, as shown below by comparing the Julia sets in Figures 2, 4–6, 8, and 10. Up to now, most toral bands have come from more elaborate constructions of elliptic functions, though an early paper by the authors [**11**] provided examples of toral bands for $\wp_\Lambda$.

THEOREM 2.1 ([**10**]). *For a lattice $\Lambda = [\lambda_1, \lambda_2]$ the following hold for every even elliptic function $f_\Lambda = f$ with period lattice Λ:*

(1) $(-1)J(f) = J(f)$ *and* $(-1)F(f) = F(f)$.
(2) *There is symmetry of $F(f)$ and $J(f)$ about each half lattice point; i.e., for any $j = 1, 2, 3$,*

$$f(c_j + tc_k) = f(c_j - tc_k), \text{ and } f(c_j + z) = f(c_j - z) \tag{2.1}$$

for $t \in \mathbb{R}, z \in \mathbb{C}$.

COROLLARY 1. *For a real lattice Λ*

(1) *$F(\wp_\Lambda)$ and $J(\wp_\Lambda)$ are symmetric about all half lattice lines parallel to the axes and about all critical points.*

(2) $\overline{F(\wp_\Lambda)} = F(\wp_\Lambda)$ *and* $\overline{J(\wp_\Lambda)} = J(\wp_\Lambda)$.

PROPOSITION 2.1. *If* $\wp_\Lambda$ *has a toral band* $T \subset F(\wp_\Lambda)$, *then* T *is the only toral band on* $\mathbb{C}/\Lambda$ *and* $\wp_\Lambda$ *maps* T *(in* $\mathbb{C}$*) either to itself or into a bounded component of* $F(\wp_\Lambda)$. *In particular, if* $T_1 \neq T_2$ *are disjoint toral bands in* $F(\wp_\Lambda) \subset \mathbb{C}$, *then* $T_2 = T_1 + \lambda$ *for some* $\lambda \in \Lambda$ *and* T_1 *is a single toral band.*

PROOF. The authors showed in [**13**, Theorem 2.15] that a toral band must contain at least 2 residue classes of critical points. There are only 3 residue classes of critical points for $\wp_\Lambda$, so at most one toral band can occur on $\mathbb{C}/\Lambda$. Suppose there are two disjoint toral bands in $\mathbb{C}$, say T_1 and T_2. Then they must project to the same set on $\mathbb{C}/\Lambda$ by the first result. Therefore $T_2 = T_1 + \lambda$ for some $\lambda \in \Lambda$. □

REMARK 2.2.

(1) Following from the definitions, every vertical toral band must intersect $\mathbb{R}$ and every horizontal toral band intersects the imaginary axis. In fact every toral band that is not horizontal intersects $\mathbb{R}$.
(2) We know from double periodicity that toral bands cannot themselves be Siegel disks or Herman rings. However, there are elliptic functions f with a toral band T for which $f(T)$ is a Siegel disk [**13**].
(3) Multiple toral bands are possible, and exist, for higher order elliptic functions [**13**].

We showed in earlier papers [**9**]–[**15**] that real and complex scalings of lattices Λ along the imaginary axis and vertical boundary of the primitive fundamental lattice region shown in Figure 1 give rise to many toral bands for $\wp_\Lambda$. We expand the result to include large regions of lattice shapes in the region in Figure 7. We start with a proposition using a function from [**8**].

DEFINITION 2.3. For a lattice generated by $[1, \tau]$, we define

$$I(\tau) = \frac{g_2^3}{27g_3^2}(\tau).$$

THEOREM 2.4. *$I(\tau)$ is meromorphic in the open upper half plane, with poles at zeroes of $g_3(\tau)$. There is a double pole at $\tau = i$, a zero of $I(\tau)$ occurs at $\tau = \exp(2\pi i/3)$, and $I(\tau) < 0$ for all τ with $|\tau| = 1$ and* $\arg(\tau) \in (2\pi/3, \pi/2)$. *Every negative real number is $I(\tau)$ for some τ on that arc.*

A version of this can be found in [**8**, Thm 3.4].

PROPOSITION 2.2. *Let $\mathfrak{U}$ denote the primitive region for lattice shape space, defined to be all $\tau \in \mathbb{C}$ satisfying*

$$\mathrm{Im}(\tau) > 0, -1/2 \leq \mathrm{Re}(\tau) < 1/2, |\tau| \geq 1, \text{ and } |\tau| > 1 \text{ if } \mathrm{Re}(\tau) > 0.$$

Then the following hold for $\Lambda = [1, \tau]$:

(1) *$g_2(\tau), g_3(\tau)$, and $J(\tau) = I(\tau)/(I(\tau) - 1) = g_2^3/\Delta$ are $\mathbb{Z}$ periodic and analytic on $\mathfrak{U}$; that is, their values are unchanged under the map $T(\tau) = \tau + 1$;*
(2) *As* $\mathrm{Im}(\tau) \to \infty$ *in $\mathfrak{U}$,* $\mathrm{Arg}(\tau) \to \pi/2$, *and* $g_2 \to 120\zeta(4) = 4\pi^4/3 = g_2^*$, *where ζ is the Riemann zeta function given by*

$$\zeta(s) = \sum_{r=1}^{\infty} r^{-s}, s > 1.$$

We also have that $\lim_{\mathrm{Im}(\tau)\to\infty} g_3 = g_3^*$ *exists and the limits satisfy*

$$(g_2^*)^3 = 27(g_3^*)^2.$$

(3) *The doubly periodic function* $\wp_{[1,\tau]}$ *converges to a simply periodic function, with period 1, and the critical values converge to* $e^* = \sqrt{g_2^*/3}$, $-1/2e^*$, *and* $-1/2e^*$.
(4) *If the lattice is generated by* $[k, k\tau]$, *then* $e^*(k\tau) = k^{-2}e^*$.

Remarks about the proof. Most of these are classical results and can be found in sources like [**8**] and [**16**]. Fixing $|\tau| = b$, in order for $\tau \in \mathfrak{U}$, we must have $\arg(\tau) \in (\arccos(-1/2b), \arccos(1/2b))$. Clearly as $\tau \to \infty$ in $\mathfrak{U}$, $\arg(\tau) \to \pi/2$ and $\mathrm{Re}(\tau)$ remains bounded; therefore $\mathrm{Im}(\tau) \to \infty$ and the lattice approaches a degenerate one-dimensional lattice generated by a single generator [1]. This is because the lattice points along the integers remain unchanged while all other lattice points "disappear" to ∞. Using [**8**, Theorem 1.2], it follows that the sums S_n defined in (1.4), converge to $2\,\zeta(n)$ for even positive integers n, and recall that $g_2 = 60S_4$ and $g_3 = 140S_6$. □

Algorithm 1. Obtaining vertical toral bands for $\wp_\Lambda$.

This method yields toral bands that are forward invariant or periodic.

(1) We first choose a lattice shape $\Lambda = [1, \tau]$, with $\tau \in \mathfrak{U}$. Labeling the critical values $\varepsilon_1 = \wp_\Lambda(1/2)$, $\varepsilon_2 = \wp_\Lambda(\tau/2)$, and $\varepsilon_3 = -\varepsilon_1 - \varepsilon_2$, by applying Proposition 2.2(2) and 2.2(3) we choose τ large enough so that ε_1 is close to its limiting value, $(2/3)\pi^2 \approx 6.5797$ and ε_2 and ε_3 are close to each other. (Note: $|\tau| = 2$ is likely large enough for this, as shown in Figure 6.)
(2) Multiplying Λ by i, we next replace the lattice Λ by $\Xi = [i, \tau i]$, whose shortest generator lies on the imaginary axis. We (re)label the critical points so that $c_2 = i/2, c_1 = i\tau/2$, and $c_3 = c_1 + c_2$. Then the critical values are: $e_1 = \wp_\Xi(c_1) = -\varepsilon_2$, $e_2 = \wp_\Xi(c_2) = -\varepsilon_1$, and $e_3 = -\varepsilon_3$; we set Ξ as our base lattice. If τ is purely imaginary, then e_1 and e_3 are positive and $e_2 < 0$; otherwise, $\tau \in \mathfrak{U}$ with $|\tau|$ large implies that $\mathrm{Im}(e_j)$ is very small for all j.
(3) We next scale Ξ by a real number κ in order to have attracting fixed points, so we use a variation of a result from [**11**] to choose a range for it. A good value can be pinpointed as follows: we have $e_1 \approx -e_2/2$, and e_2 is not far from $-e^* \approx -6.5797$, so e_3 is close to e_1. The values $e_j(\tau)$ are, for general $\tau \in \mathfrak{U}$, complex numbers with imaginary part that shrinks to 0 as $|\tau|$ increases inside $\mathfrak{U}$ by Proposition 2.2. For a fixed τ, the exact value $k = \sqrt[3]{2e_1/i\tau}$, not necessarily real, yields a superattracting fixed point for the resulting $\wp_{k\Xi}$ whose basin contains $k^{-2}e_1$ which equals the critical point $c_1^{(k)} = ik\tau/2$. For nearby real κ, the fixed point is near $c_1^{(\kappa)}$, and its immediate basin of attraction contains both $\kappa^{-2}e_1$ and $\kappa^{-2}e_3$, which are now closer to each other. Therefore a pre-image of $\kappa^{-2}e_3$ is in the basin as well, so we have a fixed toral band containing two critical points and two critical values. If τ is close to $2i$, any value $\kappa \in (1.5, 1.7)$ works well since $\kappa_0 = \sqrt[3]{2e_1/i\tau} \approx 1.5$.
(4) The resulting lattice, $\kappa\Xi = i\kappa[1, \tau]$, now has a vertical toral band since the images of the critical points $c_1^{(\kappa)}$ and $c_3^{(\kappa)}$ are in the same Fatou component.

The critical points $c_1^{(\kappa)}$ and $c_3^{(\kappa)}$ have the same real part, namely $\text{Re}(i\tau)/2$, so the toral band is parallel to the imaginary axis.

(5) As κ moves around in $\mathfrak{U}$ period doubling bifurcations occur while the toral band remains. The bifurcation values depend on τ.

Algorithm 2. Obtaining horizontal toral bands for $\wp_{k\Lambda}$.

This method yields toral bands that are nonperiodic.

(1) We start with a lattice $\Lambda = [1, \tau]$, choosing $\tau \in \mathfrak{U}$. We label $e_1 = \wp_\Lambda(1/2)$, and $e_2 = \wp_\Lambda(\tau/2)$ and $e_3 = -e_1 - e_2$ and choose τ large enough so that e_1 is close to its limiting value, $(2/3)\pi^2 \approx 6.5797$ as above. (Note: $|\tau| = 3$ seems to be big enough for this, and could be smaller.) We use Λ as our base lattice.

(2) We next scale to ensure attracting fixed points (and higher period attracting orbits), so we use a result from **[11]** to choose $k \in \mathbb{R}$ near $\kappa = \sqrt[3]{2e_1} \approx 2.36$; the value κ yields a superattracting fixed point for the resulting $\wp_{\kappa\Lambda}$. Nearby ks yield a fixed point whose attracting basin contains $k^{-2}e_1$ and kc_1, but the immediate basin of attraction needs to be a bit bigger so that one of its preimages can contain both $k^{-2}e_2$ and $k^{-2}e_3$, which are now a lot closer to each other. Often a value near $k = 2.4$ works well.

(3) The resulting lattice, $k\Lambda = k[1, \tau]$, now has a horizontal toral band containing representatives from the critical point classes $[kc_2]$ and $[kc_3]$, since the images of the critical points kc_2 and kc_3 are in the same Fatou component. The critical points kc_2 and kc_3 have the same imaginary part, namely $\text{Im}(kc_2) = \text{Im}(kc_3) = \text{Im}(k\tau)/2$, so the toral band is parallel to the real axis and therefore horizontal.

(4) We note that the number of iterations before the points $\wp_{k\Lambda}^j(kc_2)$ and $\wp_{k\Lambda}^j(kc_3)$ land in the immediate attracting basin of the fixed point can vary.

(5) As k increases (very slowly in $\mathbb{R}$), period doubling bifurcations occur while the toral band remains.

If $|\tau|$ is large enough (bigger than 2), most of the k spaces look very similar to the space in Figure 3, where the white region shows the values of k for which $\wp_{k\Lambda}$ has a toral band. We have just outlined the proof of the following.

THEOREM 2.5. *There exists an $\alpha > 1$ such that for any $|\tau| > \alpha$, the lattice $\Lambda = [1, \tau]$ can be scaled by real or imaginary complex numbers so that $\wp_{k\Lambda}$ has either a vertical or a horizontal toral band.*

There are reductions possible in every k space, and a typical k-space is shown in Figure 3, with 6-fold symmetry apparent; results on this appear in **[11]**.

REMARK 2.6. We summarize the properties of the toral bands that the algorithms produce for the Weierstrass elliptic $\wp$ function. Suppose a lattice $\Lambda = k[1, \tau]$, $\tau \in \mathfrak{U}$, with k chosen using one of the algorithms, has a toral band in $F(\wp_{k\Lambda})$.

(1) The discriminant function $\triangle = g_2^3 - 27g_3^2 \neq 0$ is small, which is equivalent to saying that the function $I(\tau) = \dfrac{g_2^3}{27g_3^2}(\tau)$, is very close to 1. However the rates at which $I(\tau)$ and $\triangle(\tau)$ approach their limits as τ increases varies, as discussed in (5); $I(\tau)$ approaches 1 earlier (than $\triangle$ shrinks) and gives toral bands.

FIGURE 3. Starting with the lattice $\Lambda = [1, 3i]$, we show k-space, where points that are white have multiple critical points attracted to a periodic orbit. This is used to study the properties of $\wp_{k\Lambda}$ for each complex number k shown, as in Figure 2.

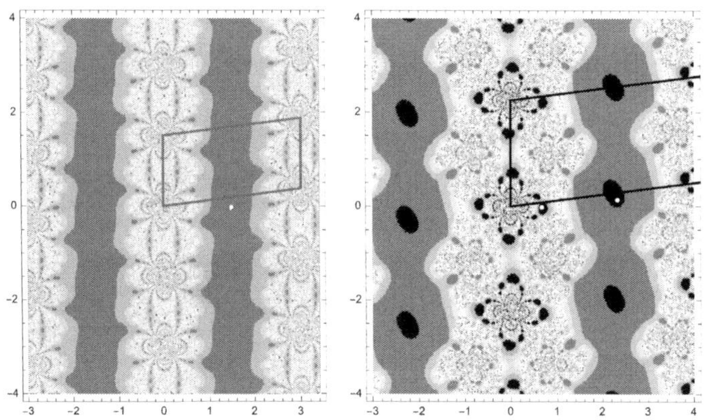

FIGURE 4. Starting with the lattice $\Lambda = [1, -.25 + 2i]$, we scale it by $k = 1.5i$ on the left and $k = 2.25i$ on the right. On the left an attracting fixed point is marked, and on the right, an attracting period 2 orbit, both are white.

(2) If k is either purely imaginary or real, then the toral band is vertical or horizontal, respectively.
(3) Since the existence of a toral band is stable when an attracting orbit exists for $\wp_{k_o\Lambda}$, then there exists a neighborhood of k_o such that for k in the neighborhood, the toral bands will persist, but will no longer be vertical or horizontal. Instead, $F(\wp_{k\Lambda})$ will contain lines parallel to a side of the resulting period parallelogram for $k\Lambda$.
(4) The labeling convention described above uses $c_1 = 1/2$ or $i\tau/2$ to denote the critical point closest to the positive real axis, $c_2 = \tau$ or $i/2$ for the

FIGURE 5. Starting with the rectangular lattice $\Lambda = [1, 2.52i]$, we scale by $k = 2.47$ and see that horizontal toral bands appear. They also have a period doubling cascade as k increases, as a small Mandelbrot set appears in k-space for this lattice.

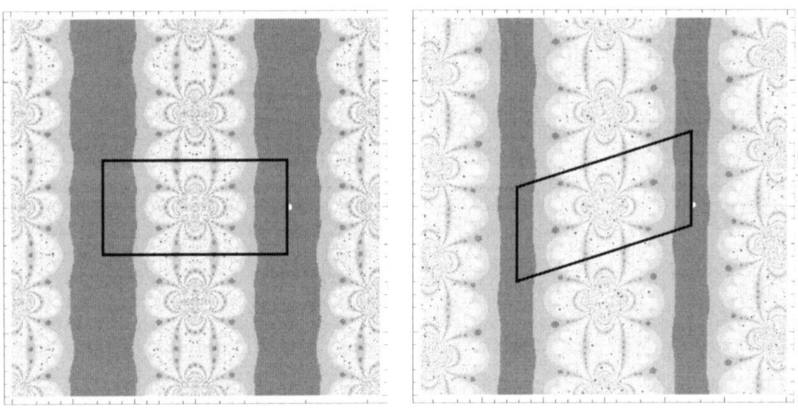

FIGURE 6. The rectangular lattice $\Lambda = [1, \tau]$, with $\tau = 2i$ is on the left and $\tau' = \kappa\tau$, with $\kappa \approx 0.955 - 0.296i$, is a small rotation of τ by $\exp(.3i)$ on the right. We use the same scaling factor (1.5) and see that vertical toral bands persist as described in Algorithm 1.

critical point closest to the positive imaginary axis, and $c_3 = c_1 + c_2$. Then for the scaled lattice $[k, k\tau]$, the toral bands constructed above always contain either kc_1 and kc_3, or kc_2 and kc_3. In the first instance the toral band is vertical and in the second instance it is horizontal. The main point is that for the scaled critical values, $e_3^{(k)}$ is extremely close to either $e_1^{(k)}$ or $e_2^{(k)}$ in each case.

(5) In the case where the critical values $e_1^{(k)}$ and $e_3^{(k)}$ lie close to each other, we set $e_1^{(k)} = \eta$ for convenience of notation to analyze Algorithm 1. Using equation (1.3), the critical values satisfy the equations $e_2^{(k)} = -2\eta - j$ and $e_3^{(k)} = \eta + j$, for some complex number j with very small modulus. Applying equation (1.3) again, we have $g_2 = 4(3\eta^2 + 3\eta j + j^2)$ and $g_3 = -4(2\eta^3 + 3\eta^2 j + \eta j^2)$. The discriminant gives

$$\begin{aligned}\triangle &= g_2^3 - 27g_3^2 \\ &= 1296\,\eta^4 j^2 + 2592\,\eta^3 j^3 + 1872\,\eta^2 j^4 + 576\,\eta j^5 + 64 j^6,\end{aligned}$$

which is not very close to 0 when η is a value greater than 3; even if $j = .1$, it is large. The term $1296\,\eta^4 j^2$ dominates, and gives a value above 1000 when $j = .1$. However, the function

$$I(\tau) = \frac{g_2^3}{27g_3^2}(\tau),$$

is very close to 1 when η is between 3 and 4 and $j \leq .5$ and this gives toral bands.

In the case where the critical values $e_2^{(k)}$ and $e_3^{(k)}$ lie close to each other (Algorithm 2), using the same notation, the critical values satisfy $e_2^{(k)} = -\eta/2 + j$ and $e_3^{(k)} = -\eta/2 - j$, for some complex number j with very small modulus. Equation (1.3) gives $g_2 = 3\eta^2 + 4j^2$ and $g_3 = \eta^3 - 4\eta j^2$. The discriminant $\triangle = g_2^3 - 27g_3^2 = 324\eta^4 j^2 - 288\eta^2 j^4 + 64 j^6$ is increasing in η for fixed j. However, the function

$$\begin{aligned}I(\tau) &= \frac{g_2^3}{27g_3^2}(\tau) \\ &= \frac{(3\eta^2 + 4j^2)^3}{27(\eta^3 - 4\eta j^2)^2},\end{aligned}$$

is very close to 1 when j is reasonably small, again resulting in toral bands.

3. Failure of existence of toral bands for $\wp_\Lambda$ for certain lattice shapes

It is now well-known that every lattice shape admits scalars for which some or all of the critical points land on poles [**12**]. In this section we give arguments that toral bands are rare or possibly even nonexistent in the setting where $\Lambda = k[1, e^{i\theta}]$, with $\theta \in [\pi/2, 2\pi/3] = \pi[.5, .\overline{6}]$, and $k \in \mathbb{C} \setminus \{0\}$. This curve of lattice shapes makes up the lower boundary of the region in Figure 7. The endpoints of the arc correspond to square and triangular lattices, and it has already been shown, in [**5, 10–12, 20**] that $F(\wp_\Lambda)$ for Λ square or triangular contains no toral bands. This is because the critical values are maximally spread out and no amount of scaling will bring two of them into the same Fatou component. In this section we address

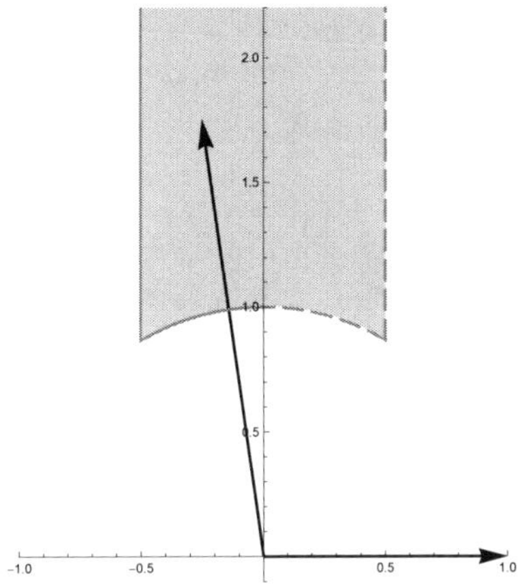

FIGURE 7. *The primitive region* $\mathfrak{U}$ *for lattices.* When $|\tau| = 1$, only the lattices with generators $\Lambda = [1, e^{i\theta}]$, with $\theta \in [\pi/2, 2\pi/3]$ are needed. The endpoints of the interval give square ($\theta = \pi/2$) and triangular ($\theta = 2\pi/3$) lattices. The dashed curves are not in the region.

the algorithms given in the previous section, but apply them to lattices when $|\tau|$ is 1.

3.1. Medium rhombic lattices. A rhombic lattice is one where there is a period parallelogram with all four sides of the same length. A medium rhombic lattice has the additional property that the length of the shortest diagonal is longer than one of its sides. Some notable cases occur when $\tau = i$, and the lattice is square so both diagonals have the same length (longer than the side of the square), and the triangular lattice, where they are equal since the lattice is comprised of 2 equilateral triangles. The other lattice shapes in this category are made by deforming a square lattice to a triangular one by increasing one angle. The next result follows from a variety of classical results given in Duval [**8**], for example.

If Λ is a real lattice and is also medium rhombic, then it can be written as $\Lambda = [a + ib, a - ib]$, $a, b \in (0, \infty)$. If $a < b$, then we say Λ is *vertical*, and otherwise it is *horizontal*, as in the first case the longer diagonal is parallel to the imaginary axis, and in the other the longer diagonal lies along the real axis.

PROPOSITION 3.1. *If* $|\tau| = 1$, *and* $\tau = \exp(i\theta)$, *with* $\theta \in [\pi/2, 2\pi/3]$, *then the following hold for* $\Lambda = [1, \tau]$:

(1) Λ *is a medium rhombic lattice .*
(2) Λ *is similar to a real lattice* Λ_h, *which is a horizontal medium rhombic lattice.*
(3) $\Lambda_h(g_2, g_3)$ *has* $g_2 < 0$ *and* $g_3 < 0$; *if* $\theta \in (\pi/2, 2\pi/3)$ *then* $\Lambda_h = k[1, e^{i\theta}] = [a + ib, a - ib]$, *where* $k = ie^{-i\theta/2}$ *and* $a = \sin(\theta/2)$, $b = \cos(\theta/2)$.
(4) Λ *is also similar to a real lattice* Λ_v, *which is a vertical medium rhombic lattice.*
(5) $\Lambda_v(g_2, g_3)$ *has* $g_2 < 0$ *and* $g_3 > 0$; *if* $\theta \in (\pi/2, 2\pi/3)$ *then* $\Lambda_v = [b + ia, b - ia]$, *where* a *and* b *are as above.*

(6) *For either* Λ_h *or* Λ_v, *if* z *lies on a vertical line passing through any real lattice point or any real half-lattice point, then* $\wp_\Lambda(z) \in (-\infty, e_3]$.

(7) *Defining* $I(\tau) = \dfrac{g_2^3}{27g_3^2}(\tau)$, $I(\tau) < 0$ *for all* τ *with* $|\tau| = 1$, $\arg(\tau) \in (2\pi/3, \pi/2)$. *Also every negative real number is* $I(\tau)$ *for some* τ *in that arc.*

Conversely, if a real lattice $\Lambda = \Lambda(g_2, g_3)$ *has* $g_2 < 0$ *and* $g_3 < 0$, *then* Λ *is similar to a lattice with* $\Lambda = [1, \tau]$ *with* $|\tau| = 1$.

PROOF. Most of these results are found in Duval [**8**]. The proof (3) and (4) follows from the homogeneity equations. □

The dynamics of $\wp_\Lambda$ change as a medium rhombic lattice Λ rotates and stretches by scalars, but some invariants are unaffected. In particular, the next result shows that scaling a medium rhombic lattice leaves the rather large angles between the e_js the same.

PROPOSITION 3.2. *Let* $|\tau| = 1$, *and* $\tau = \exp(i\theta)$, *with* $\theta \in [\pi/2, 2\pi/3]$, *so that* $\Lambda = [1, \tau]$ *is medium rhombic. Then using* $k = \exp(-i\theta/2)$, *we have the following:*

- $k\Lambda$ *is a real lattice so for all* $z \in \mathbb{C}$, $\overline{\wp_{k\Lambda}(z)} = \wp_{k\Lambda}(\overline{z})$.
- *Denoting* $c_1 = 1/2$, $c_2 = \tau/2$ *and* $c_3 = c_1 + c_2$ *as critical points of* $\wp_\Lambda$, *we set* r_1, r_2, *and* r_3 *as the corresponding critical points for* $k\Lambda$; *we have that* $r_3 \in \mathbb{R}$.
- *Denoting* $e_1 = \wp_\Lambda(1/2)$, $e_2 = \wp_\Lambda(\tau/2)$ *and* $e_3 = \wp_\Lambda(c_3)$ *as the three critical values of* $\wp_\Lambda$, *we set* s_1, s_2, *and* s_3 *as the corresponding critical values for* $k\Lambda$; *we have that* $s_3 \in \mathbb{R}$, *and* $s_2 = \overline{s_1}$;
- $e_1 = k^2 s_1$, $e_2 = k^2 s_2$, *so* $\arg(e_1/e_2) = \arg(s_1/s_2)$; *also* $e_2 = \exp(2\theta)\overline{e_1}$.

PROOF. The statements follow from the homogeneity equations and the elementary observation that $\overline{k} = k^{-1}$. Under our hypotheses we have these identities: $\overline{k\Lambda} = k\Lambda$, both being the same real lattice, so $\overline{\wp_{k\Lambda}(kz)} = \overline{k^2\wp_\Lambda(z)}$. We note that $k\Lambda = \overline{k\Lambda}$ implies that $\Lambda = k^{-2}\overline{\Lambda}$, so we use:

$$s_2 = \overline{s_1} = \overline{(k^{-2})e_1} = k^2\overline{e_1}.$$

Also, $s_2 = k^{-2}e_2$ so the result follows. □

<u>Failure of the algorithm when Λ is medium rhombic lattice.</u> The first step of Algorithms 1 and 2 fails when Λ is medium rhombic. Step 1 instructs us to move through the region $\mathfrak{U}$ until $I(\tau)$ is quite close to 1. This is the starting point to look for attracting dynamics. However, when $|\tau| = 1$ in $\mathfrak{U}$, as we move counterclockwise with $\tau = \exp(i\theta)$, from $\theta = \pi/2$ to $\theta = 3\pi/2$, $I(\tau)$ ranges from 0 to $-\infty$, hitting all the negative real numbers along the way.

We can use the result from [**11**] to obtain infinitely many values of k for which the resulting lattice $k\Lambda$ yields a superattracting or attracting fixed point and attracting orbits of higher period, but this is not enough. Both the critical points and the critical values are too separated to allow for more than one to appear in a single Fatou component.

We conjecture that these lattices, occurring at the bottom of the primary region of lattice shapes cannot have toral bands, and we give some partial results in that direction. We start by using the result from [**11**, Lemma 7.2] to show that given a medium rhombic lattice we can scale it by $k = \sqrt[3]{e_j/c_j}$, $j = 1, 2$ or 3, choosing

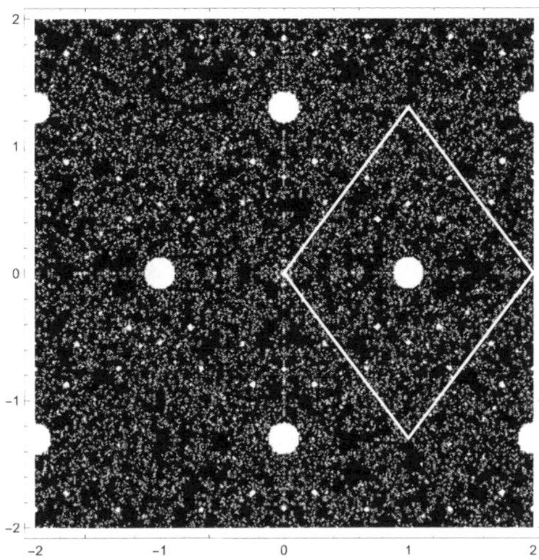

FIGURE 8. $J(\wp_\Xi)$ is quite thick (in black), with no toral bands for a superattracting fixed point at ≈ 1, for $\Xi = k[1, \exp(7\pi i/12)]$, a medium rhombic lattice.

the cube root so that we have a real lattice with a superattracting fixed point. For large regions in τ space, this lands us near or among parameters giving toral bands, but the next example shows this does not happen in this setting.

EXAMPLE 3.1. We show a Julia and Fatou set for a medium rhombic lattice of the form $\Xi = k[1, \exp(7\pi i/12)] = k[1, \tau]$. We note that τ is halfway along the lower boundary of the region shown in Figure 7. Using [**11**, Lemma 7.2], we obtain that for $(g_2, g_3) \approx (-19.8033, 23.7727)$ we have not changed τ but the resulting map $\wp_\Xi$ has a real superattracting fixed point; $J(\wp_\Xi)$ is shown in Figure 8. In fact the results in [**12**] show that there are infinitely many values of k with this property, and infinitely many ℓs such that $\Xi = \ell[1, \tau]$ satisfies $J(\wp_\Xi) = \hat{\mathbb{C}}$. For these, the techniques given in [**12**] and some discussion below show that there are no toral bands.

We finish with a result in the direction of showing that we cannot have toral bands for medium rhombic lattices. While we state the lemma for one scalar, there are infinitely many that work.

PROPOSITION 3.3. *If $\Lambda = [1, \tau]$ is medium rhombic and has critical points for $\wp_\Lambda$ labelled by $c_1 = 1/2, c_2 = \tau/2$, and $c_3 = c_1 + c_2$, then scaling by: $k = \sqrt[3]{e_3/(2c_3)}$ gives a lattice with $e_3^{(k)} = \wp_{k\Lambda}(kc_3)$ a pole. Moreover, $\arg(k) = -\arg(\tau)/2$, so the lattice $k\Lambda$ is real, $\wp^n_{k\Lambda}(kc_j) \in \mathbb{R}$ for $n \geq 2$ and there are no toral bands.*

PROOF. Results very similar to this have been proved in [**10**] and [**20**] and the proof follows from the homogeneity equations. Since we know that $e_3^{(k)} = 2(kc_3)$, a pole, it follows that the real parts of $e_3^{(k)}$ and $e_3^{(k)}$ are critical points $(-kc_3)$. Therefore $\wp_{k\Lambda}(e_1^{(k)}) = \wp_{k\Lambda}(e_2^{(k)}) \in \mathbb{R}$, using Proposition 3.1(6); hence there can be no toral bands. This is because there are no nonrepelling real cycles and the critical orbits remain on $\mathbb{R}$. □

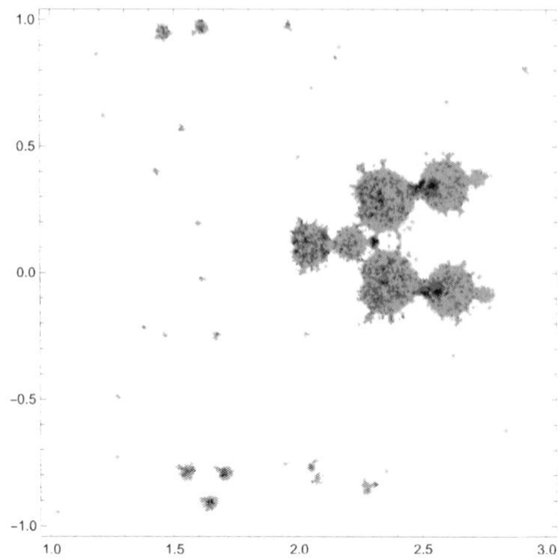

FIGURE 9. We scale the medium rhombic lattice $\Lambda = [1, \tau]$, $\tau \approx \exp(.633\pi i)$, by $k = a + ib$. The values of k for which $\wp_\Lambda$ has an attracting orbit are colored gray. The darker the shade of gray, the more critical values are attracted to periodic orbits.

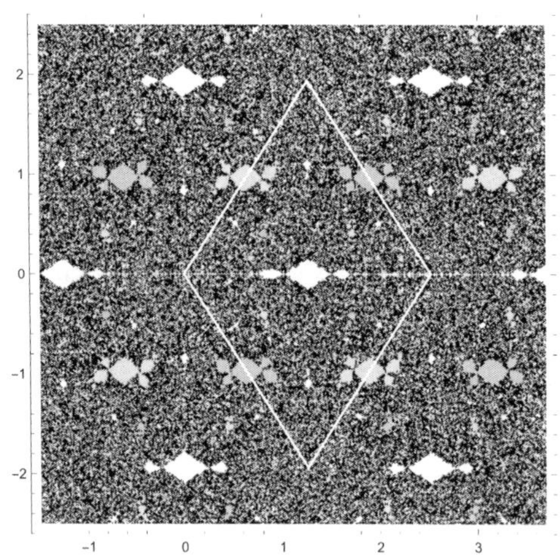

FIGURE 10. We show $J(\wp_\Lambda)$, for a medium rhombic lattice $\Lambda = k[1, \tau]$, $\tau \approx \exp(.633\pi i)$. Points in the Julia set are the darkest. $F(\wp_\Lambda)$ consists of 3 disjoint basins of attraction: one for an attracting real period 4 orbit, and two others for disjoint period 3 attracting orbits, complex conjugates of each other, shown in varying shades of light gray. A fundamental period parallelogram is white.

The proof details use that the scaled lattice $k\Lambda$ is vertical rhombic, and therefore yield the result that $J(\wp_{k\Lambda}) = \hat{\mathbb{C}}$, but the techniques appear in [**10**] and [**20**] so we do not give them here. The most general statement is that scaling the lattice in

Proposition 3.3 by any k of the form $k = \sqrt[3]{e_j/(2mc_j)}$, $j = 1, 2, 3$, $m \in \mathbb{N}$ gives a pole for the critical value e_j.

There is a parameter space for each medium rhombic lattice, since there is a k-space for every value of τ. Small Mandelbrot sets appear, suggesting the existence of attracting periodic orbits and bifurcations, which can be proved in many settings. What seems to be the difference is that the spread of the critical values and critical points prevents toral bands from occurring. In Figure 9 we show a typical parameter space for a medium rhombic lattice of the form $\Lambda = [1, \tau]$, with $\tau \approx \exp(.633\pi i)$. We see that the attracting orbits can act independently for the three different critical values. In Figure 10 we see, for the map $\wp(z)$ with $(g_2, g_3) \approx (-2.512, 4.483)$ that three distinct attracting orbits can occur, one each attracting a critical value. The large discriminants for the lattices and invariants are what causes the spread of the critical values, and this is what changes as we move up along the imaginary axis in the region $\mathfrak{U}$.

References

[1] I. N. Baker, J. Kotus, and Lü Yinian, *Iterates of meromorphic functions. I*, Ergodic Theory Dynam. Systems **11** (1991), no. 2, 241–248, DOI 10.1017/S014338570000612X. MR1116639

[2] I. N. Baker, J. Kotus, and Lü Yinian, *Iterates of meromorphic functions. IV. Critically finite functions*, Results Math. **22** (1992), no. 3-4, 651–656, DOI 10.1007/BF03323112. MR1189754

[3] Joseph Bak and Donald J. Newman, *Complex analysis*, Undergraduate Texts in Mathematics, Springer-Verlag, New York-Berlin, 1982. MR671250

[4] Walter Bergweiler, *Iteration of meromorphic functions*, Bull. Amer. Math. Soc. (N.S.) **29** (1993), no. 2, 151–188, DOI 10.1090/S0273-0979-1993-00432-4. MR1216719

[5] Joshua J. Clemons, *Connectivity of Julia sets for Weierstrass elliptic functions on square lattices*, Proc. Amer. Math. Soc. **140** (2012), no. 6, 1963–1972, DOI 10.1090/S0002-9939-2011-11079-7. MR2888184

[6] Robert L. Devaney and Linda Keen, *Dynamics of tangent*, Dynamical systems (College Park, MD, 1986), Lecture Notes in Math., vol. 1342, Springer, Berlin, 1988, pp. 105–111, DOI 10.1007/BFb0082826. MR970550

[7] Robert L. Devaney and Linda Keen, *Dynamics of meromorphic maps: maps with polynomial Schwarzian derivative*, Ann. Sci. École Norm. Sup. (4) **22** (1989), no. 1, 55–79. MR985854

[8] Patrick Du Val, *Elliptic functions and elliptic curves*, London Mathematical Society Lecture Note Series, No. 9, Cambridge University Press, London-New York, 1973. MR0379512

[9] Jane Hawkins, *Stability of Cantor Julia sets in the space of iterated elliptic functions*, Dynamical systems and random processes, Contemp. Math., vol. 736, Amer. Math. Soc., [Providence], RI, [2019] ©2019, pp. 69–95, DOI 10.1090/conm/736/14837. MR4011907

[10] Jane Hawkins and Lorelei Koss, *Ergodic properties and Julia sets of Weierstrass elliptic functions*, Monatsh. Math. **137** (2002), no. 4, 273–300, DOI 10.1007/s00605-002-0504-1. MR1947915

[11] Jane Hawkins and Lorelei Koss, *Parametrized dynamics of the Weierstrass elliptic function*, Conform. Geom. Dyn. **8** (2004), 1–35, DOI 10.1090/S1088-4173-04-00103-1. MR2060376

[12] Jane Hawkins and Lorelei Koss, *Connectivity properties of Julia sets of Weierstrass elliptic functions*, Topology Appl. **152** (2005), no. 1-2, 107–137, DOI 10.1016/j.topol.2004.08.018. MR2160809

[13] Jane Hawkins and Lorelei Koss, *Single and double toral band Fatou components in meromorphic dynamics*, Conform. Geom. Dyn. **27** (2023), 118–144, DOI 10.1090/ecgd/380. MR4550316

[14] Jane Hawkins and Mónica Moreno Rocha, *Dynamics and Julia set of iterated elliptic functions*, New York J. Math. **24** (2018), 947–979. MR3874958

[15] Jane Hawkins and Mónica Moreno Rocha, *Unbounded Fatou components for elliptic functions over square lattices*, Houston J. Math. **47** (2021), no. 2, 353–374. MR4447977

[16] G. Jones and D. Singerman, *Complex Functions: An Algebraic and Geometric Viewpoint*, Cambridge Univ. Press, (1997).

[17] L. Koss, *A fundamental dichotomy for Julia sets of a family of elliptic functions*, Proc. Amer. Math. Soc. **137** (2009), no. 11, 3927–3938, DOI 10.1090/S0002-9939-09-09967-5. MR2529903

[18] L. Koss, *Cantor Julia sets in a family of even elliptic functions*, J. Difference Equ. Appl. **16** (2010), no. 5-6, 675–688, DOI 10.1080/10236190903217796. MR2675599

[19] Lorelei Koss, *Elliptic functions with disconnected Julia sets*, Internat. J. Bifur. Chaos Appl. Sci. Engrg. **26** (2016), no. 6, 1650095, 10, DOI 10.1142/S0218127416500954. MR3516961

[20] Lorelei Koss and Katie Roy, *Dynamics of vertical real rhombic Weierstrass elliptic functions*, Involve **10** (2017), no. 3, 361–378, DOI 10.2140/involve.2017.10.361. MR3583871

[21] Janina Kotus and Mariusz Urbański, *Hausdorff dimension and Hausdorff measures of Julia sets of elliptic functions*, Bull. London Math. Soc. **35** (2003), no. 2, 269–275, DOI 10.1112/S0024609302001686. MR1952406

[22] Mónica Moreno Rocha, *Herman rings of elliptic functions*, Arnold Math. J. **6** (2020), no. 3-4, 551–570, DOI 10.1007/s40598-020-00167-3. MR4181724

[23] P. J. Rippon and G. M. Stallard, *Iteration of a class of hyperbolic meromorphic functions*, Proc. Amer. Math. Soc. **127** (1999), no. 11, 3251–3258, DOI 10.1090/S0002-9939-99-04942-4. MR1610785

Department of Mathematics, University of North Carolina at Chapel Hill, CB #3250, Chapel Hill, North Carolina 27599-3250

Email address: jmh@math.unc.edu

Department of Mathematics and Computer Science, P.O Box 1773, Carlisle, Pennsylvania 17013

Email address: koss@dickinson.edu

Current address: Department of Mathematics, Colby College, 5845 Mayflower Hill, Waterville, Maine 04901

Contemporary Mathematics
Volume **797**, 2024
https://doi.org/10.1090/conm/797/15956

Fractal functions using weak contraction theory in some function spaces and generalized α-fractal functions

Manuj Verma and Amit Priyadarshi

ABSTRACT. In this paper, we demonstrate the existence of a new class of fractal functions in standard function spaces such as the Lebesgue function space and the bounded function space. We also provide the existence of the general version of the so-called 'α-fractal function'. We emphasize the fact that our construction provides some new classes of fractal functions that are not obtained by the previously known classes of α-fractal functions. We also discuss the stability and the fractal dimension of the newly constructed fractal functions. Lastly, we plot approximate graphs of some newly constructed fractal functions.

1. Introduction

The iterated function system (IFS) is one of the most important tools in fractal geometry for generating fractals. Generally, fractals are irregular in nature and have complex structures. Due to these properties, they are very close to nature. There are several examples of fractals in nature, such as trees, peaks of mountains, coastlines, clouds, and many more. Using the idea of the IFS theory and the Banach contraction principle, Barnsley [**3**] introduced the concept of fractal interpolation functions (FIFs). An FIF is a function that interpolates the given data set, and the graph of this function is an attractor for some IFS. In general, the FIFs are nowhere differentiable functions, whereas the classical interpolation functions are piecewise smooth. This makes the FIFs better for approximation than the classical interpolation functions. The FIFs are widely applicable in approximation theory, image compression, modeling of natural objects, etc. (see [**1**, **8**, **10**]). Some authors have attempted to figure out the fractal dimensions of the graphs of the FIFs (see [**3**, **5**, **6**, **13**, **15**, **24**]). For more details on fractals, fractal dimensions, and the IFS theory, we refer the readers to see [**4**, **9**].

In 1991, Massopust [**15**] generalized the theory of Barnsley [**3**] in higher dimensions and defined the vector-valued FIFs. He also determined the box dimension of its graph. After that, Wang and Yu [**25**] constructed FIFs with variable parameters and discussed their stability, smoothness and sensitivity. In 2015, Barnsley and Massopust [**5**] defined bilinear FIFs and computed the box dimension of their graphs. Motivated by the work of Barnsley [**3**], Massopust [**14**, **16**] constructed

2020 *Mathematics Subject Classification.* Primary 28A80; Secondary 33C47, 41A30.
Key words and phrases. Iterated function systems, fractal interpolation functions, Hausdorff dimension, box dimension, Rakotch contraction.

bivariate FIFs and non-stationary FIFs. In 2005, Navascués [**18**] defined the parametric form of FIFs, namely 'α-fractal functions'. After that, Jha and Verma [**13**] determined some dimensional results for the α-fractal functions. Inspired by the work of Massopust [**16**], Navascués and Verma [**19**] defined the nonstationary version of α-fractal surfaces and discussed some dimensional results for them.

In the previous paragraph, we discussed some generalizations of Barnsley's FIF [**3**]. In all these generalizations, the authors used the Banach contraction principle and considered finite data sets. Using Banach fixed point theory and countable IFS, some authors extended Barnsley's work [**3**] for countable data sets (see [**7, 23**]). In 2018, Ri [**22**] constructed FIFs for finite data sets using some weak contraction theory instead of the Banach contraction theory. By doing so, Ri [**22**] generalized Barnsley's theory [**3**] and produced some novel FIFs that could not have been generated by any earlier FIFs theory. After that, Pacurar [**20**] used weak contraction theory and countable IFS to construct FIFs for a countable data set.

Motivated by the work of Ri [**22**], using the Rakotch contraction theory and the IFS theory, we prove the existence of FIFs in some standard spaces such as Lebesgue function space and bounded function space. After that, we extend the work of Navascués [**18**] and construct a more general version of α-fractal functions. We provide the stability of the general version of the α-fractal functions. We also determine the exact value of the fractal dimension of the newly constructed fractal function. Due to this construction, we get some new types of fractal functions. We also plot the approximate graphs for a few of them.

The structure of the paper is as follows: In the following section, we provide the necessary definitions and preliminary results for the subsequent sections. In Section 3, we provide the existence of new type of fractal functions in some standard function spaces. In the next section, we generalize the concept of α-fractal function and provide the stability and dimensional results for the generalized α-fractal functions. In the last section, we give examples of some newly constructed fractal functions and plot their approximate graphs.

2. Preliminaries

Let (X, d) be a complete metric space, and we denote the family of all nonempty compact subsets of X by $K(X)$. We define the Hausdorff metric on $K(X)$ by

$$\mathcal{D}(A, B) = \inf\{\delta > 0 : A \subset B_\delta \text{ and } B \subset A_\delta\},$$

where A_δ and B_δ denote the δ-neighborhoods of the sets A and B, respectively. Then it is well-known that $(K(X), \mathcal{D})$ is a complete metric space.

A map $f : (X, d) \to (X, d)$ is called a contraction if there exists a constant $s < 1$ such that

$$d(f(x), f(y)) \leq s d(x, y) \ \forall\ x, y \in X.$$

DEFINITION 2.1. Let (X, d) be a metric space and let A be a subset of X. The Hausdorff dimension of A is defined by

$$\dim_H A = \inf \Big\{ \beta > 0 : \text{for each } \epsilon > 0, \text{ there is a countable cover } \{U_i\}_{i=1}^{\infty} \text{ of } A \text{ with} \sum_{i=1}^{\infty} |U_i|^{\beta} < \epsilon \Big\},$$

where $|U_i|$ denotes the diameter of the set U_i.

DEFINITION 2.2. Let (X, d) be a metric space. Let A be a bounded nonempty subset of X. The box dimension of A is defined by

$$\dim_B A = \lim_{\delta \to 0} \frac{\log \mathcal{N}_\delta(A)}{-\log \delta},$$

provided the limit exists, where $\mathcal{N}_\delta(A)$ is the least number of sets of diameter $\delta > 0$ that can cover the set A. If this limit does not exist, then limsup and liminf are known as the upper box dimension and the lower box dimension, respectively.

DEFINITION 2.3. The system $\mathcal{I} = \big\{(X, d); f_1, f_2, \ldots, f_N\big\}$ is called an iterated function system (IFS), if f_i is a continuous self-map on X for each $i \in \{1, 2, \ldots, N\}$.

DEFINITION 2.4. Let (X, d) be a metric space. A map $f : X \to X$ is said to be a ϕ-contraction if there exists a map $\phi : \mathbb{R}^+ \to \mathbb{R}^+$ such that

$$d(f(x), f(y)) \leq \phi(d(x, y)) \ \forall x, y \in X,$$

where $\mathbb{R}^+$ denotes the set of all nonnegative real numbers.

DEFINITION 2.5. A map $f : (X, d) \to (X, d)$ is said to be a Rakotch contraction if f is a ϕ-contraction such that $\phi(t) < t$ for $t > 0$ and the function $\alpha(t) = \frac{\phi(t)}{t}$ is a decreasing function for $t > 0$.

REMARK 2.6. If we choose $\phi(t) = st$ for some $s < 1$, then $\phi(t) < t$ for $t > 0$ and $\frac{\phi(t)}{t}$ is a constant function. Thus each contraction is a Rakotch contraction. But the converse is not true. Let $X = [0, \infty)$ and $f : X \to X$ be such that $f(x) = \frac{x}{1+x}$ or $\frac{1}{1+x}$ for all $x \in X$. Let $x, y \in X$. Then we have

$$|f(x) - f(y)| \leq \frac{|x - y|}{1 + |x - y|} = \phi(|x - y|),$$

where $\phi : \mathbb{R}^+ \to \mathbb{R}^+$ is such that $\phi(t) = \frac{t}{1+t}$. Thus, the map f is a Rakotch contraction but it is not a contraction.

NOTE 2.7 ([**11**, Theorem 1(a),1(g)]). A map $f : X \to X$ is a Rakotch contraction if and only if there exists an increasing function $\phi : \mathbb{R}^+ \to \mathbb{R}^+$ such that

$$d(f(x), f(y)) \leq \phi(d(x, y)) \ \forall \ x, y \in X,$$

where $\phi(t) < t \ \forall \ t > 0$ and the function $\alpha(t) = \frac{\phi(t)}{t}$ is a decreasing function for $t > 0$.

THEOREM 2.8 ([**21**], p. 463). *Let (X, d) be a complete metric space and let $f : X \to X$ be a Rakotch contraction. Then f has a unique fixed point x_* in X. Moreover, for any $x_0 \in X$, the sequence $x_n = f(x_{n-1})$ converges to x_* in X.*

NOTE 2.9 ([**12**, Theorem 1]). The function $f : X \to X$ is a Rakotch contraction if and only if the function f is a ϕ-contraction such that ϕ is strictly increasing, concave and $\phi(t) < t$ for all $t > 0$.

2.1. New type of fractal functions. Motivated by the idea of Barnsley's FIFs [**3**], Ri [**22**] constructed new types of FIFs using the IFS theory and weak contraction theory. Here we recall the construction of Ri's FIFs [**22**].

Let us consider the data set $\{(x_n, y_n) : n \in \{1, 2, \ldots, N+1\}\}$ such that $x_1 < x_2 < \cdots < x_{N+1}$. Set $J = [x_1, x_{N+1}]$ and $I = \{1, 2, \ldots, N\}$. Let $J_n = [x_n, x_{n+1}]$ for all $n \in I$. Let K be some closed interval in $\mathbb{R}$ such that $\{y_1, y_2, \ldots, y_{N+1}\} \subseteq K$. For each $n \in I$, let $P_n : J \to J_n$ be the map given by

$$P_n(x) = a_n x + d_n \quad \text{that satisfies} \quad P_n(x_1) = x_n \text{ and } P_n(x_{N+1}) = x_{n+1}.$$

One can easily see that $a_n = \frac{x_{n+1} - x_n}{x_{N+1} - x_1}$ and $d_n = x_n - \frac{x_{n+1} - x_n}{x_{N+1} - x_1} x_1$. For each $n \in I$, let us define a continuous map $F_n : J \times K \to K$ such that F_n is a Rakotch contraction in the second variable such that

$$|F_n(x, y) - F_n(x, y*)| \leq \phi(|y - y*|)$$

and

$$F_n(x_1, y_1) - y_n, \quad F_n(x_{N+1}, y_{N+1}) = y_{n+1}.$$

Using P_n and F_n, we define a map $W_n : J \times K \to J \times K$ such that

$$W_n(x, y) = (P_n(x), F_n(x, y)),$$

for every $n \in I$. Then, $\mathcal{I} = \{J \times K; W_n : n \in I\}$ is an IFS. Using the above construction and the weak contraction theory, Ri proved that there exists a unique continuous function h on J such that the function h interpolates the data set and satisfies the following self-referential equation

$$h(P_n(x)) = F_n(x, h(x))$$

for all $x \in J$ and he also proved that the graph of the function h is the attractor of the IFS $\mathcal{I}$.

In the above construction of FIFs, if we take $\phi(t) = \max_{n \in I}\{s_n\}(t)$, where $0 < s_n < 1$ such that $|F_n(x, y) - F_n(x, y^*)| \leq s_n|y - y^*|$, then we get FIFs obtained by Barnsley. But if we consider $\phi(t) = \frac{t}{1+t}$, then we get some new type of FIFs which are not obtained by the theory of Barnsley [**3**]. Thus using the weak contraction theory, we get a more general class of FIFs.

3. Fractal functions in Lebesgue function spaces and bounded function spaces

In this section, we define some function spaces and construct new types of FIFs in those spaces.

For each $q \in [1, \infty)$, we define the space of all q-integrable functions on the interval $[a, b]$ with $a < b$ as follows

$$\mathcal{L}^q[a, b] = \{f : [a, b] \to \mathbb{R} \mid f \text{ is a measurable function and } \int_a^b |f(x)|^q dx < \infty\}.$$

The space $\mathcal{L}^q[a, b]$ is a Banach space with respect to the norm $\|.\|_q$, where $\|.\|_q$ is defined as follows

$$\|f\|_q = \left(\int_a^b |f(x)|^q dx\right)^{\frac{1}{q}}.$$

Now, we provide a construction of the new fractal functions in $\mathcal{L}^q$ space.

Let us consider the data set $\{(x_n, y_n) : n \in \{1, 2, \dots, N+1\}\}$ such that $x_1 < x_2 < \cdots, x_{N+1}$. Set $J = [x_1, x_{N+1}]$ and $I = \{1, 2, \dots, N\}$. Let $J_n = [x_n, x_{n+1}]$ for all $n \in I$. Let K be some closed interval in $\mathbb{R}$ such that $\{y_1, y_2, \dots, y_{N+1}\} \subseteq K$. For each $n \in I$, let $P_n : J \to J_n$ be the map defined as in Section 2. For each $n \in I$, we define a map $F_n : J \times K \to K$ as follows

$$F_n(x, y) = q_n(y)g_n(x) + b_n(x),$$

where $q_n : K \to K$ is a Rakotch contraction map such that

$$|q_n(y) - q_n(y*)| \leq \phi_n(|y - y*|),$$

$g_n : J \to \mathbb{R}$ is a map such that $\|g_n\|_\infty < 1$ and $b_n : J \to \mathbb{R}$ is such that $b_n(x_1) = y_n - q_n(y_1)g_n(x_1)$ and $b_n(x_{N+1}) = y_{n+1} - q_n(y_{N+1})g_n(x_{N+1})$. Now, we define a map $W_n : J \times K \to J \times K$ such that

$$W_n(x, y) = (P_n(x), F_n(x, y)).$$

Then, $\mathcal{I} = \{J \times K; W_n : n \in I\}$ is an IFS.

NOTE 3.1. We are given that $q_n : K \to K$ is a Rakotch contraction map such that

$$|q_n(y) - q_n(y*)| \leq \phi_n(|y - y*|).$$

By Note 2.9, we can take ϕ_n such that ϕ_n is strictly increasing, concave and $\phi_n(t) < t$ for all $t > 0$.

THEOREM 3.2. *Assume the assumptions and notations as in the previous paragraph. Let $b_n, q_n \in \mathcal{L}^q(J)$ for each $n \in I$. Let $(\phi_n(t))^s \leq \phi_n(t^s)$ for all $t \geq 0, s > 0$ and $n \in I$, and $\sum_{n\in I} \phi_n(t) < t$ for all $t > 0$, where ϕ_n is defined as in Note 3.1. Then, there exists a unique function $h \in \mathcal{L}^q(J)$ such that h interpolates the data set and satisfies*

$$h(P_n(x)) = q_n(h(x))g_n(x) + b_n(x)$$

for all $x \in J$. Moreover, the graph $G(h)$ of the function h satisfies

$$G(h) = \cup_{n\in I} W_n(G(h)).$$

PROOF. We define a closed subset $\mathcal{L}^q(J)^* := \{f \in \mathcal{L}^q(J) : f(x_1) = y_1, f(x_{N+1}) = y_{N+1}\}$ of $\mathcal{L}^q(J)$. Since $(\mathcal{L}^q(J), \|.\|_q)$ is a Banach space, we have $\mathcal{L}^q(J)^*$ is a complete metric space with respect to the metric d_q induced by the norm $\|.\|_q$. Now, we define the Read–Bajraktarevic (RB) operator $\mathcal{T} : \mathcal{L}^q(J)^* \to \mathcal{L}^q(J)^*$ as follows

$$\mathcal{T}g(x) = q_n(g(P_n^{-1}(x)))g_n(P_n^{-1}(x)) + b_n(P_n^{-1}(x)),$$

for all $x \in J_n$ and $n \in I$. Since $b_n, q_n, g \in \mathcal{L}^q(J)$ and $\|g_n\|_\infty < 1$, we have $\mathcal{T}g \in \mathcal{L}^q(J)$. Using the fact that $b_n(x_1) = y_n - q_n(y_1)g_n(x_1)$, $b_n(x_{N+1}) = y_{n+1} - q_n(y_{N+1})g_n(x_{N+1})$, $g(x_1) = y_1$ and $g(x_{N+1}) = y_{N+1}$, one can easily get $\mathcal{T}g(x_n) = y_n$ for all $n \in I$. Therefore, for any $g \in \mathcal{L}^q(J)^*$, we have $\mathcal{T}g \in \mathcal{L}^q(J)^*$ and $\mathcal{T}g$ interpolates the data set. This implies that the map $\mathcal{T}$ is well defined. Now, we show that the map $\mathcal{T}$ is a Rakotch contraction. Let $f, g \in \mathcal{L}^q(J)^*$ and $f \neq g$. Then we have

$$\begin{aligned}
d_q(\mathcal{T}f,\mathcal{T}g) &= \|\mathcal{T}f-\mathcal{T}g\|_q \\
&= \left(\int_J |\mathcal{T}f(x)-\mathcal{T}g(x)|^q dx\right)^{\frac{1}{q}} \\
&\le \sum_{n\in I} \|g_n\|_\infty \left(\int_{J_n} |q_n(f(P_n^{-1}(x))) - q_n(g(P_n^{-1}(x)))|^q dx\right)^{\frac{1}{q}}.
\end{aligned}$$

Using the fact that the map q_n is a Rakotch contraction, $\|g_n\|_\infty < 1$ and $(\phi_n(t))^s \le \phi_n(t^s)$ for all $t \ge 0, s > 0$ and $n \in I$, we obtain

$$\begin{aligned}
d_q(\mathcal{T}f,\mathcal{T}g) &\le \sum_{n\in I} \left(\int_{J_n} (\phi_n(|f(P_n^{-1}(x)) - g(P_n^{-1}(x))|))^q dx\right)^{\frac{1}{q}} \\
&\le \sum_{n\in I} \left(\int_{J_n} \phi_n(|f(P_n^{-1}(x)) - g(P_n^{-1}(x))|^q) dx\right)^{\frac{1}{q}}
\end{aligned}$$

Using the concavity of ϕ_n, we have

$$\begin{aligned}
d_q(\mathcal{T}f,\mathcal{T}g) &\le \sum_{n\in I} \left(\phi_n\left(\int_{J_n} |f(P_n^{-1}(x)) - g(P_n^{-1}(x))|^q dx\right)\right)^{\frac{1}{q}} \\
&\le \sum_{n\in I} \phi_n\left(\int_{J_n} |f(P_n^{-1}(x)) - g(P_n^{-1}(x))|^q dx\right)^{\frac{1}{q}} \\
&\le \sum_{n\in I} \phi_n\left(\int_{J} |(f-g)(x)|^q dx\right)^{\frac{1}{q}} \\
&= \sum_{n\in I} \phi_n(d_q(f,g)).
\end{aligned}$$

We define a map $\Phi : \mathbb{R}^+ \to \mathbb{R}^+$ by $\Phi(t) = \sum_{n\in I} \phi_n(t)$. Since $\sum_{n\in I} \phi_n(t) < t$ for all $t > 0$, we have $\Phi(t) < t$ for all $t > 0$. And also, it is not difficult to see that, Φ is strictly increasing and concave. Thus, we have

$$d_q(\mathcal{T}f,\mathcal{T}g) \le \Phi(d_q(f,g)).$$

This implies that $\mathcal{T}$ is a Rakotch contraction map. By Theorem 2.8, the map $\mathcal{T}$ has a unique fixed point $h \in \mathcal{L}^q(J)^*$, that is, $\mathcal{T}h = h$. Since the image of h under $\mathcal{T}$ is h, the function h interpolates the data set. Since $\mathcal{T}h = h$, the function h satisfies

$$h(P_n(x)) = q_n(h(x))g_n(x) + b_n(x)$$

for all $x \in J$. Next, we show that the graph $G(h)$ of the function h satisfies $G(h) = \cup_{n\in I} W_n(G(h))$.

$$\begin{aligned}
\cup_{n\in I} W_n(G(h)) &= \cup_{n\in I} W_n\{(x,h(x)) : x \in J\} \\
&= \cup_{n\in I}\{(P_n(x), F_n(x,h(x))) : x \in J\} \\
&= \cup_{n\in I}\{(P_n(x), q_n(h(x))g_n(x) + b_n(x)) : x \in J\} \\
&= \cup_{n\in I}\{(P_n(x), h(P_n(x))) : x \in J\} = G(h).
\end{aligned}$$

This completes the proof. □

REMARK 3.3. Let μ_q be the invariant measure corresponding to the IFS $\{J; P_n : n \in I\}$ with the probability vector $q = (q_1, q_2, \dots, q_N)$. Then, μ satisfies the following equation

$$\mu_q = \sum_{n \in I} q_n \mu_q \circ P_n^{-1}$$

and $\mu_q(J) = 1$. It is not difficult to see that the IFS $\{J; P_n : n \in I\}$ satisfies the open set condition (OSC) with open set $O = (x_1, x_{N+1})$. The overlapping set C of J and the dynamical boundary ∂J of J are defined as

$$C = \cup_{n \neq m \in I} P_n(J) \cap P_m(J) \quad \text{and} \quad \partial J = \overline{\cup_{k=1}^{\infty} \mathcal{F}^{-k}(C) \cap J},$$

where $\mathcal{F}^{-1}(C) = \cup_{n \in I} P_n^{-1}(C)$ (see [**2**]). Thus, $C = \{x_2, x_3, \dots, x_N\}$ and $\partial J = \{x_1, x_{N+1}\}$. By [**2**, Theorem 2.2], we have

$$\mu_q(C) = 0 \quad \text{and} \quad \mu_q(\partial J) = 0$$

for any probability vector $q = (q_1, q_2, \dots, q_N)$. Since $\sum_{n \in I} a_n = 1$, where a_n is the contraction ratio of P_n, we have $a = (a_1, a_2, \dots, a_N)$ is a probability vector. Then, $\mu_a = \sum_{n \in I} a_n \mu_a \circ P_n^{-1}$. By using $\mu_a(C) = 0$ and $\mu_a(\partial J) = 0$, we have

$$\mu_a[x_n, x_{n+1}] = \sum_{m \in I} a_m \mu_a \circ P_m^{-1}[x_n, x_{n+1}] = a_n = \frac{x_{n+1} - x_n}{x_{N+1} - x_1} = L_J^1([x_n, x_{n+1}]),$$

where L_J^1 denotes the normalized Lebesgue measure on the interval J. In fact, for this choice of probability vector, $\mu_a = L_J^1$ (see [**17**, Theorem 2.2]). Thus, in the previous theorem, we are taking this invariant measure corresponding to the probability vector $(a_1, a_2, \dots, a_N)$. But for different choice of probability vector $(q_1, q_2, \dots, q_N) \neq (a_1, a_2, \dots, a_N)$, we get different invariant measure supported on the interval J. One can also prove the previous theorem by using the same idea and considering the different invariant measure associated with the IFS $\{J; P_n : n \in I\}$.

Now, we define another space. The space of all bounded functions on the interval $[a, b]$ with $a < b$ is denoted by $\mathcal{B}([a, b])$ and defined as

$$\mathcal{B}([a, b]) = \{f : [a, b] \to \mathbb{R}; \sup_{x \in [a,b]} |f(x)| < \infty\}.$$

The space $\mathcal{B}([a, b])$ is a Banach space with respect to the sup norm $\|.\|_\infty$. Now, our aim is to provide a construction of a new fractal function in the bounded function space.

Let us consider the data set $\{(x_n, y_n) : n \in \{1, 2, \dots, N+1\}\}$ such that $x_1 < x_2 < \dots < x_{N+1}$. Set $J = [x_1, x_{N+1}]$ and $I = \{1, 2, \dots, N\}$. Let $J_n = [x_n, x_{n+1}]$ for all $n \in I$. For each $n \in I$, let $P_n : J \to J_n$ be the map defined as in the previous section.

THEOREM 3.4. *Let the data set and P_n be defined as above. For each $n \in I$, let $F_n : J \times K \to K$ be a bounded map such that F_n is a Rakotch contraction in the second variable such that $|F_n(x, y) - F_n(x, y*)| \leq \phi_n(|y - y*|)$ and satisfies $F_n(x_1, y_1) = y_n$, $F_n(x_{N+1}, y_{N+1}) = y_{n+1}$. Then the fractal interpolation function $h \in \mathcal{B}(J)$.*

PROOF. First, we define a set $\mathcal{B}(J)^* = \{f \in \mathcal{B}(J) : f(x_1) = y_1, f(x_{N+1}) = y_{N+1}\}$. It is not difficult to see that $\mathcal{B}(J)^*$ is a complete metric space with respect to the metric d_∞ induced by sup norm $\|.\|_\infty$. Now, we define the RB operator $\mathcal{T} : \mathcal{B}(J)^* \to \mathcal{B}(J)^*$ by

$$\mathcal{T}g(x) = F_n(P_n^{-1}(x), g(P_n^{-1}(x)))$$

for all $x \in J_n$ and $n \in I$. Clearly, the operator $\mathcal{T}$ is well-defined and for each $g \in \mathcal{B}(J)^*$, $\mathcal{T}g(x_n) = y_n$ for all $n \in \{1, 2, \ldots, N+1\}$. Let $f, g \in \mathcal{B}(J)^*$. Then we have

$$\begin{aligned}
d_\infty(\mathcal{T}f, \mathcal{T}g) &= \|\mathcal{T}f - \mathcal{T}g\|_\infty \\
&= \max_{n\in I}\left(\sup_{x\in J_n} |\mathcal{T}f(x) - \mathcal{T}g(x)|\right) \\
&= \max_{n\in I}\left(\sup_{x\in J_n} |F_n(P_n^{-1}(x), f(P_n^{-1}(x))) - F_n(P_n^{-1}(x), g(P_n^{-1}(x)))|\right) \\
&\le \max_{n\in I}\left(\sup_{x\in J_n} \phi_n(|f(P_n^{-1}(x)) - g(P_n^{-1}(x))|)\right).
\end{aligned}$$

Using the increasing property of ϕ_n, we have

$$\begin{aligned}
d_\infty(\mathcal{T}f, \mathcal{T}g) &\le \max_{n\in I} \phi_n\left(\sup_{x\in J_n} |f(P_n^{-1}(x)) - g(P_n^{-1}(x))|\right) \\
&= \max_{n\in I} \phi_n(\|f - g\|_\infty) \\
&= \max_{n\in I} \phi_n(d_\infty(f, g)).
\end{aligned}$$

We define a mapping $\Phi : \mathbb{R}^+ \to \mathbb{R}^+$ by $\Phi(t) = \max_{n\in I} \phi_n(t)$. Thus, it is easy to see that Φ is an increasing function, $\Phi(t) < t$ for all $t > 0$ and $\frac{\Phi(t)}{t}$ is a decreasing function for $t > 0$. Thus, we get

$$d_\infty(\mathcal{T}f, \mathcal{T}g) \le \Phi(\|f - g\|_\infty).$$

This implies that the map $\mathcal{T}$ is a Rakotch contraction. Theorem 2.8 yields that there exists a unique function $h \in \mathcal{B}(J)^*$ such that $\mathcal{T}h = h$. Thus, $h(x_n) = \mathcal{T}h(x_n) = y_n$ for all $n \in \{1, 2, \ldots, N+1\}$. This implies that h interpolates the data set. Also, h satisfies the equation $h(x) = F_n(P_n^{-1}(x), h(P_n^{-1}(x)))$ for all $x \in J_n$ and $n \in I$. Now, we show that $G(h)$ is the attractor of some IFS. For each $n \in I$, we define $W_n : J \times \mathbb{R} \to J \times \mathbb{R}$ such that

$$W_n(x, y) = (P_n(x), F_n(x, y)).$$

Then $\{J \times \mathbb{R}; W_n : n \in I\}$ is an IFS. We will show that $G(h) = \cup_{n\in I} W_n(G(h))$. We prove our claim by using the self-referential equation satisfied by the function h.

$$\begin{aligned}
\cup_{n\in I} W_n(G(h)) &= \cup_{n\in I} W_n\{(x, h(x)) : x \in J\} \\
&= \cup_{n\in I}\{(P_n(x), F_n(x, h(x))) : x \in J\} \\
&= \cup_{n\in I}\{(P_n(x), h(P_n(x))) : x \in J\} = G(h).
\end{aligned}$$

Thus, the function h is a fractal function and $h \in \mathcal{B}(J)$. This completes the proof. □

4. Generalized version of α-fractal functions, their stability and fractal dimensions

Motivated by the theory of Barnsley's FIFs [**3**], Navascués [**18**] described a family of FIFs, widely known as α-fractal functions corresponding to a continuous function. First, we recall the construction given by Navascués [**18**].

Let $J = [c, d]$ with $c < d$ and let $\mathcal{C}(J)$ denote the space of all continuous real valued functions on J. Let $f \in \mathcal{C}(J)$ be a given function. Let $\Delta = \{x_1, x_2, \ldots, x_{N+1}\}$ be a partition of J such that $c = x_1 < x_2 < \cdots < x_{N+1} = d$. Let $I = \{1, 2, \ldots, N\}$. Set $J_n = [x_n, x_{n+1}]$ for $n \in I$. Let $\alpha = (\alpha_1, \alpha_2, \ldots, \alpha_N)$ be a scaling vector such that $|\alpha_n| < 1$ for all $n \in I$. Let us consider a continuous function $b : J \to \mathbb{R}$ such that $b \neq f$ and $b(x_n) = f(x_n)$ for $n = 1, N+1$. The function b is called the base function. For each $n \in I$, let $P_n : J \to J_n$ be the map given by

$$P_n(x) = a_n x + d_n \quad \text{satisfying} \quad P_n(x_1) = x_n \text{ and } P_n(x_{N+1}) = x_{n+1}.$$

Using the above Navascués [**18**] defined an IFS as follows. For each $n \in I$, let $W_n : J \times \mathbb{R} \to J \times \mathbb{R}$ be defined as

$$W_n(x, y) = (P_n(x), F_n(x, y)),$$

where $F_n : J \times \mathbb{R} \to \mathbb{R}$ is a continuous function such that

$$F_n(x, y) = \alpha_n y + f(P_n(x)) - \alpha_n b(x).$$

Then, $\mathcal{I} = \{J \times \mathbb{R}; W_n : n \in I\}$ is an IFS.

With the help of the above construction, Navascués [**18**] proved that there exists a unique continuous function $f^\alpha_{\Delta,b} : J \to \mathbb{R}$ such that $f^\alpha_{\Delta,b}(x_n) = f(x_n)$ for all $n \in \{1, 2, \ldots, N+1\}$ and $f^\alpha_{\Delta,b}$ satisfies

$$f^\alpha_{\Delta,b}(P_n(x)) = f(P_n(x)) + \alpha_n (f^\alpha_{\Delta,b} - b)(x)$$

for all $x \in J$ and $n \in I$. Moreover, the graph $G(f^\alpha_{\Delta,b})$ of the function $f^\alpha_{\Delta,b}$ satisfies

$$G(f^\alpha_{\Delta,b}) = \cup_{n\in I} W_n(G(f^\alpha_{\Delta,b})).$$

Thus, the function $G(f^\alpha_{\Delta,b})$ is a fractal function and is named as α-fractal function. This function depends on the parameters α, b and the partition Δ.

Next, we describe the construction of generalized version of the above construction by using the Rakotch contraction theory and the IFS theory. We also emphasize the fact that by our construction, we can obtain some new type of fractal functions which are not obtained by the above construction.

Let $J, \mathcal{C}(J), \Delta, f, I, P_n$ and J_n be defined as in the previous construction. Let K be a closed interval in $\mathbb{R}$ such that $f(J) \subseteq K$. For each $n \in I$, we define a continuous mapping $F_n : J \times K \to K$ by

$$F_n(x, y) = q_n(x, y) + f(P_n(x)) - q_n(x, b(x)),$$

where f is the given continuous map, $b : J \to K$ is the base function defined as above and $q_n : J \times K \to \mathbb{R}$ is a continuous map such that q_n is a Rakotch contraction in the second variable map satisfying

$$|q_n(x, y) - q_n(x, y^*)| \leq \phi_n(|y - y^*|).$$

For each $n \in I$, we define a map $W_n : J \times K \to J \times K$ as follows

$$W_n(x, y) = (P_n(x), F_n(x, y)).$$

Then, $\mathcal{I} = \{J \times K; W_n : n \in I\}$ is an IFS.

THEOREM 4.1. *Let the IFS $\mathcal{I}$ be defined as above. Then there exists a unique continuous function $f^{\phi}_{\Delta,b} : J \to K$ such that $f^{\phi}_{\Delta,b}(x_n) = f(x_n)$ for all $n \in \{1, 2, \ldots, N+1\}$ and the function $f^{\phi}_{\Delta,b}$ satisfies the following self-referential equation*

$$f^{\phi}_{\Delta,b}(P_n(x)) = f(P_n(x)) + q_n(x, f^{\phi}_{\Delta,b}(x)) - q_n(x, b(x)).$$

for all $x \in J$ and $n \in I$. Furthermore, the graph $G(f^{\phi}_{\Delta,b})$ is the attractor of the IFS $\mathcal{I}$.

PROOF. First, we define a complete metric space as follows

$$\mathcal{C}(J)^* = \{g : J \to K \mid g \text{ is a continuous map and } g(x_1) = f(x_1),\ g(x_{N+1}) = f(x_{N+1})\}.$$

It is well known that $(\mathcal{C}(J)^*, d_\infty)$ is a complete metric space, where d_∞ is the metric induced by the sup norm. Now, we define the RB operator $\mathcal{T} : \mathcal{C}(J)^* \to \mathcal{C}(J)^*$ by

$$\mathcal{T}g(x) = q_n(P_n^{-1}(x), g(P_n^{-1}(x))) + f(x) - q_n(P_n^{-1}(x), b(P_n^{-1}(x)))$$

for all $x \in J_n$ and $n \in I$. It is not difficult to see that the map $\mathcal{T}$ is well defined and for each $g \in \mathcal{C}(J)^*$, $\mathcal{T}g(x_n) = f(x_n)$ for all $n \subset \{1, 2, \ldots, N+1\}$. Now, we will show that the operator $\mathcal{T}$ is a Rakotch contraction. Let $g_1, g_2 \in \mathcal{C}(J)^*$. Then, we have

$$\begin{aligned}
d_\infty(\mathcal{T}g_1, \mathcal{T}g_2) &= \|\mathcal{T}g_1 - \mathcal{T}g_2\|_\infty \\
&= \max_{n \in I}\left(\sup_{x \in J_n} |\mathcal{T}g_1(x) - \mathcal{T}g_2(x)|\right) \\
&= \max_{n \in I}\left(\sup_{x \in J_n} |q_n(P_n^{-1}(x), g_1(P_n^{-1}(x))) - q_n(P_n^{-1}(x), g_2(P_n^{-1}(x)))|\right) \\
&\le \max_{n \in I}\left(\sup_{x \in J_n} \phi_n(|g_1(P_n^{-1}(x)) - g_2(P_n^{-1}(x))|)\right) \\
&\le \max_{n \in I} \phi_n(d_\infty(g_1, g_2)) \\
&= \Phi(d_\infty(g_1, g_2)),
\end{aligned}$$

where $\Phi : \mathbb{R}^+ \to \mathbb{R}^+$ is given by $\Phi(t) = \max_{n \in I} \phi_n(t)$. Using the definition of ϕ_n, it is clear that Φ is an increasing function, $\Phi(t) < t$ for all $t > 0$ and $\frac{\Phi(t)}{t}$ is a decreasing function for $t > 0$. This implies that the operator $\mathcal{T}$ is a Rakotch contraction. Thus, by Theorem 2.8, there exists a unique function $f^{\phi}_{\Delta,b} \in \mathcal{C}(J)^*$ such that $\mathcal{T}f^{\phi}_{\Delta,b} = f^{\phi}_{\Delta,b}$. The function $f^{\phi}_{\Delta,b}$ depends on the comparison function vector $\phi = (\phi_1, \phi_2, \ldots, \phi_N)$, the partition Δ and the base function b. Since $\mathcal{T}f^{\phi}_{\Delta,b} = f^{\phi}_{\Delta,b}$, we have $f^{\phi}_{\Delta,b}(x_n) = f(x_n)$ for all $n \in \{1, 2, \ldots, N+1\}$ and the function $f^{\phi}_{\Delta,b}$ satisfies the following self-referential equation

$$f^{\phi}_{\Delta,b}(P_n(x)) = f(P_n(x)) + q_n(x, f^{\phi}_{\Delta,b}(x)) - q_n(x, b(x)).$$

for all $x \in J$ and $n \in I$. Using the above self-referential equation satisfied by $f^{\phi}_{\Delta,b}$, one can prove that

$$G(f^{\phi}_{\Delta,b}) = \cup_{n \in I} W_n(G(f^{\phi}_{\Delta,b})).$$

Thus, the proof is done. □

REMARK 4.2. In the above construction, if we consider $q_n(x,y) = \alpha_n y$ for all $n \in I$, where $|\alpha_n| < 1$. Then, by our construction, we get the same α-fractal function that was introduced by Navascués [**18**]. If we take $q_n(x,y) = \alpha_n(x)y$ for all $n \in I$, where $\alpha_n : J \to \mathbb{R}$ is a continuous map such that $\sup_{x\in J} |\alpha_n(x)| < 1$, then we get variable type α-fractal function. On the other hand, if we consider $q_n(x,y) = \alpha_n(x)\frac{y}{1+y}$ or $\alpha_n(x)\frac{1}{1+y}$ or $\frac{y}{1+y}$ or $\frac{1}{1+y}$, where $\alpha_n : J \to \mathbb{R}$ is a continuous map such that $\sup_{x\in J} |\alpha_n(x)| < 1$, then we obtain some new type of fractal functions which are not obtained by the previous ones.

Motivated by the work of Wang and Yu [**25**], we now discuss about the stability of the FIFs constructed above. For this, we consider two IFSs by taking the same partition Δ and the same set of Rakotch contraction maps q_n's but different given functions f, $\tilde{f}$ and base functions b, $\tilde{b}$.

Let J, Δ, J_n, I, P_n be the same as defined in the previous construction. Let f and $\tilde{f}$ be two continuous functions on J. Let b and $\tilde{b}$ be two affine continuous functions on J satisfying end point conditions $b(x_1) = f(x_1)$, $b(x_{N+1}) = f(x_{N+1})$, $\tilde{b}(x_1) = \tilde{f}(x_1)$ and $\tilde{b}(x_{N+1}) = \tilde{f}(x_{N+1})$. Let K be some closed interval in $\mathbb{R}$ such that $f(J) \subset K$ and $\tilde{f}(J) \subset K$. For each $n \in I$, let $q_n : J \times K \to K$ be a continuous map which is a Rakotch contraction in the second variable. Then, we define two mapping $F_n : J \times K \to K$ and $\tilde{F}_n : J \times K \to K$ by

$$F_n(x,y) = \alpha_n q_n(x,y) + f(P_n(x)) - \alpha_n q_n(x, b(x))$$

and

$$\tilde{F}_n(x,y) = \alpha_n q_n(x,y) + \tilde{f}(P_n(x)) - \alpha_n q_n(x, \tilde{b}(x)),$$

where $|\alpha_n| < 1$ for all $n \in I$. From the above, for each $n \in I$, we define two mappings $W_n : J \times K \to J \times K$ and $\tilde{W}_n : J \times K \to J \times K$ as follows

$$W_n(x,y) = (P_n(x), F_n(x,y)),\ \tilde{W}_n(x,y) = (P_n(x), \tilde{F}_n(x,y)).$$

Thus, from the help of the above, we get two IFSs $\mathcal{I} = \{J \times K; W_n : n \in I\}$ and $\tilde{\mathcal{I}} = \{J \times K; \tilde{W}_n : n \in I\}$. Further, using Theorem 4.1 corresponding to the IFSs $\mathcal{I}$ and $\tilde{\mathcal{I}}$, we get unique fractal interpolation functions $f^{\phi}_{\Delta,b}$ and $\tilde{f}^{\phi}_{\Delta,\tilde{b}}$, respectively. Theorem 4.3 describes the relation between $f^{\phi}_{\Delta,b}$ and $\tilde{f}^{\phi}_{\Delta,\tilde{b}}$.

THEOREM 4.3. *Let $\mathcal{I}$ and $\tilde{\mathcal{I}}$ be two IFSs defined as above. Let $f^{\phi}_{\Delta,b}$ and $\tilde{f}^{\phi}_{\Delta,\tilde{b}}$ be the unique FIFs associated with the IFSs $\mathcal{I}$ and $\tilde{\mathcal{I}}$, respectively. Then*

$$\|f^{\phi}_{\Delta,b} - \tilde{f}^{\phi}_{\Delta,\tilde{b}}\|_\infty \leq \frac{1 + \|\alpha\|_\infty}{1 - \|\alpha\|_\infty} \|f - \tilde{f}\|_\infty,$$

where $\|\alpha\|_\infty = \max\{|\alpha_n| : n \in I\}$.

PROOF. We are given that $f^{\phi}_{\Delta,b}$ and $\tilde{f}^{\phi}_{\Delta,\tilde{b}}$ are the unique FIFs associated with the IFSs $\mathcal{I}$ and $\tilde{\mathcal{I}}$, respectively. Thus, these fractal functions satisfy the following self-referential equations:

$$f^{\phi}_{\Delta,b}(P_n(x)) = f(P_n(x)) + \alpha_n q_n(x, f^{\phi}_{\Delta,b}(x)) - \alpha_n q_n(x, b(x)),$$
$$\tilde{f}^{\phi}_{\Delta,\tilde{b}}(P_n(x)) = \tilde{f}(P_n(x)) + \alpha_n q_n(x, \tilde{f}^{\phi}_{\Delta,\tilde{b}}(x)) - \alpha_n q_n(x, \tilde{b}(x))$$

for all $x \in J$ and $n \in I$. This implies that

$$\begin{aligned}|f^{\phi}_{\Delta,b}(P_n(x)) - \tilde{f}^{\phi}_{\Delta,\tilde{b}}(P_n(x))| &= |f(P_n(x)) + \alpha_n q_n(x, f^{\phi}_{\Delta,b}(x)) - \alpha_n q_n(x, b(x)) \\ &\quad - \tilde{f}(P_n(x)) - \alpha_n q_n(x, \tilde{f}^{\phi}_{\Delta,\tilde{b}}(x)) + \alpha_n q_n(x, \tilde{b}(x))| \\ &\leq |f(P_n(x)) - \tilde{f}(P_n(x))| \\ &\quad + \|\alpha\|_\infty |q_n(x, f^{\phi}_{\Delta,b}(x)) - q_n(x, \tilde{f}^{\phi}_{\Delta,\tilde{b}}(x))| \\ &\quad + \|\alpha\|_\infty |q_n(x, b(x)) - q_n(x, \tilde{b}(x))|.\end{aligned}$$

Using the fact that the map q_n is a Rakotch contraction in the second variable, we have

$$\begin{aligned}|f^{\phi}_{\Delta,b}(P_n(x)) - \tilde{f}^{\phi}_{\Delta,\tilde{b}}(P_n(x))| \leq& \|f - \tilde{f}\|_\infty + \|\alpha\|_\infty \phi_n(|f^{\phi}_{\Delta,b}(x) - \tilde{f}^{\phi}_{\Delta,\tilde{b}}(x)|) \\ &+ \|\alpha\|_\infty \phi_n(|b(x) - \tilde{b}(x)|).\end{aligned}$$

Since ϕ_n is an increasing function, $\phi_n(t) < t$ for all $t > 0$ and $b, \tilde{b}$ are affine continuous function on J with end point conditions, we have

$$\begin{aligned}\phi_n(|f^{\phi}_{\Delta,b}(x) - \tilde{f}^{\phi}_{\Delta,\tilde{b}}(x)|) \leq \phi_n(\|f^{\phi}_{\Delta,b} - \tilde{f}^{\phi}_{\Delta,\tilde{b}}\|_\infty) \leq \|f^{\phi}_{\Delta,b} - \tilde{f}^{\psi}_{\Delta,\tilde{b}}\|_\infty, \\ \phi_n(|b(x) - \tilde{b}(x)|) \leq \phi_n(\|f - \tilde{f}\|_\infty) \leq \|f - \tilde{f}\|_\infty.\end{aligned}$$

Thus, using the above inequality we get

$$|f^{\phi}_{\Delta,b}(P_n(x)) - \tilde{f}^{\phi}_{\Delta,\tilde{b}}(P_n(x))| \leq \|f - \tilde{f}\|_\infty + \|\alpha\|_\infty \|f^{\phi}_{\Delta,b} - \tilde{f}^{\phi}_{\Delta,\tilde{b}}\|_\infty + \|\alpha\|_\infty \|f - \tilde{f}\|_\infty.$$

This implies that

$$\|f^{\phi}_{\Delta,b} - \tilde{f}^{\phi}_{\Delta,\tilde{b}}\|_\infty \leq \frac{1 + \|\alpha\|_\infty}{1 - \|\alpha\|_\infty} \|f - \tilde{f}\|_\infty.$$

Thus, the proof is done. □

Next, we determine a dimensional result for the constructed generalized α-fractal functions.

For the sake of simplicity, in Theorem 4.4, we assume that $J = [0, 1]$ and the partition $\Delta = \{0, \frac{1}{N}, \frac{2}{N}, \ldots, 1\}$, where $N > 1$ is a fixed natural number.

In this case, we have $I = \{1, 2, \ldots, N\}$ and $J_n = [\frac{n-1}{N}, \frac{n}{N}]\ \forall\ n \in I$. For each $n \in I$, the map $P_n : J \to J_n$ is given by $P_n(x) = \frac{1}{N}x + \frac{1}{N}(n - 1)$. Let f be a given continuous map on J and b be a base function defined as above. Let K be a closed interval of $\mathbb{R}$ such that $f(J) \subseteq K$. For each $n \in I$, we define a map $F_n : J \times K \to K$ as follows

$$F_n(x, y) = \alpha_n q_n(y) + f(P_n(x)) - \alpha_n q_n(b(x)),$$

where $q_n : K \to K$ is a Rakotch contraction map and $\|\alpha\|_\infty = \max\{|\alpha_n| : n \in I\} < 1$. Then, by Theorem 4.1, there exists a unique continuous function $f^{\phi}_{\Delta,b} : J \to K$ interpolating the data set $\{(\frac{n-1}{N}, f(\frac{n-1}{N})) : n \in 1, 2, \ldots, N + 1\}$ and satisfying

$$f^{\phi}_{\Delta,b}(P_n(x)) = f(P_n(x)) + \alpha_n q_n(f^{\phi}_{\Delta,b}(x)) - \alpha_n q_n(b(x)).$$

In Theorem 4.4, we determine the exact value of the fractal dimension of the graph of the fractal function $f^{\phi}_{\Delta,b}$.

THEOREM 4.4. *Let $f : J \to \mathbb{R}$ be the given map such that f is a Lipschitz map. Let the base function $b : J \to \mathbb{R}$ be also a Lipschitz function. If $\|\alpha\|_\infty N^2 < 1$, then*

$$\dim_H(G(f^{\phi}_{\Delta,b})) = \dim_B(G(f^{\phi}_{\Delta,b})) = 1,$$

where $G(f^{\phi}_{\Delta,b})$ denotes the graph of the function $f^{\phi}_{\Delta,b}$.

PROOF. Let $x, y \in J$. Using the self-referential equation satisfied by $f^{\phi}_{\Delta,b}$, we have

$$\begin{aligned}|f^{\phi}_{\Delta,b}(P_n(x)) - f^{\phi}_{\Delta,b}(P_n(y))| \leq & |f(P_n(x)) - f(P_n(y))| \\ & + |\alpha_n||q_n(f^{\phi}_{\Delta,b}(x)) - q_n(f^{\phi}_{\Delta,b}(y))| \\ & + |\alpha_n||q_n(b(x)) - q_n(b(y))|.\end{aligned}$$

Using the fact that the map q_n is a Rakotch contraction, $\phi_n(t) < t$ for all $t > 0$, and f and b are Lipschitz maps with Lipschitz constants L_f and L_b, respectively, we get

$$\begin{aligned}|f^{\phi}_{\Delta,b}(P_n(x)) - f^{\phi}_{\Delta,b}(P_n(y))| \leq & L_f|P_n(x) - P_n(y)| + |\alpha_n|\phi_n(|f^{\phi}_{\Delta,b}(x) - f^{\phi}_{\Delta,b}(y)|) \\ & + |\alpha_n|\phi_n(|b(x) - b(y)|) \\ \leq & L_f\frac{1}{N}|x - y| + \|\alpha\|_\infty|f^{\phi}_{\Delta,b}(x) - f^{\phi}_{\Delta,b}(y)| \\ & + \|\alpha\|_\infty L_b|x - y|.\end{aligned}$$

Let $0 < \delta < \frac{1}{N}$ and let m be the smallest natural number greater than or equal to $\frac{1}{N\delta}$. Let $\mathcal{N}_\delta(f^{\phi}_{\Delta,b}, J)$ denote the number of squares of δ-mesh that intersect with $G(f^{\phi}_{\Delta,b})$ on the domain J. Since $J = \cup_{n\in I} P_n(J)$, we have

$$\mathcal{N}_\delta(f^{\phi}_{\Delta,b}, J) \leq \sum_{n\in I} \mathcal{N}_\delta(f^{\phi}_{\Delta,b}, P_n(J)).$$

Here, we define the notation $R_{f,D}[t_1, t_2]$ as follows

$$R_{f,D}[t_1, t_2] := \sup_{x,y\in[t_1,t_2]} |f(x) - f(y)|,$$

where $[t_1, t_2] \subseteq D \subset \mathbb{R}$ and D is a subset of domain of the function f.

Now, for $i \in \{0, 1, \ldots, m-1\}$, we calculate

$$\begin{aligned}& R_{f^{\phi}_{\Delta,b}, P_n(J)}\left[\frac{n-1}{N} + i\delta, \frac{n-1}{N} + (i+1)\delta\right] \\ & = \sup_{\substack{x,y\in J, \\ P_n(x),P_n(y)\in[\frac{n-1}{N}+i\delta, \frac{n-1}{N}+(i+1)\delta]}} |f^{\phi}_{\Delta,b}(P_n(x)) - f^{\phi}_{\Delta,b}(P_n(y))|\end{aligned}$$

By the previous inequalities, we have

$$\begin{aligned}
&R_{f^{\phi}_{\Delta,b},P_n(J)}\left[\frac{n-1}{N}+i\delta,\frac{n-1}{N}+(i+1)\delta\right]\\
&\quad\le L_f\delta+\|\alpha\|_{\infty}L_bN\delta+\|\alpha\|_{\infty}\sup_{\substack{x,y\in J,\\ P_n(x),P_n(y)\in[\frac{n-1}{N}+i\delta,\frac{n-1}{N}+(i+1)\delta]}}|f^{\phi}_{\Delta,b}(x)-f^{\phi}_{\Delta,b}(y)|\\
&\quad= L_f\delta+\|\alpha\|_{\infty}L_bN\delta+\|\alpha\|_{\infty}\sup_{\substack{x,y\in J,\\ x,y\in[iN\delta,(i+1)N\delta]}}|f^{\phi}_{\Delta,b}(x)-f^{\phi}_{\Delta,b}(y)|\\
&\quad= L_f\delta+\|\alpha\|_{\infty}L_bN\delta+\|\alpha\|_{\infty}R_{f^{\phi}_{\Delta,b},J}[iN\delta,(i+1)N\delta].
\end{aligned}$$

By using [**9**, Proposition 11.1], we have

$$\mathcal{N}_{\delta}(f^{\phi}_{\Delta,b},P_n(J))\le 2m+\delta^{-1}\sum_{i=0}^{m-1}R_{f^{\phi}_{\Delta,b},P_n(J)}\left[\frac{n-1}{N}+i\delta,\frac{n-1}{N}+(i+1)\delta\right].$$

With the help of previous inequalities, we get

$$\mathcal{N}_{\delta}(f^{\phi}_{\Delta,b},P_n(J))\le 2m+mL_f+\|\alpha\|_{\infty}NmL_b+\|\alpha\|_{\infty}\delta^{-1}\sum_{i=0}^{m-1}R_{f^{\phi}_{\Delta,b},J}[iN\delta,(i+1)N\delta].$$

Again, [**9**, Proposition 11.1] yields that

$$(N\delta)^{-1}\sum_{i=0}^{m-1}R_{f^{\phi}_{\Delta,b},J}[iN\delta,(i+1)N\delta]\le\mathcal{N}_{N\delta}(f^{\phi}_{\Delta,b},J).$$

By considering the above two inequalities, we get

$$\mathcal{N}_{\delta}(f^{\phi}_{\Delta,b},P_n(J))\le 2m+mL_f+\|\alpha\|_{\infty}NmL_b+\|\alpha\|_{\infty}N\mathcal{N}_{N\delta}(f^{\phi}_{\Delta,b},J).$$

Using the fact that $\mathcal{N}_{\delta}(f^{\phi}_{\Delta,b},J)\le\sum_{n\in I}\mathcal{N}_{\delta}(f^{\phi}_{\Delta,b},P_n(J))$ and $\delta<N\delta$, we have

$$\begin{aligned}
\mathcal{N}_{\delta}(f^{\phi}_{\Delta,b},J)&\le\sum_{n=1}^{N}\mathcal{N}_{\delta}(f^{\phi}_{\Delta,b},P_n(J))\\
&\le 2Nm+mNL_f+\|\alpha\|_{\infty}N^2mL_b+\|\alpha\|_{\infty}N^2\mathcal{N}_{N\delta}(f^{\phi}_{\Delta,b},J)\\
&\le 2Nm+mNL_f+\|\alpha\|_{\infty}N^2mL_b+\|\alpha\|_{\infty}N^2\mathcal{N}_{\delta}(f^{\phi}_{\Delta,b},J)
\end{aligned}$$

Since $\|\alpha\|_{\infty}N^2<1$, we have

$$\mathcal{N}_{\delta}(f^{\phi}_{\Delta,b},J)\le\frac{1}{(1-\|\alpha\|_{\infty}N^2)}(2+L_f+\|\alpha\|_{\infty}NL_b)mN.$$

Since $m\le 1+\frac{1}{N\delta}$ and $0<\delta<\frac{1}{N}$, we have $mN\le\frac{2}{\delta}$. Thus, the above inequality implies that

$$\mathcal{N}_{\delta}(f^{\phi}_{\Delta,b},J)\le\frac{1}{(1-\|\alpha\|_{\infty}N^2)}(4+2L_f+2\|\alpha\|_{\infty}NL_b)\delta^{-1}.$$

Thus, from the above inequality, we get

$$\overline{\dim}_B(G(f^{\phi}_{\Delta,b}))=\limsup_{\delta\to 0}\frac{\log\mathcal{N}_{\delta}(f^{\phi}_{\Delta,b},J)}{-\log\delta}\le 1.$$

Since $1\le\dim_H(G(f^{\phi}_{\Delta,b}))\le\overline{\dim}_B(G(f^{\phi}_{\Delta,b}))$, we get

$$\dim_H(G(f^{\phi}_{\Delta,b}))=\dim_B(G(f^{\phi}_{\Delta,b}))=1.$$

This completes the proof. □

5. Graphs of some new type of fractal functions

In this section, we plot some approximate graphs of the fractal interpolation functions, which are obtained by the Rakotch contraction theory as mentioned in Section 4.

EXAMPLE 5.1. Let $J = [0,1]$ and $f : [0,1] \to \mathbb{R}$ be the function given by $f(x) = x$. We consider the partition $\Delta = \{0, 0.2, 0.4, 0.6, 0.8, 1\}$. Take the base function $b : [0,1] \to \mathbb{R}$ such that $b(x) = x^{20} + x(1-x)$. Then, it is clear that $b(0) = f(0)$ and $b(1) = f(1)$. In this case, we take $K = [0,\infty)$, $I = \{1,2,\ldots,5\}$ and $J_n = [0.2(n-1), 0.2n]$ for all $n \in I$. Clearly, $f([0,1]) \subset [0,\infty)$. Here, the map $P_n : J \to J_n$ is defined as $P_n(x) = 0.2x + 0.2(n-1)$ for all $n \in I$. For each $n \in I$, we define a map $q_n : J \times K \to K$ as follows

$$q_n(x,y) = \alpha_n \frac{y}{1+y},$$

where $\alpha_1 = 0.2, \alpha_2 = 0.80, \alpha_3 = 0.95, \alpha_4 = 0.98$ and $\alpha_5 = 0.99$. Clearly, for each $n \in I$, the map q_n is a Rakotch contraction map. Thus, the map $F_n : J \times K \to K$ is defined as follows

$$F_n(x,y) = \alpha_n \frac{y}{1+y} + P_n(x) - \alpha_n \frac{x^{20} + x(1-x)}{1 + x^{20} + x(1-x)}.$$

Thus, by Theorem 4.1, there exists a unique fractal function $f^{\phi}_{\Delta,b} : J \to K$ such that $f^{\phi}_{\Delta,b}(0.2(n-1)) = f(0.2(n-1))$ for all $n \in \{1,2,\ldots,6\}$. In Figure 1, we plot the approximate graph of the fractal interpolation function $f^{\phi}_{\Delta,b}$.

EXAMPLE 5.2. Consider $J = [0,1]$ and the map $f : [0,1] \to \mathbb{R}$ is defined by $f(x) = x^5 + 1$. Take the partition $\Delta = \{0, 0.2, 0.4, 0.6, 0.8, 1\}$ and the base function $b : [0,1] \to \mathbb{R}$ such that $b(x) = x^{10} + 20x^2(1 - x^{10}) + 1$. Clearly, the base function b satisfies the end point condition. Consider K, I, J_n and P_n to be the same as in the previous example. For each $n \in I$, we define a mapping $q_n : J \times K \to K$ by

$$q_n(x,y) = \frac{1}{1+y}.$$

It is not difficult to see that the map q_n is a Rakotch contraction with $\phi_n(t) = \frac{t}{1+t}$. Therefore, the map $F_n : J \times K \to K$ is defined by

$$F_n(x,y) = \frac{1}{1+y} + (P_n(x))^5 + 1 - \frac{1}{x^{10} + 20x^2(1-x^{10}) + 2},$$

for each $n \in I$. Clearly, the above construction satisfies all the assumptions of Theorem 4.1. Thus, Theorem 4.1 yields that there exists a unique fractal function $f^{\phi}_{\Delta,b} : J \to K$ such that $f^{\phi}_{\Delta,b}(0.2(n-1)) = f(0.2(n-1))$ for all $n \in \{1,2,\ldots,6\}$. We plot the approximate graph of the fractal function $f^{\phi}_{\Delta,b}$ in Figure 2.

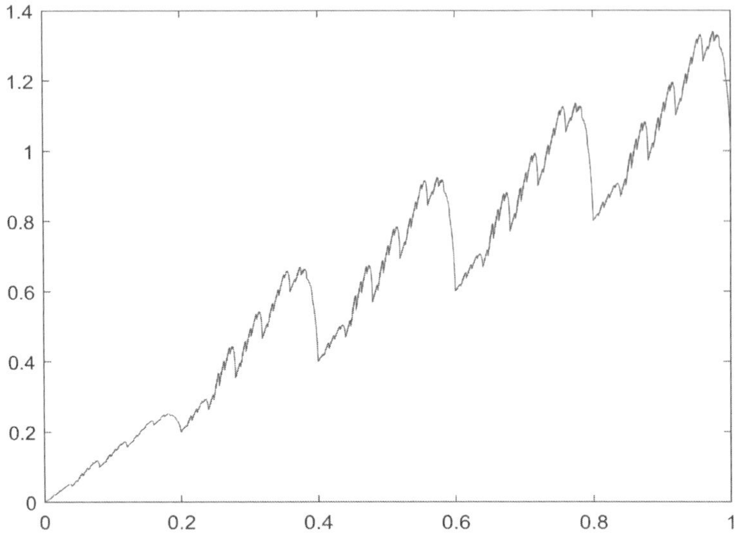

FIGURE 1. Fractal function corresponding to data set $\{(0,0),(0.2,0.2),(0.4,0.4),(0.6,0.6),(0.8,0.8),(1,1)\}$ and $\phi_n(t) = \frac{t}{1+t}$.

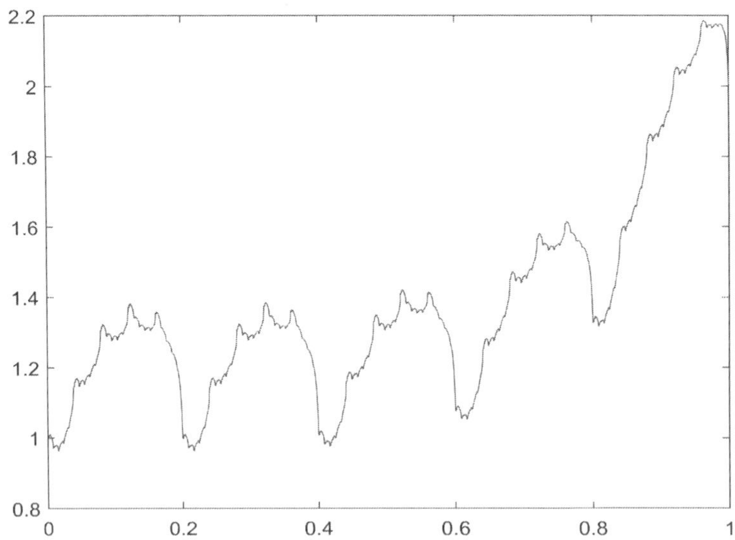

FIGURE 2. Fractal function corresponding to data set $\{(0,1),(0.2,1.003),(0.4,1.0102),(0.6,1.0778),(0.8,1.3277),(1,2)\}$ and $\phi_n(t) = \frac{t}{1+t}$.

References

[1] Vishal Agrawal and Tanmoy Som, *$\mathcal{L}^p$-approximation using fractal functions on the Sierpiński gasket*, Results Math. **77** (2022), no. 2, Paper No. 74, 17, DOI 10.1007/s00025-021-01565-5. MR4382684

[2] Christoph Bandt, Michael Barnsley, Markus Hegland, and Andrew Vince, *Old wine in fractal bottles I: Orthogonal expansions on self-referential spaces via fractal transformations*, Chaos Solitons Fractals **91** (2016), 478–489, DOI 10.1016/j.chaos.2016.07.007. MR3551732

[3] Michael F. Barnsley, *Fractal functions and interpolation*, Constr. Approx. **2** (1986), no. 4, 303–329, DOI 10.1007/BF01893434. MR892158

[4] Michael Barnsley, *Fractals everywhere*, Academic Press, Inc., Boston, MA, 1988. MR977274

[5] Michael F. Barnsley and Peter R. Massopust, *Bilinear fractal interpolation and box dimension*, J. Approx. Theory **192** (2015), 362–378, DOI 10.1016/j.jat.2014.10.014. MR3313490

[6] Subhash Chandra and Syed Abbas, *On fractal dimensions of fractal functions using function spaces*, Bull. Aust. Math. Soc. **106** (2022), no. 3, 470–480, DOI 10.1017/S0004972722000685. MR4510139

[7] Subhash Chandra, Syed Abbas, and Saurabh Verma, *Bernstein super fractal interpolation function for countable data systems*, Numer. Algorithms **92** (2023), no. 4, 2457–2481, DOI 10.1007/s11075-022-01398-5. MR4562899

[8] V. Drakopoulos, P. Bouboulis, and S. Theodoridis, *Image compression using affine fractal interpolation on rectangular lattices*, Fractals **14** (2006), no. 4, 259–269, DOI 10.1142/S0218348X06003271. MR2282778

[9] Kenneth Falconer, *Fractal geometry*, 2nd ed., John Wiley & Sons, Inc., Hoboken, NJ, 2003. Mathematical foundations and applications, DOI 10.1002/0470013850. MR2118797

[10] Yuval Fisher (ed.), *Fractal image compression*, Springer-Verlag, New York, 1995. Theory and application, DOI 10.1007/978-1-4612-2472-3. MR1313035

[11] Jacek R. Jachymski, *Equivalence of some contractivity properties over metrical structures*, Proc. Amer. Math. Soc. **125** (1997), no. 8, 2327–2335, DOI 10.1090/S0002-9939-97-03853-7. MR1389524

[12] Jacek Jachymski and Izabela Jóźwik, *Nonlinear contractive conditions: a comparison and related problems*, Fixed point theory and its applications, Banach Center Publ., vol. 77, Polish Acad. Sci. Inst. Math., Warsaw, 2007, pp. 123–146, DOI 10.4064/bc77-0-10. MR2338580

[13] S. Jha and S. Verma, *Dimensional analysis of α-fractal functions*, Results Math. **76** (2021), no. 4, Paper No. 186, 24, DOI 10.1007/s00025-021-01495-2. MR4303614

[14] Peter R. Massopust, *Fractal surfaces*, J. Math. Anal. Appl. **151** (1990), no. 1, 275–290, DOI 10.1016/0022-247X(90)90257-G. MR1069462

[15] Peter R. Massopust, *Vector-valued fractal interpolation functions and their box dimension*, Aequationes Math. **42** (1991), no. 1, 1–22, DOI 10.1007/BF01818475. MR1112180

[16] P. R. Massopust, *Non-stationary fractal interpolation*, Mathematics 7(8) (2019), Art. 666 (14pp).

[17] Manuel Morán and José-Manuel Rey, *Singularity of self-similar measures with respect to Hausdorff measures*, Trans. Amer. Math. Soc. **350** (1998), no. 6, 2297–2310, DOI 10.1090/S0002-9947-98-02218-1. MR1475691

[18] M. A. Navascués, *Fractal polynomial interpolation*, Z. Anal. Anwendungen **24** (2005), no. 2, 401–418, DOI 10.4171/ZAA/1248. MR2174271

[19] M. A. Navascués and S. Verma, *Non-stationary α-fractal surfaces*, Mediterr. J. Math. **20** (2023), no. 1, Paper No. 48, 18, DOI 10.1007/s00009-022-02242-9. MR4527888

[20] Cristina Maria Pacurar, *A countable fractal interpolation scheme involving Rakotch contractions*, Results Math. **76** (2021), no. 3, Paper No. 161, 19, DOI 10.1007/s00025-021-01470-x. MR4287311

[21] E. Rakotch, *A note on contractive mappings*, Proc. Amer. Math. Soc. **13** (1962), 459–465, DOI 10.2307/2034961. MR148046

[22] SongIl Ri, *A new idea to construct the fractal interpolation function*, Indag. Math. (N.S.) **29** (2018), no. 3, 962–971, DOI 10.1016/j.indag.2018.03.001. MR3800662

[23] Nicolae-Adrian Secelean, *The fractal interpolation for countable systems of data*, Univ. Beograd. Publ. Elektrotehn. Fak. Ser. Mat. **14** (2003), 11–19 (2004), DOI 10.2298/PETF0314011S. MR2076305

[24] S. Verma and P. R. Massopust, *Dimension preserving approximation*, Aequationes Math. **96** (2022), no. 6, 1233–1247, DOI 10.1007/s00010-022-00893-3. MR4514041

[25] Hong-Yong Wang and Jia-Shan Yu, *Fractal interpolation functions with variable parameters and their analytical properties*, J. Approx. Theory **175** (2013), 1–18, DOI 10.1016/j.jat.2013.07.008. MR3101056

Department of Mathematics, Indian Institute of Technology Delhi, New Delhi, India 110016

Email address: mathmanuj@gmail.com

Department of Mathematics, Indian Institute of Technology Delhi, New Delhi, India 110016

Email address: priyadarshi@maths.iitd.ac.in

Contemporary Mathematics
Volume **797**, 2024
https://doi.org/10.1090/conm/797/15957

Unmating of expanding Thurston maps with Julia set $\mathbb{S}^2$

Mary Wilkerson

Abstract. Every expanding Thurston map f without periodic critical points is known to have an iterate f^n which is the topological mating of two polynomials. This has been examined by Kameyama and Meyer; the latter has offered an explicit construction for finding two polynomials in the unmating of the iterate. Initializing Meyer's algorithm depends on an invariant Jordan curve through the postcritical set of f–but we propose adjustments to this unmating algorithm for Kameyama's more general case where there exists a curve which is fully f-invariant up to homotopy and not necessarily simple. When f is a critically pre-periodic expanding Thurston map, extending the algorithm to accommodate non-Jordan curves in this manner may allow us to unmate without iterates.

1. Introduction

Some rational maps contain dynamics similar to those of polynomials. In the 1980s, Douady and Hubbard explored this phenomenon with the introduction of *mating*—which is now the name for a collection of operations that combine two polynomials into a new map with shared dynamics. The classic approach starts with two monic degree d postcritically finite polynomials acting on disjoint copies of their respective filled Julia sets: if appropriate conditions are satisfied, the domains are identified along their Julia set boundaries, yielding a topological sphere. The polynomial mappings then descend to a branched covering of this sphere called the *topological mating*, which may be Thurston-equivalent to a rational map on $\widehat{\mathbb{C}}$.

In the event that the two polynomials are critically preperiodic, their Julia sets are dendrites. This yields that there is an equator-like curve containing a full set of equivalence class representatives for the equivalence relation $\approx$ generating this quotient space. An immediate consequence is that the topological mating in this case is then semiconjugate to the d-fold map on $\mathbb{S}^1$.

Closed curves on $\mathbb{S}^2$ with similar properties to this "equator" can be very useful in investigating the construction and decomposition of certain rational maps: namely those exhibiting behavior similar to polynomials within their dynamics. For example, [**4**] and [**16**] both utilize the Thurston pullback algorithm along with equator-like structures to develop approximations to the associated rational map. Conversely, curves with properties resembling those of an equator can be used to decompose or "unmate" branched coverings arising as matings. In [**6**],

2020 *Mathematics Subject Classification.* Primary 37F20; Secondary 37F10.
Key words and phrases. Mating, unmating, Thurston maps, rational maps, invariant curve.

Kameyama demonstrates that any expanding Thurston maps which is *orientedly* $\mathbb{S}^1$*-parametrizable* (See Subsection 2.5) has an iterate which is Thurston equivalent to an essential mating. In [**8**], Meyer offers the explicit construction of an $\mathbb{S}^1$-parameterization. This is elaborated upon in the subsequent papers [**7**] and [**9**], where the $\mathbb{S}^1$-parameterization is used to unmate a rational map.

Meyer notes that a *pseudo-equator* is a sufficient condition for an expanding postcritically finite rational map with Julia set the Riemann sphere to be a mating, but that it is not necessary: some such rational maps which arise as matings do not have a pseudo-equator. [**7**]. Further, the unmating algorithm described in [**9**] does not apply when an initial curve is not Jordan. Our goal is to explore and address these issues.

We start with prerequisite topics in Section 2. This section includes preliminaries on expanding Thurston maps and matings to discuss the objects we wish to decompose; as well as laminations, oriented curves, and tilings to highlight tools that will be used to develop their decomposition. We detail the current literature on decomposition of rational maps and unmating in Section 3. The main results are developed in Section 4, where we discuss accommodations to extend the reach of Meyer's unmating algorithm. As the arguments presented here are intended to build upon the constructions given in [**6**], [**7**], [**8**], and [**9**], we will use corresponding results that are still applicable freely, but also emphasize required adjustments where necessary and highlight implications of our changes. We then conclude with an example and discussion of further avenues of exploration.

2. Preliminaries

2.1. Notation. We utilize many notational conventions adopted in [**6**], [**7**], [**8**], and [**9**] for ease of reference. Some items are highlighted here.

We will use $\mathbb{S}^2$ to refer to the two-sphere, but $\widehat{\mathbb{C}}$ to refer to the two-sphere when endowed with a Riemannian metric and structure.

We occasionally identify $\mathbb{S}^1$ with $\partial\mathbb{D} = \{z \in \mathbb{C} : |z| = 1\}$ where convenient. Most frequently $\mathbb{S}^1$ will be used to parametrize closed curves however, and our preference in these instances will be to identify $\mathbb{R}/\mathbb{Z}$ with $\mathbb{S}^1$ via the map $t \mapsto e^{2\pi it}$ and use parameters in the interval $[0,1)$. The "d-fold map on $\mathbb{S}^1$" is then taken to be the map $q_d(t) := dt \bmod 1$ applied to this interval.

When the word *equator* is used in this text, it is taken to represent a curve separating hemispheres of $\mathbb{S}^2$. (This is mentioned as a point of disambiguation, as this word has a distinct mathematical interpretation in [**9**].)

The primary setting of this paper assumes that f is taken to be an expanding Thurston map which is critically preperiodic. We will denote the degree of f as d and the postcritical set of f as post_f or possibly $\mathbf{V}^0$, depending on context.

If a homotopy $H : X \times [0,1] \to Y$ is an isotopy on $X \times [0,1)$, we will refer to H as a *pseudo-isotopy*. We may use H_t to refer to the function $H(\cdot,t)$, and generally construct H_0 as an identity map. If H_t is constant on a set, we will refer to H as a homotopy *relative to* that set, and say that the homotopy is *supported* on its complement.

We use $\bigvee$ to indicate the join of two equivalence relations.

2.2. Expanding Thurston maps. We begin with some fundamentals for expansion.

DEFINITION 2.1. *Thurston maps* are orientation-preserving postcritically finite branched coverings of $\mathbb{S}^2$. If $f, g : \mathbb{S}^2 \to \mathbb{S}^2$ are two such maps with postcritical sets post_f and post_g, we say that f and g are *Thurston equivalent* if and only if there exist orientation-preserving homeomorphisms $h, h' : (\mathbb{S}^2, \text{post}_f) \to (\mathbb{S}^2, \text{post}_g)$ such that $g \circ h' = h \circ f$ and h is isotopic to h' relative to post_f.

In particular, polynomials and rational maps on $\widehat{\mathbb{C}}$ that are postcritically finite are Thurston maps. The Thurston maps we will focus on are rational maps with no periodic critical points. Such rational maps have Julia set equal to $\widehat{\mathbb{C}}$, and are also *expanding* on $\mathbb{S}^2$ as described in [**1**]:

DEFINITION 2.2. Suppose that f is a Thurston map with postcritical set post_f, and that $\mathcal{C} \subseteq \mathbb{S}^2$ is a Jordan curve containing post_f. Note that we may generate a cellular decomposition of $\mathbb{S}^2$ by marking post_f as our 0-cell vertices, the arcs between these on $\mathcal{C}$ as 1-cell open edges, and the two components of $\mathbb{S}^2 \setminus \mathcal{C}$ as 2-cell open tiles. Further, we denote elements of $f^{-n}(\text{post}_f)$ as *n-vertices*, the closures of components of $f^{-n}(\mathcal{C} \setminus \text{post}_f)$ as *n-edges*, closures of components of $f^{-n}(\mathbb{S}^2 \setminus \mathcal{C})$ as *n-tiles*.

We say that f is *expanding* if as $n \to \infty$ the n-tile diameters shrink to 0 uniformly with respect to some metric generating the topology on $\mathbb{S}^2$.

While the property of expansion is independent of the metric used, much of our discussion hinges on usage of the *visual metric* in particular:

DEFINITION 2.3. Let $f : \mathbb{S}^2 \to \mathbb{S}^2$ be an expanding Thurston map, $\mathcal{C} \subset \mathbb{S}^2$ be a Jordan curve containing post_f, and $x, y \in \mathbb{S}^2$. If $x \neq y$, we assign $m(x, y)$ to be the largest $n \in \mathbb{N}$ for which there exist intersecting n-tiles X and Y with $x \in X$ and $y \in Y$. Otherwise, we set $m(x, y) := \infty$.

A metric ϱ on $\mathbb{S}^2$ is then called a *visual metric* (for f) if there exists a constant $\Lambda > 1$ such that

$$\frac{1}{C}\varrho(x, y) \leq \Lambda^{-m(x,y)} \leq C\varrho(x, y) \tag{2.1}$$

for all $x, y \in \mathbb{S}^2$, where the constant C is independent of both x and y. We call this Λ the *expansion factor* of ϱ. [**1**]

Bonk and Meyer note that visual metrics are guaranteed to exist for expanding Thurston maps, and that each visual metric induces the given topology on $\mathbb{S}^2$. Further, the $\mathcal{C}$ that we start with is not important: if we develop a different $\tilde{\mathcal{C}}$ containing post_f, we may obtain a similar statement as in (2.1), except with different $\tilde{m}$ and $\tilde{C}$. The expansion factor Λ, however, will remain the same for this curve. [**1**, Proposition 8.3]

Sets composed of n-tiles then have a natural relationship with balls in the visual metric:

LEMMA 2.4. *Let f be an expanding Thurston map, and $\mathcal{C} \subset \mathbb{S}^2$ be a Jordan curve containing post_f. Let ϱ be a visual metric for f corresponding to this choice of $\mathcal{C}$ with expansion factor $\Lambda > 1$. For $x \in \mathbb{S}^2, n \in N_0$, set*

$U^n(x) := \bigcup\{Y \in \mathbf{X}^n : Y$ intersects an n-tile X containing $x\}$.

Then there are constants $K \geq 1$ and $n_0 \in \mathbb{N}_0$ with the following property.

For all $x \in \mathbb{S}^2$ and all $n \in \mathbb{Z}$, $B_\varrho(x, r/K) \subset U^n(x) \subset B_\varrho(x, Kr)$, where $r = \Lambda^{-n}$.
[**1**, Lemma 8.10]

2.3. Matings. The work in this paper will emphasize matings of monic polynomials which are critically preperiodic in $\mathbb{C}$. Any such polynomial has an associated Julia set which is a connected and locally connected dendrite–that is, the filled Julia set K has no interior and is the same as the Julia set J for the polynomial.

When K is connected, there exists a unique conformal isomorphism $\phi : \hat{\mathbb{C}} \smallsetminus \overline{\mathbb{D}} \to \hat{\mathbb{C}} \smallsetminus K$ for which $\phi(z^d) = f \circ \phi(z)$ and $\phi(z)/z \to 1$ as $|z| \to \infty$. We then let $R(t)$ denote the set of points forming the *external ray of angle t*; where the collection of points is given by $R(t) := \{\phi(re^{2\pi it}) \mid r \in (1,\infty)\}$. Since any filled Julia set K discussed here will be locally connected, ϕ extends continuously to $\partial\mathbb{D}$ and thus external rays of angle t are said to have a *landing point*, which is given by $\gamma(t) = \lim_{r\to 1^+} \phi(re^{2\pi it})$. The map γ is called the *Carathéodory semiconjugacy*, as we have that $\gamma(dt) = f(\gamma(t))$. In essence, this semiconjugacy relates the mapping behavior of f on its Julia set to that of the d-fold map on the circle.

Let $\widetilde{\mathbb{C}}$ be the compactification of $\mathbb{C}$ formed by union with the circle at infinity, $\widetilde{\mathbb{C}} = \mathbb{C} \cup \{\infty \cdot e^{2\pi i\theta} | \theta \in \mathbb{R}/\mathbb{Z}\}$. Then, we may take the *closed external ray of angle t*, $\overline{R(t)}$, to be the closure of $R(t)$ in $\widetilde{\mathbb{C}}$. This closure is formed by including the landing point on the Julia set, $\gamma(t)$, and the limit point on the circle at infinity, $\lim_{r\to\infty} \phi(re^{2\pi it})$.

We may now introduce three fundamental constructions: the formal, topological, and essential matings.

Definition 2.5. Let $P_1 : \widetilde{\mathbb{C}}_1 \to \widetilde{\mathbb{C}}_1$ and $P_2 : \widetilde{\mathbb{C}}_2 \to \widetilde{\mathbb{C}}_2$ be postcritically finite monic degree d polynomials taken on two disjoint copies of $\widetilde{\mathbb{C}}$, and let $\sim_f$ be the equivalence relation which identifies $\infty \cdot e^{2\pi it}$ on $\widetilde{\mathbb{C}}_1$ with $\infty \cdot e^{-2\pi it}$ on $\widetilde{\mathbb{C}}_2$ for all $t \in \mathbb{R}/\mathbb{Z}$. Then, the quotient space $\widetilde{\mathbb{C}}_1 \bigsqcup \widetilde{\mathbb{C}}_2 / \sim_f$ may be identified with $\mathbb{S}^2$. The map which descends to this quotient space is the *formal mating* of P_1 and P_2.

It should be noted that the formal mating is well defined and yields a continuous branched covering of $\mathbb{S}^2$ to itself.

Definition 2.6. Let $\mathbb{S}^2$ denote the domain of the formal mating as above, and let $\sim_t$ be an equivalence relation on $\mathbb{S}^2$ where any points sharing a closed external ray are contained in the same equivalence class . Allowing the formal mating to descend to the quotient space $\mathbb{S}^2 / \sim_t$ yields the *topological mating* $P_1 \perp\!\!\!\perp P_2$.

Note that a given ray equivalence class of $\sim_t$ may contain points from several external rays because pairs of closed external rays meet where identified by $\sim_f$, and also because multiple external rays may land at multiply accessible points of J_1 or J_2. This equivalence relation generates a quotient space that can alternately be viewed as resulting from gluing the filled Julia sets of P_1 and P_2 together along the boundaries. Since the polynomials we examine are critically preperiodic and possess (filled) Julia sets that are dendrites, $\mathbb{S}^2 / \sim_t$ has the potential to be a very peculiar space. However, by Moore's theorem if the equivalence classes of $\sim_t$ are closed, and no equivalence class of $\sim_t$ separates $\mathbb{S}^2$, then $\mathbb{S}^2 / \sim_t$ is homeomorphic to $\mathbb{S}^2$.

Moore's theorem is a powerful result, but it may be difficult to visualize the mapping behavior of the topological mating on $\mathbb{S}^2$. We will frequently make use of the *essential* (or *degenerate*) *mating* $P_1 \perp\!\!\!\perp_e P_2$ instead, as it is somewhat of an intermediary between the formal and topological mating. The intuition behind its development in [**12**] is this: occasionally, the formal mating is not Thurston equivalent to a rational map due to Levy cycles which yield Thurston obstructions. In essence, we collapse only the equivalence classes of $\sim_t$ which are needed to fix this problem, and make minor adjustments so that the map descending to the quotient remains a branched covering, as described below.

DEFINITION 2.7. Allow $h : \mathbb{S}^2 \to \mathbb{S}^2$ to denote the formal mating of P_1 and P_2, and suppose that no point on the critical orbit of h is in an equivalence class of $\sim_t$ which forms a closed loop.

Let $\{\tau_1, \ldots, \tau_m\}$ be the set of equivalence classes of $\sim_t$ that (a) contain at least one point on the critical orbit of h and (b) have an equivalence class of $\sim_t$ containing ≥ 2 elements of P_h in their forward orbit under h. for some n. Define $\sim_e$ as the refinement of $\sim_t$ whose only nontrivial equivalence classes are $\{\tau_1, \ldots, \tau_m\}$.

Let $\{V_1, \ldots, V_n\}$ be a collection of open tubular neighborhoods (respectively) of $\{\tau_1, \ldots, \tau_n\}$, selected so that distinct V_i are disjoint and that each V_i contains precisely the points on the critical orbit that τ_i does. Then, define $\{U_1, \ldots, U_m\}$ to be the collection of connected components of $h^{-1}(\bigcup_{i=1}^{n} V_i)$ which do not intersect any τ_i.

Let $\pi : \mathbb{S}^2 \to \mathbb{S}^2/\sim_e$ *be* the natural projection. We define $g : \mathbb{S}^2/\sim_e \to \mathbb{S}^2/\sim_e$ as follows: On each $\overline{U_i}$, define g to be a local homeomorphism which is equal to $\pi \circ h \circ \pi^{-1}$ on ∂U_i. Set $g := \pi \circ h \circ \pi^{-1}$ on the remainder of $\mathbb{S}^2/\sim_e$. The map g is the *essential mating* of P_1 and P_2.

Note that since none of the τ_i are loops, $\mathbb{S}^2/\sim_e$ is homeomorphic to $\mathbb{S}^2$. (We will thus simplify notation by treating g as a self-map of $\mathbb{S}^2$.) As h maps equivalence classes to equivalence classes, $\pi \circ h \circ \pi^{-1}$ is well defined on $\mathbb{S}^2$. With the alteration of this composition on each U_i though, we ensure that g does not map arcs to points. The result is a map which is Thurston equivalent to the topological and possibly geometric matings while retaining much of the relative simplicity in structure of the formal mating. In fact, when $\sim_t$ never identifies postcritical points of the formal mating, the formal and essential matings are the same map.

2.4. Laminations. Recall that a set of the form $\widetilde{\mathbb{C}}_1 \sqcup \widetilde{\mathbb{C}}_2/\sim_f \cong \mathbb{S}^2$ served as the domain of the formal mating. The canonical equator of this space contains a set of equivalence class representatives of $\sim_t$. Further, restricting to this set, the topological mating is semiconjugate to the d-fold map on $\mathbb{S}^1$. As equivalence relations like this will be crucial to our discussion, we introduce a common tool for geometric representation of equivalence relations on $\mathbb{S}^1$: the geodesic lamination.

DEFINITION 2.8. Let L be a set of geodesics on the closed hyperbolic disk $\overline{\mathbb{D}}$ which are disjoint, except possibly at their endpoints. We will call these geodesics *leaves*, and will also allow for the inclusion of degenerate leaves in L, which are single points on ∂D. If $\bigcup L$ is closed, we call L a *lamination* on $\overline{\mathbb{D}}$.

The closures of components of $\mathbb{D} \setminus \bigcup L$ are called *gaps*. If a gap has finite intersection with $\partial\mathbb{D}$, this gap is an *ideal polygon*.

Laminations were introduced by Thurston as a tool for discussing the geometry and dynamics of polynomials, although 2-sided analogues of laminations have also been used to discuss matings and other maps [**3**], [**11**], [**13**], [**14**]. To draw a parallel to previous ideas, consider the Julia set J of a postcritically finite polynomial of degree d: If any external rays land at a multiply accessible point of J, we mark the corresponding angles for these rays on $\partial\mathbb{D}$, and take the boundary of the convex hull between the points to determine a collection of leaves. The set of all such leaves defines a lamination which, when collapsed along its leaves and filled-in ideal polygons, yields a topological model of the structure of J.

REMARK 2.9. While geodesic laminations are not given significant further attention within this manuscript, we have introduced them here as they are referenced heavily in the discussion of equivalence relations in [**7**]. For the convenience of the reader following along, we note that in that discussion, the text uses a somewhat nonstandard definition for "leaves": these are taken to be the convex hull of a finite collection of points on the boundary of $\mathbb{D}$.

2.5. Oriented curves and 2-color tilings. Like laminations, tilings are another structure frequently used to represent matings and their approximations. In particular, we will examine 2-color tilings that are associated with oriented curves on $\mathbb{S}^2$.

DEFINITION 2.10. We say that a closed curve $\mathcal{C}:\mathbb{S}^1\to\mathbb{S}^2$ with at most finitely many self-intersections is *oriented* if $\mathcal{C}$ can be deformed to a simple closed curve by a small perturbation–that is, there exists a continuous map $h:\mathbb{S}^1\times[0,1]\to\mathbb{S}^2$ such that $h(\mathbb{S}^1,0)=\mathcal{C}$, and $h(\theta,t)\neq h(\theta',t)$ whenever $\theta\neq\theta'$. A key implication here is that while such a curve may self-intersect, it is not allowed to "self-cross."

Kameyama refers to such distorted Jordan curves $h(\mathbb{S}^1,t), t\neq 0$ as *unlacings* of $\mathcal{C}$. These are also analagous to Meyer's concept of *geometric representation* as in [**8**, Section 6].

If desired, unlacings can be constructed so as to be pseudo-isotopic to the original $\mathcal{C}$ [**6**, Definition 6.7].

A subtle point of note is that while $\mathcal{C}$ may have an *orientation* implied by its parameterization, the distinction of being *oriented* as above is meant to emphasize how arcs of $\mathcal{C}$ can be viewed as oriented with respect to the components of $\mathbb{S}^2\smallsetminus\mathcal{C}$. We may view an oriented curve $\mathcal{C}$ as generating a cellular decomposition of $\mathbb{S}^2$ by marking points of self intersection (and possibly finitely many other point along $\mathcal{C}$) as 0-cell vertices, arcs of $\mathcal{C}$ between these points as 1-cell edges, and the connected components of $\mathbb{S}^2\smallsetminus\mathcal{C}$ as 2-cell tiles. If these marked points are elements of post_f for some Thurston map f, this is similar to the cellular decomposition discussed in Definition 2.2—except here there are more than two tiles. We will denote the sets of closed vertices, edges, and tiles respectively as the set of 0-*vertices* $\mathbf{V}^0$, set of 0-*edges* $\mathbf{E}^0$, and set of 0-*tiles* $\mathbf{X}^0$. (We similarly refer to the sets of pullbacks of these objects by f^n respectively as $\mathbf{V}^n$, $\mathbf{E}^n$, and $\mathbf{X}^n$. To emphasize, the n in this context is meant to specify an iterate rather than a dimension.)

If $\mathcal{C}$ is an oriented curve, then the 1-cell arcs surrounding a given tile are either all positively oriented to the tile (in which case we will color it white), or all negatively oriented to the tile (in which case we will color it black). There are no other options for tiles: note that an unlacing of $\mathcal{C}$ is colored similarly, the pseudo-isotopy deforming the unlacing to $\mathcal{C}$ induces this "checkerboard" coloring, where every edge separates a white tile from a black one.

Any time that we refer to a tiling of $\mathbb{S}^2$ associated with an oriented curve $\mathcal{C}$, we will assume that the tiling was constructed in the above manner. Given such a tiling, we will define connections at vertices of the tiling as follows:

DEFINITION 2.11. Let v be any self intersection point of $\mathcal{C}$, $\mathcal{C}'$ an unlacing of $\mathcal{C}$ for which $h : \mathbb{S}^2 \times [0,1] \to \mathbb{S}^2$ is a pseudo-isotopy. Further, let $U(v)$ be some small neighborhood containing v and each other point $u \in \mathcal{C}'$ for which $h(u,1) = v$. (We assume however that no other marked points of $\mathcal{C}$ are contained in this set.) If the deformation of any connected component of $U(v) \setminus \mathcal{C}'$ via h has multiple connected components, the (same-color) tiles containing these components are said to be *connected at* v.

The intuition is to make note of which tiles an unlacing "connects" at the vertex v. With additional labeling of structures, this satisfies the conditions of a connection per [**8**].

3. Unmating

We have previously noted that the topological mating is semi-conjugate to the d-fold map on $\mathbb{S}^1$. Much of the decomposition theory of matings centers on whether such a semiconjugacy—or at least a curve with similar behavior under the d-fold map—exists for a given map f. We reproduce a number of definitions from [**6**] to clarify the properties of this curve which are desired, and to give context for later results on the decomposition of expanding Thurston maps via algorithms detailed by Meyer.

3.1. Fundamental results.

DEFINITION 3.1. Let f be an expanding degree d Thurston map, and let P_f^a denote the set of postcritical points which eventually map forward to a critically periodic cycle for f. Recall that $q_d : \mathbb{S}^1 \to \mathbb{S}^1$ denotes the d-fold map given by $q_d(t) = td \bmod 1$ on the circle $\mathbb{S}^1 \cong \mathbb{R}/\mathbb{Z} \cong [0,1)$.

We say that J is $\mathbb{S}^1$*-parametrizable* if there exists a continuous surjection $\phi : \mathbb{S}^1 \to J$ so that $f \circ \phi = \phi \circ q_N$ for some $N \in \mathbb{N}$.

A closed curve $\mathcal{C} : \mathbb{S}^1 \to \mathbb{S}^2 \setminus P_f^a$ is said to be *fully f-invariant up to homotopy* if there exists a closed curve $\mathcal{C}_1 : \mathbb{S}^1 \to f^{-1}(\mathcal{C})$ and homotopy $h : \mathbb{S}^1 \times [0,1] \to \mathbb{S}^2 \setminus P_f^a$ from $\mathcal{C}$ to $\mathcal{C}_1$ with $(\mathrm{post}_f \setminus P_f^a)$ fixed so that $f \circ \mathcal{C}_1 = \mathcal{C} \circ q_d$. In other words, we should be able to continuously deform $\mathcal{C}$ to its pullback relative to the postcritical set.

We now summarize a key result of [**6**] in Theorem 3.2:

THEOREM 3.2 (Kameyama). *Suppose f is an expanding Thurston map. If $\mathcal{C} : \mathbb{S}^1 \to \mathbb{S}^2 \setminus P_f^a$ is an oriented closed curve which is fully f-invariant up to homotopy, then J is orientedly $\mathbb{S}^1$-parametrizable and some iterate of f is Thurston equivalent to a degenerate mating of polynomials.*

While Kameyama investigates the more general case of expanding Thurston maps with suitable $\mathcal{C}$, in this paper we emphasize where f is expanding and has only strictly preperiodic critical points. The above result simplifies in this setting: when $P_f^a = \varnothing$ we have that J is the sphere, and the arguments of [**6**] show that f itself (rather than a higher iterate) is Thurston equivalent to a degenerate mating.

In a series of papers [**8**],[**7**],[**9**], Meyer adds substantial detail to Kameyama's result. In the setting where f has only strictly preperiodic critical points, he constructs a Jordan curve which is fully f^n-invariant up to homotopy, giving explicit instructions for using this curve to obtain $\mathbb{S}^1$-parameterization, and also demonstrating how to use the $\mathbb{S}^1$-parameterization to find labels for the polynomials in the unmating. We briefly summarize the unmating approach below, giving key sections for relevant discussion of each "step":

(1) Suppose that $f : \mathbb{S}^2 \to \mathbb{S}^2$ is an expanding Thurston map without periodic critical points. A Jordan curve $\mathcal{C} \subseteq \mathbb{S}^2$ containing post_f is selected. For a sufficiently large $n \in \mathbb{N}$, there exists a Jordan curve γ^0 which is invariant for $F = f^n$ and isotopic to $\mathcal{C}$ relative to post_f, thus F can be viewed as the subdivision map for a 2-tile subdivision rule on $\mathbb{S}^2$ as in [**1**, Section 15] and [**2**]. Connections at vertices are investigated to develop a pseudo-isotopy H^0 which deforms γ^0 to $\gamma^1 = F^{-1}(\gamma^0)$ relative to post_f. See [**8**, Sections 6–8].
(2) The connections at vertices are marked to assist with appropriately parameterizing γ in a later step. See [**8**, Definition 6.4].
(3) Successive lifts via F of the pseudo-isotopy H^0 are used to develop a sequence of closed curves, γ^n. This sequence converges uniformly (with respect to the visual metric for F) to a curve γ. See [**8**, Sections 3–4].
(4) A re-parameterization $\gamma^0 : \mathbb{S}^1 \to \mathbb{S}^2$ is assigned to γ^0, so that F is topologically semiconjugate to the dn-fold map on $\mathbb{S}^1$ via γ^0. With this assigned parameterization, γ^0 is an oriented curve which is fully F-invariant up to homotopy. The successive lifts of γ^0 induce a parameterization of γ, which allows it to serve as an $\mathbb{S}^1$-parameterization of $\mathbb{S}^2$. See [**8**, Sec 4], as well as [**9**, Section 9] for implementation.
(5) We consider an equivalence relation $\sim$ on $\mathbb{S}^1$ given by $s \sim t \iff \gamma(s) = \gamma(t)$, along with several other equivalence relations induced by γ^0. The equivalence relation $\sim$ can be interpreted in terms of limiting operations on the $\sim_n$ relations, as well as in terms of a two-sided lamination. See [**7**, Sections 5–7].
(6) The two "sides" of this lamination are generated by polynomials. Further, we can use the finite data from self intersections of γ^1 to generate critical portraits of the specific polynomials in the mating. See [**7**, Section 7], as well as [**9**, Section 9] for implementation.

While Kameyama and Meyer's results are very similar, there are a few key differences that are of interest to us. First, the unmating process is much more explicitly constructive than Kameyama's–to the tune of well over a hundred pages of additional detail! This comes at the cost that the initial setting considered by Meyer is more restrictive: the oriented fully f-invariant up to homotopy curve that Meyer initializes with is Jordan. Indeed, when a map possesses a *pseudo-equator* (roughly speaking, a Jordan γ^0 and the pseudo-isotopy noted in step (1) above), it is a sufficient condition for F to be a mating–but Meyer notes that this criterion is not a necessary one. For example, any topological mating with no postcritical-to-postcritical identifications has a pseudo-equator [**15**], but there are examples in [**8**] and [**9**] of matings that do not have one. Of note is that both of these examples *do* still possess curves which are fully f invariant up to homotopy. Kameyama's construction, which effectively take the equator curve in a degenerate mating (as in [**6**, Proposition 6.6])—does not necessitate that $\mathcal{C}$ is Jordan. In the event that $\sim$ contains at least one equivalence class with two postcritical points from the same polynomial, the equator curve is "pinched" in the resulting quotient space for the degenerate mating, and is no longer a Jordan curve.

3.2. An example. We consider the case of the example from [**8**, Section 10]: Combinatorial data for the two polynomials and their mating are noted in Figures 1 and 2. In Figure 1, $f_{5/12}$ and $f_{1/12}$ reference two Misiurewicz polynomials that are parametrized by angles of external rays landing at the respective critical values for each polynomial. The critical orbit portraits are as follows:

$$f_{5/12}:\ c_1 \xrightarrow{2} p_1 \longrightarrow p_3 \longrightarrow p_0 \circlearrowleft$$

$$f_{1/12}:\ c_2 \xrightarrow{2} p_2^* \longrightarrow p_3^* \longrightarrow p_0^* \rightleftarrows p_0'$$

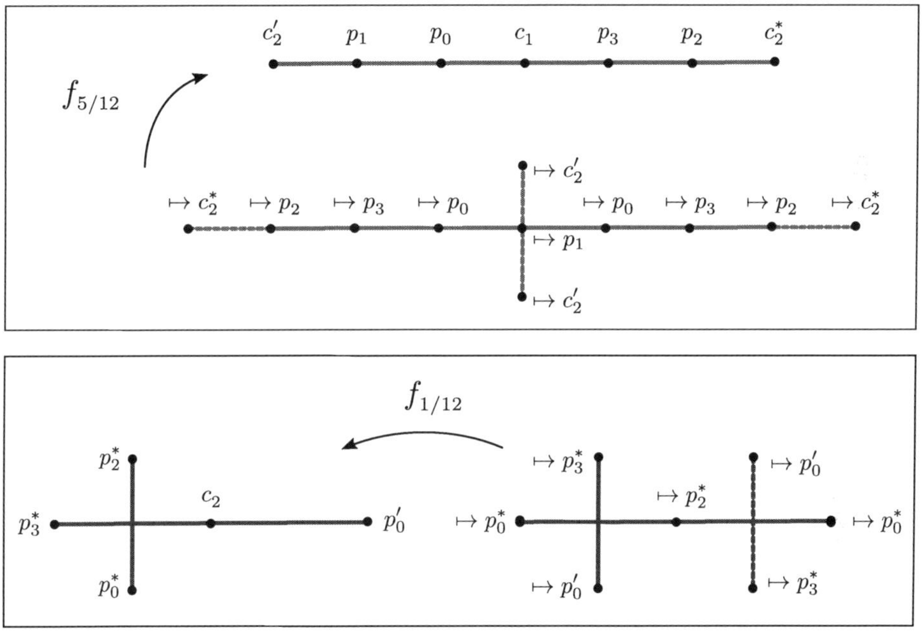

FIGURE 1. Combinatorial mapping data for two polynomials whose mating has no pseudo-equator. The Hubbard trees for $f_{5/12}$ and $f_{1/12}$ are subsets of the above trees; additional edges and vertices have been marked to emphasize where postcritical identification will occur in the essential mating of these polynomials..

The graph structures noted in Figure 1 contain Hubbard trees for the polynomials $f_{5/12}$ and $f_{1/12}$, which give a combinatorial description of the mapping behavior of these polynomials. The postcritical points here are labeled to emphasize postcritical identifications forced by $\sim_e$: any points sharing a letter and subscript are collapsed in the topological and essential matings. We then default to using the relevant letter and subscript (sans other markings) to demonstrate the mapping behavior of the essential mating in Figure 2.

Of note in Figure 2: the dashed lines indicate a curve isotopic to $\mathbb{S}^1/\sim_e$ where $\mathbb{S}^1$ is the equator of the formal mating; along with its pullback. This is clearly no longer a Jordan curve, but there is a parameterization for the pullback of this curve which is orientation preserving and pseudo-isotopic to the original relative to the postcritical set. (The dashed lines have been drawn slightly offset from the postcritical set to emphasize this.)

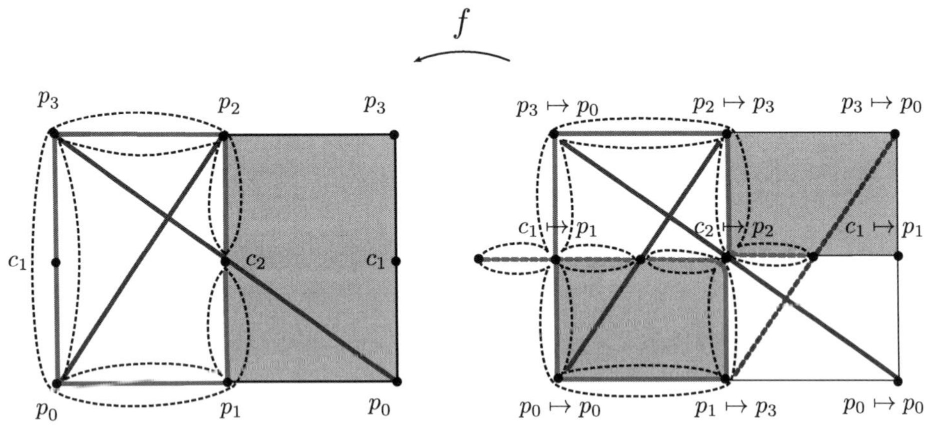

FIGURE 2. Meyer's example of a map with no pseudo-equator. The finite subdivision rule has been overlaid with 1-skeletons containing Hubbard trees for polynomials $f_{1/12}$ and $f_{5/12}$. This map is Thurston-equivalent to the essential mating of these polynomials.

4. Revisiting the unmating construction

Let f be an expanding Thurston map with no periodic critical points, so that J is the sphere.

An early step in Meyer's unmating process is the construction of a pseudo-isotopy which deforms a Jordan curve $\mathcal{C}$ containing post_f into its pullback by an iterate of f. We present an outline of Meyer's unmating algorithm with considerations for changes to be made in light of initializing with a fully f-invariant up to homotopy curve which is not necessarily simple. We begin by highlighting an immediate point of concern, and using this as a tool to discuss our approach through the rest of this section.

One potential problem to reconcile is that the Jordan curve $\mathcal{C}$ is used to generate much of the topological structure on $\mathbb{S}^2$ in which we discuss notions of convergence. However, as noted earlier in Subsection 2.2, the visual metric ϱ and expansion constant Λ that are associated with this $\mathcal{C}$ are not constrained to only work with $\mathcal{C}$: any Jordan curve through post_f can be shown to have a similar relationship (up to a scaling factor) with ϱ and Λ [**1**, Proposition 8.3]. With this in mind, many arguments posed involving open neighborhoods still apply because these sets are still open in our setting; however they may not relate to the surrounding structures in the initially intended manner. For example, consider the following:

LEMMA 4.1 (existence of neighborhoods between balls [**8**, Lemma 2.3]). *Let ϱ be a visual metric for f with expansion factor Λ. Then there exist $\varepsilon_0 > 0$ and some constant $K \geq 1$ such that the following is true:*

Suppose $0 < \varepsilon < \varepsilon_0$ and let $\mathcal{N}_\varrho(\mathbf{V}^1, \varepsilon)$ be the ε-neighborhood of $\mathbf{V}^1$. Then there exists some neighborhood V^1 of $\mathbf{V}^1$ such that $\mathcal{N}_\varrho(\mathbf{V}^1, \frac{\varepsilon}{K}) \subset V^1 \subset \mathcal{N}_\varrho(\mathbf{V}^1, \varepsilon)$.

Further, the set $V^{n+1} := F^{-n}(V^1)$ satisfies
$\mathcal{N}_\varrho(\mathbf{V}^{n+1}, \Lambda^{-n}\frac{\varepsilon}{K}) \subset V^{n+1} \subset \mathcal{N}_\varrho(\mathbf{V}^{n+1}, \Lambda^{-n}\varepsilon)$ for all $n \in \mathbb{N}$.

The above lemma is significantly dependent on [**1**, Lemma 8.10], which we have reproduced here as Lemma 2.4. The $U^n(x)$ set referenced in Lemma 2.4 is a union of n-tiles associated with a Jordan curve $\mathcal{C}$ through post_f. In Meyer's setting, the invariant curve γ^0 was being used with an iterate of f for this $\mathcal{C}$, and this curve serves as the 1-skeleton for a 2-tile finite subdivision rule on $\mathbb{S}$. In other words, the n-tiles subdivide into $(n+1)$-tiles and the open set int $U^n(x)$ can be expressed neatly in terms of n-tiles of γ^0.

In our setting, we work with f directly, and there may not be a Jordan curve through post_f which is invariant with regards to this map. We do assume, however that a Jordan $\mathcal{C}$ is chosen, and that the relevant ϱ and expansion constant Λ have been selected for our setting. Even though $\mathcal{C}$ doesn't generate a finite subdivision rule, we can still pull back the curve to generate n-tiles–so, $U^n(x)$ is still a meaningful object in this setting. However, our algorithm doesn't use $\mathcal{C}$ or any other Jordan curve as γ^0. The set $U^n(x)$ then will have little to do with our choice of γ^0 or its n-tiles, aside from living in the same metric space generated by ϱ.

With that said, Lemma 2.4 is only needed to obtain an open set of a particular size for Lemma 4.1, and we do not in this instance require the combinatorial constraints of a finite subdivision rule—so existing in the same metric space is enough. Regardless of setting, int $U^n(x)$ can still function as the desired open set, and so Lemma 4.1 still holds.

We largely follow the constructions of Meyer outlined in the previous section, and for brevity assume that minor adjustments like the above are taken. Our emphasis will be on places where the analagous constructions differ more significantly—however, arguments and supporting lemmas will be summarized or reproduced to provide context. To assist the reader, we have tried to conform to the notational conventions used in these papers as closely as possible. Subsections here roughly align with the steps enumerated in Section 3.

4.1. Initializing with non-Jordan curves. We begin by stating a key lemma from Meyer's construction to initialize our unmating.

LEMMA 4.2. *(Isotopic Schönflies theorem* [**8**, Theorem 5.1]*) Let $\gamma, \sigma \subset \mathbb{D}$ be two Jordan arcs with common endpoints $p, q \in \overline{\mathbb{D}}$. Then, there is an isotopy of $\overline{\mathbb{D}}$ relative to $\partial\mathbb{D} \cup \{p, q\}$ that deforms γ to σ.*

THEOREM 4.3. *Suppose that f is an expanding degree d Thurston map whose Julia set is $\mathbb{S}^2$, and let $\gamma^0 : \mathbb{S}^1 \to \mathbb{S}^2$ be an oriented closed curve which is fully f-invariant up to homotopy. Then, there exists a pseudo-isotopy $H_0 : \mathbb{S}^2 \times [0,1] \to \mathbb{S}^2$ satisfying the following properties:*

1. *H^0 is a pseudo-isotopy relative to $V_0 = \text{post}_f$.*
2. *The set of all 0-edges is deformed by H^0 to the set of all 1-edges.*
3. *Only finitely many points of γ^0 are deformed by H_0 to any individual 1-vertex v.*

(4) *Let* $\varepsilon_0 > 0$ *be the constant from Lemma* 4.1, $0 < \varepsilon < min\{\varepsilon_0, 1/2\}$ *and* V^1 *be a neighborhood of* $\mathbf{V}^1$ *also as in Lemma* 4.1. *We will further require that* $H^0 : \mathbb{S}^2 \times [1-\varepsilon, 1] \to \mathbb{S}^2$ *is supported on* V^1.

PROOF. Without loss of generality, we may assume that γ^0 has at most finitely many self-intersections occurring only at points in post_f [**6**, Theorem 3.6]. As such, we may assume that $\gamma^1 = f^{-1}(\gamma^0)$ is an oriented closed curve with at most finitely many self-intersections as well, and construct H^0 as follows.

Since γ^0 is oriented, a small perturbation of this curve yields a Jordan curve $\gamma^{0'}$ which differs from γ^0 only on small neighborhoods surrounding its points of self-intersection. We obtain a similar result for γ^1, where a small perturbation on neighborhoods $U_1, \ldots, U_n$ of intersection points yields the Jordan curve $\gamma^{1'}$. We may take $\gamma^{1'}$ to be an "unlacing" of γ^1 which is homeomorphic to $\gamma^{0'}$ and which traverses the elements of post_f in the same cyclic order. Repeated application of Lemma 4.2 to corresponding boundary arcs of $\gamma^{0'}$ and $\gamma^{1'}$ yields an isotopy $H^a : \mathbb{S}^2 \times [0, 1-\varepsilon] \to \mathbb{S}^2$ that deforms $\gamma^{0'}$ into $\gamma^{1'}$. We may concatenate this isotopy with a pseudo-isotopy $H^b : \mathbb{S}^2 \times [1-\varepsilon] \to \mathbb{S}^2$ that fixes post_f, but deforms the Jordan curve $\gamma^{1'}$ into γ^1. We will call this concatenation the pseudo-isotopy H^0.

Criteria (1) and (2) follow immediately from the construction. Note that the cycle of postcritical points visited by γ^0 is a subsequence of those visited by γ^1, which includes repeat visits to elements of post_f that are included at the very end of the pseudo-isotopy H^0. As Lemma 4.2 is the machinery behind the pseudo-isotopy H^b, (3) follows. We may guarantee (4) with careful selection of both ε and a maximum allowable diameter for the sets $U_1, \ldots, U_n$. □

REMARK 4.4. It should be noted that the original construction of H^0 requires a 5th property regarding $f : H_1^0(\gamma^0) \to \gamma^0$ being a d-fold cover, however we start with a stronger condition in light of γ^0 being fully f-invariant up to homotopy: this implies by definition that $f \circ \gamma^1 = \gamma^0 \circ q_d$.

REMARK 4.5. Per the proof of [**8**, Lemma 3.11], we expect this d-fold cover property exactly when f is orientation preserving. In light of this, applying the above construction with a Jordan γ^0 would satisfy conditions for what Meyer calls a *pseudo-equator*.

REMARK 4.6. If a map has a pseudo-equator it is equivalent to a mating, but at the comparable point in Meyer's construction $F = f^n$ is used to pull back γ^0 to obtain γ^1, where we instead use f. This demonstrates that the iterate f^n is a mating instead of f.

4.2. Marking connections at vertices. As the H^0 of Theorem 4.3 initially assumes γ^0 to be fully f-invariant up to homotopy, its relationship with q_d yields that our parameterizations for γ^0 and γ^1 are implied. With this extra information, we have less to construct than in [**8**]: for Meyer there may be multiple oriented Eulerian paths along the set of points in $F^{-1}(\gamma^0)$—and one must be constructed and selected before constructing any sort of pseudo-isotopy. Meyer's tile-centered construction thus eventually allows for the discovery of shared matings—however, these typically reflect an iterate of f rather than f itself.

With that said, as long as the Eulerian paths determined by γ^0 and γ^1 are understood up to homotopy type, we can recover an $\mathbb{S}^1$-parameterization for f via the methods described here with or without the assumed parameterizations. We just need to know the sequence of postcritical points visited along γ^0 (in order, and including return visits), can be viewed as a subsequence of the postcritical points visited in order along the pseudo-isotopic pullback curve γ^1.

To do this requires an understanding of connections–and we must be particularly careful because γ^0 is not necessarily assumed to be Jordan. This means we need to note connections of both γ^0 *and* γ^1 instead of just γ^1. To help with this, we will adopt a new convention regarding *marking* vertices at connections, which is singling out an arc on a curve in order to "keep track" of where we expect a relevant postcritical point to be. (In figures of a geometric representation, Meyer typically denotes this with a dot on the relevant arc; we may name or label vertices and/or arcs as needed.) Meyer marks a single arc at each connection in his version of γ^1; we mark *all* arcs in the geometric representation of γ^0. As elements of post_f are possible intersection points of γ^0, we expect that multiple parameters may yield a visit to the same point—and we "keep track" of the parameters accordingly.

Note that considerations determining which vertices are connected and how markings are constructed, etc. do not differ significantly from the unmating approach—we simply have more markings to keep track of.

4.3. The limit of approximating curves γ^n. We now develop a sequence of approximations to the desired $\mathbb{S}^1$-parameterization by taking lifts of our pseudo-isotopy. To do so, we recall a fundamental result for lifts of pseudo-isotopies:

LEMMA 4.7 (Lifts of pseudo-isotopies [**8**, Lemma 3.4]). *Let $H : \mathbb{S}^2 \times [0,1] \to \mathbb{S}^2$ be a pseudo-isotopy relative to $\text{post}_f = \mathbf{V}^0$. Then H can be lifted uniquely by f to a pseudo-isotopy $\tilde{H}_1$ relative to $f^{-1}(\mathbf{V}^0) = \mathbf{V}^1$. This means that $f(\tilde{H}_1(x,t)) = H(f(x),t)$ for all $x \in \mathbb{S}^2$ and all $t \in [0,1]$.*

Further, let H_n be the lift of H_0 by an iterate f^n. Then diam $H_n :=$ max $_{x \in \mathbb{S}^2}$ diam$\{H_n(x,t) : t \in [0,1]\}$, and we have that diam $H_n \leq C\Lambda^{-n}$.

(Diameter here is measured with respect to the fixed visual metric ϱ with expansion factor $\Lambda > 1$. The constant C is independent of n.)

The argument supporting this lemma is independent of the choice of curve $\mathcal{C}$ that ϱ and Λ are associated with, and also independent of the initial pseudo-isotopy H. Using the H^0 of Section 4.3, we may now repeatedly apply Lemma 4.7 to obtain a sequence of unique pseudo-isotopies $\{H^n$ relative to $\mathbf{V}^n\}$ deforming γ^n into γ^{n+1}.

We obtain the following analogue of [**8**, Lemma 3.5], which states how the properties of H^0 noted in Theorem 4.3 convey to the pseudo-isotopies $\{H^n$ relative to $\mathbf{V}^n\}$.

LEMMA 4.8. *Let H^0 be a pseudo-isotopy as in Theorem 4.3, and H^n denote the lift of H^0 by f^n. The lifts H^n possess the following properties:*

(1) *H^0 is a pseudo-isotopy relative to $\mathbf{V}^n = f^{-n}(\text{post}_f)$.*
(2) *The set of all n-edges is deformed by H^n to the set of all $(n+1)$-edges.*
(3) *Only finitely many points of γ^n are deformed by H^n to any individual $(n+1)$-vertex v.*
(4) *Let V^1_ε be the neighborhood of $\mathbf{V}^1$ as in Theorem 4.3. Then H^n is supported on the set $\mathbf{V}^{n+1} := f^{-n}(\mathbf{V}^1)$, which is a neighborhood of $\mathbf{V}^{n+1}$.*

PROOF. This largely follows in an identical manner to the analogous Lemma in [**8**, Lemma 3.5]. Statement (2) requires a small amount of additional bookkeeping since it is possible for more than one 0-edge in our setting to start at the same point:

Meyer uses the fact that lifts of paths with distinct starting points have distinct endpoints, and uses this to count lifts of edges. We cannot precisely replicate this argument as some distinct 0-edges share initial points. Sub-arcs contained in these 0-edges could be constructed to have unique initial points though–and so we can still appropriately index the necessary lifts to guarantee an appropriate number of $(n+1)$-edges. □

4.4. Parameterization of γ. In this subsection we outline the procedures in [**8**, Section 4], [**9**, Section 9] for context, as they are largely unaltered in our setting.

Note that if we know which values of t yield the points in $\mathbf{V}^n$ that we may pull back to obtain the parameters for the elements of $\mathbf{V}^{n+1}$, and so on as these have been appropriately marked on γ^{n+1}. Taking the γ^n as any orientation-preserving homeomorphism on the remaining arcs, we have that $\gamma : \mathbb{S}^1 \to \mathbb{S}^2$ yields an oriented $\mathbb{S}^1$-parameterization of $\mathbb{S}^2$ when defined so that $\gamma(t) := \lim_{n\to\infty} \gamma^n(t)$.

The beauty in the argument is that we obtain this parameterization starting with finite data–we need only start with parameters for the elements in post_f. The parameters for $\mathbf{V}^0 = \text{post}_f$ separate $\mathbb{S}^1$ into a collection of sub-arcs. Since we expect $f \circ \gamma(t) = \gamma(d \cdot t)$, we know which sub-arcs map onto others, and can develop an edge transition matrix to represent this transition. An extension of the Perron–Frobenius theorem guarantees a unique positive eigenvalue equal to the spectral radius of this matrix, which in this setting is the degree of f. This eigenvalue is simple, and has an eigenvector with all positive values, representing respective lengths of each of the arcs on $\mathbb{S}^1$ (after a suitable scaling for overall unit length).

Knowing the lengths of the respective arcs now means that given a parameter t for any postcritical point p, we have two ways of obtaining the parameter for $f(\gamma(t))$: we can add the appropriate lengths of arcs to t to find the new parameter, or we can multiply t by d to obtain the new parameter. This allows us to set up an equation to solve for t.

With the parameterizations above for γ and each of the γ^n, we have that the sequence of curves $\{\gamma^n\}$ converge uniformly to γ in the visual metric ϱ. Further, if desired, we could construct a pseudo-isotopy from γ^0 to γ by modifying a concatenation of the sequence of homotopies $\{H^n\}$.

REMARK 4.9. In our setting, we begin with γ^0 an oriented curve which is fully f-invariant; implying a given parameterization. However, we did not need to actually know this parameterization and could recover the parameters for the postcritical points without it using the above procedure. The only crucial difference in application is that with a self-intersecting γ^0, postcritical points may have been represented with multiple parameters–however we have offset this by our choice to make additional markings on γ^0 to compensate. The application of techniques in this section is otherwise identical.

4.5. Laminations and equivalence relations. At this point, γ and the sequence of γ^n are used to generate several equivalence relations on $\mathbb{S}^1$:

- The equivalence relations $\overset{n,w}{\sim}$ and $\overset{n,b}{\sim}$: Consider a self intersection point v of γ^n. Let $\gamma^{n'}$ be an unlacing of γ^n, and $U(v)$ be a small neighborhood of v as in Definition 2.11. We now mark each arc of $U(v) \cap \gamma^{n'}$, assuming that these marks correspond to angles in $\mathbb{S}^1$ which γ^n sends to v. If any of the marked points lie on the boundary of the same white component of $U(v)-$

γ, we say that their corresponding angles are in the same equivalence class of the equivalence relation $\overset{n,w}{\sim}$. (We may similarly define an equivalence relation for points lying on the boundary of the same black component, $\overset{n,w}{\sim}$.)

- The closure of $\overset{\infty}{\sim}$: We define the equivalence relations $\overset{n}{\sim}$ so that $s \overset{n}{\sim} t$ if and only if $\gamma^n(s) = \gamma^n(t)$. (As the γ^n are inductively defined, in a sense these can be interpreted as inductively defined from $\overset{0}{\sim}$.) We then take $\overset{\infty}{\sim} := \bigvee \overset{n}{\sim}$. The closure of this is an equivalence relation.
- The equivalence relation $\sim$: We define $\sim$ so that $s \sim t$ if and only if $\gamma(s) = \gamma(t)$.
- The equivalence relation $\overset{w}{\sim} \vee \overset{b}{\sim}$: We proceed similarly to above. Note that we have already defined the equivalence relations $\overset{n,w}{\sim}$ and $\overset{n,b}{\sim}$, and that these similarly can be interpreted as inductively defined from $\overset{0,w}{\sim}$ and $\overset{0,b}{\sim}$ respectively. We take $\overset{w}{\sim} := \bigvee \overset{n,w}{\sim}$, and similar for $\overset{b}{\sim}$. $\overset{w}{\sim} \vee \overset{b}{\sim}$ is an equivalence relation.
- The equivalence relation $\approx := \overset{w}{\approx} \vee \overset{b}{\approx}$: The equivalence relation $\overset{1,w}{\sim}$ is taken to generate a *critical portrait*, which defines the white polynomial P_w. We take $\overset{w}{\approx}$ to be the equivalence relation induced by the Carathéodory semiconjugacy σ_w, for P_w which we discussed in Section 2.3. In other words, $s \sim t$ if and only if $\sigma_w(s) = \sigma_w(t)$. We similarly define $\overset{b}{\approx}$.

Aside from a differing $\overset{0,w}{\sim}$ and $\overset{0,b}{\sim}$ in our setting, we take our equivalence relations to be defined and/or induced in the same respective manner.

In the setting of [**7**], all of these equivalence relations (except for the $\overset{n,w}{\sim}$ and $\overset{n,b}{\sim}$) are noted to be equal. We first have that $\overset{\infty}{\sim} = \sim$ as a result of a metric argument in [**7**, Theorem 4.7]. We have that $\sim = \overset{w}{\sim} \vee \overset{b}{\sim}$ via a similar metric argument along with an exercise in operations on equivalence relations. [**7**, Proposition 7.12] guarantees that $\overset{w}{\sim} \vee \overset{b}{\sim} = \approx$ by noting that $\overset{w}{\sim} = \overset{w}{\approx}$ via an argument regarding the polynomial P_w and Poirier's Theorem, and similar for the corresponding black equivalence relations. Further, $z^d / \approx$ is noted to generate a map topologically conjugate to the topological mating. In other words, this links Meyer's $\mathbb{S}^1$-parameterization to the topological mating of the two polynomials P_w and P_b.

Many of the arguments here rely on discussion of the underlying metric, and/or are exercises in properties of operations on equivalence relations. Since we are in the same metric space and define our equivalence relations similarly, these arguments do not change in a significant manner. The primary point of concern is that our $\overset{0,w}{\sim}$ and $\overset{0,b}{\sim}$, which induce the rest of these equivalence relations, are defined differently than in [**7**]. Meyer uses the nontrivial equivalence classes of $\overset{1,w}{\sim}$ to generate a *critical portrait* which yields the polynomial P_w; this is a critical step in defining $\overset{w}{\approx}$ and asserting that this equivalence relation is equal to $\overset{w}{\sim}$.

It thus remains to discuss what a *critical portrait* is, and to show that our $\overset{1,w}{\sim}$ and $\overset{1,b}{\sim}$ relations can be used to define them.

4.6. Critical portraits.

We define the following per [**10**]:

DEFINITION 4.10. We define that a set $A \subset \mathbb{S}^1$ is a *degree d preargument set* if $da = \{da : a \in A\}$ is a singleton set.

Further, consider a pair of families $\mathcal{F} = \{\mathcal{F}_1, ..., \mathcal{F}_n\}$ (associated with critically periodic orbits) and $\mathcal{J} = \{\mathcal{J}_1, ..., \mathcal{J}_m\}$ (associated with critically preperiodic orbits) of rational degree d preargument sets. We say that $\Theta = (\mathcal{F}, \mathcal{J})$ is a *degree d critical portrait* if the following conditions are fulfilled:

(c1) $d - 1 = \sum(|\mathcal{F}_k| - 1) + \sum(|\mathcal{J}_k| - 1)$,
(c2) $\mathcal{J}$ is weakly unlinked to $\mathcal{F}$ on the right,
(c3) each family is hierarchic, (That is, whenever a degree d preargument set is mapped into from another preargument set via some $a \mapsto d^i a \in A$, then $d^i a \in A$ is the "preferred" or representative element of A that is mapped to by any preargument set.)
(c4) for any a that participates in $\mathcal{F}$, some periodic forward iterate $d^i a$ also participates in $\mathcal{F}$
(c5) no a that participates in $\mathcal{J}$ is periodic.
(c6) Suppose a and t are periodic with the same period and the same symbol sequences. If a participates in $\mathcal{F}$, then $a = t$.
(c7) Let $a \in \mathcal{J}_l$ and $t \in \mathcal{J}_k$.Take $i \geq 0$.If the left symbol sequence of $d^i a$ equals the left symbol sequence of t then $d^i a = t$.

Poirier's theorem notes that critical portraits determine centered monic post-critically finite polynomials.

Meyer's choice of Jordan γ^0 yields that any point of intersection on γ^1 is automatically a critical point, since a point of intersection does not map homomorphically to any arc of γ^0. The parameters associated with that point automatically generate appropriate preargument sets, which are used to form critical portraits for the relevant polynomials.

We cannot simply choose any point of intersection in γ^1 to yield a preimage set associated with a critical point though, as some of our intersection points on γ^1 may map homeomorphically onto intersection points of γ^0. Further, as some critical values are marked by multiple parameters, we need to ensure that we actually obtain proper preargument sets that do not contain too many elements. Therefore:

DEFINITION 4.11. We form markings of selected vertices of $\mathbf{V}^1$ as follows.

- Note the elements of $\mathbf{V}^1$ which do not map homeomorphically onto vertices of $\mathbf{V}^0$ =post$_f$. These vertices $c_1, \ldots, c_k$ form our critical set.
- Consider an unlacing $\gamma^{1'}$ of γ^1 in a small neighborhood $U(c)$ of each critical point c, and examine the components of $U(c) \setminus \gamma^{1'}$. If any component does not map homeomorphically onto its image, note the color of this component, and partition the critical points by color.
- Start with the white critical points. Select a critical point (again, call it c for now) that is not in the forward orbit of any other critical point of the same color. Select any marked parameter α for which $\gamma(\alpha) = c$. The marking for this critical point is now the collection of values in the set $q_d^{-1}(q_d(\alpha))$.
- Iterate α in q_d until we find a parameter for the next white critical point in the forward orbit. Apply $q_d^{-1} \circ q_d$ to this critical parameter to obtain a marking for this new critical point. Repeat this process to continue marking critical points until we have exhausted all of the critical points in the forward orbit of c.

- If there are remaining critical points that eventually map into this critical orbit, select the parameter associated with this critical point that eventually iterates in q_d to a previously noted parameter. Similarly apply $q_d^{-1} \circ q_d$ to this critical parameter to find a set of values to mark the critical point.
- Repeat this process until all white critical points are exhausted. Then, apply this method to the set of black critical points.

The collection of white markings (or pre-image sets) yields the *critical portrait* for the white polynomial P_w, and similar for the black polynomial P_b.

THEOREM 4.12. *The above procedure generates a critical portrait for both the black and white polynomials.*

PROOF. As our case involves pre-periodic critical points only; the criteria for a degree d preargument set satisfying the conditions of a critical portrait are significantly simplified: any condition involving $\mathcal{F}$ is trivially satisfied because $\mathcal{F} = \varnothing$ here. The considerations for Rieman–Hurwitz are satisfied by an argument similar to the one that Meyer presents in the proof of [**7**, Lemma 5.7]. That this serves as a critical portrait follows mostly from construction; the one-by-one development of critical markings is intended to select preferred elements to preserve the hiearchic structure required of the remaining considerations for critical markings. □

4.7. An example, revisited. We revisit the function f of our earlier example, this time demonstrating the steps noted above.

(1) We begin with an oriented curve γ^0 which is fully f-invariant up to homotopy. As noted previously, this implies an existing parameterization for γ^0 and its pullback γ^1, but the construction only necessitates an understanding of the homotopy types of these curves, as conveyed in Figure 3.

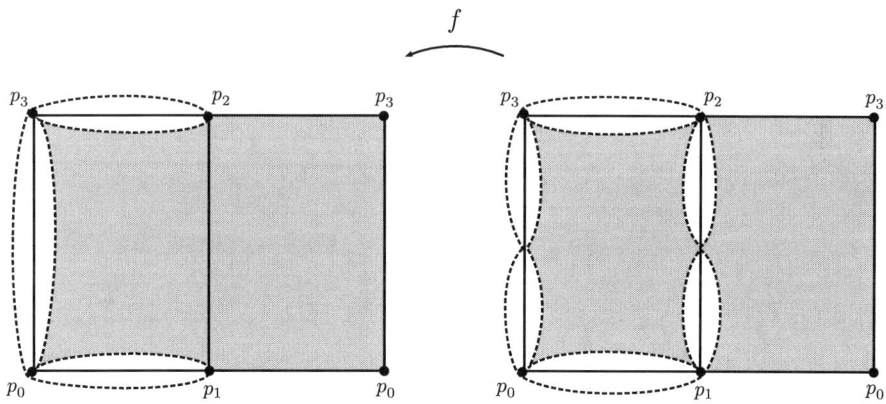

FIGURE 3. The dashed lines above note an oriented curve γ^0 with its marked postcritical set, as well as the curve $\gamma^1 := f^{-1}(\gamma^0)$.

(2) We consider an unlacing of γ^0 and γ^1 and mark these curves accordingly, as in Figure 4. Note that at each point of self-intersection on γ^0 all vertices are marked. Since γ^0 is fully f-invariant up to homotopy, we have that $f \circ \gamma^1 = \gamma^0 \circ q_2$, which induces a marking on γ^1. We construct the pseudo-isotopy H^0 of Theorem 4.3 to respect this induced marking.

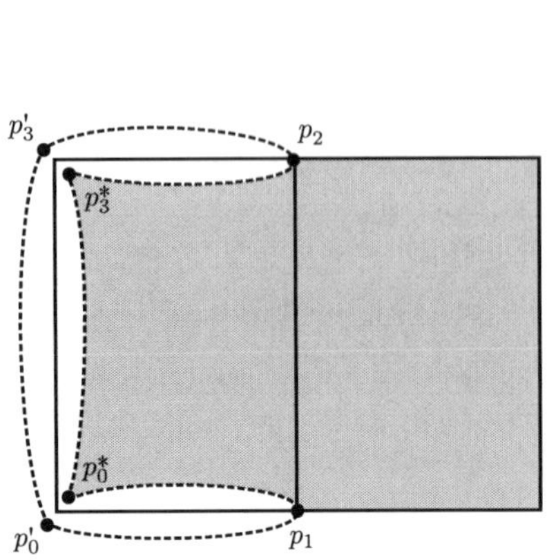

FIGURE 4. We select a marking of curves.

One may note that this is not the same oriented fully f-invariant curve of Kameyama's construction for the mating which is initially pictured in Figure 2. This was done intentionally to highlight how the marking conventions allow us to "keep track of postcritical points", as the later parameterizations of γ^n fix values of t at these locations. It should be noted that the procedure could be repeated with this curve to obtain the same polynomial decomposition, though.

(3) We may repeatedly lift pseudo-isotopies to obtain a sequence H^n and corresponding collection of curves γ^n. We take there to be some limit curve γ whose parameterization we develop in the next step.

(4) The parameterization of γ is induced by a finite amount of information, namely we need to recover the parameters at each element of post_f. To do so, we mark the edges of γ^0, and note locations of their pullbacks via f, as in Figure 5.

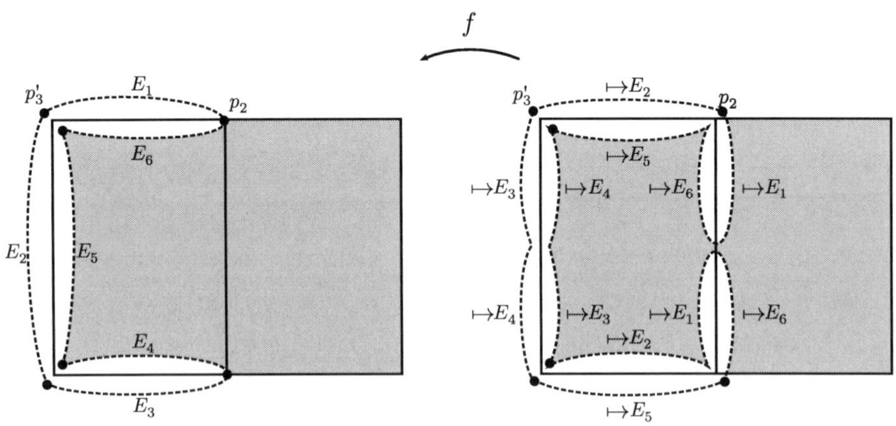

FIGURE 5. We label 0-edges E_i, as well as pullbacks.

We may then develop an edge transition matrix whose entries a_{ij} are determined by counting how many 1-edges which map to E_j are contained in the H_1 deformation of each edge E_i. For instance, we can see that H^0 deforms edge E_6 into an arc which contains sub-arcs mapping to E_1, E_5, and E_6—and this deformation is responsible for each entry on the last row of the matrix below.

$$\begin{bmatrix} 0 & 1 & 0 & 0 & 0 & 0 \\ 0 & 0 & 1 & 1 & 0 & 0 \\ 0 & 0 & 0 & 0 & 1 & 0 \\ 1 & 1 & 0 & 0 & 0 & 1 \\ 0 & 0 & 1 & 1 & 0 & 0 \\ 1 & 0 & 0 & 0 & 1 & 1 \end{bmatrix}.$$

The spectral radius of this matrix matches the degree of f, which is 2. This eigenvalue corresponds to a unique real eigenvector, $[1\ 2\ 1\ 3\ 2\ 3]^T$. If we scale the entries of this vector so that the components sum to 1, each ith component l_i represents the length of the interval $e_i \in \mathbb{S}^1$ which maps to E_i: For instance, the parameters mapping to p_2 and $f(p_2) = p_3'$ are $l_1 = 1/12$ units apart.

We can then use the lengths we've computed along with the mapping behavior of f to compute parameters corresponding to each of the markings representing elements of post_f. Suppose p_a and $p_b := f(p_a)$ are two postcritical points with corresponding parameters $t_a, t_b \in \mathbb{S}^1$. Then not only is $t_b = d \cdot t_a$, we also can find t_b by adding the sum $l(t_a, t_b)$ of lengths of subintervals between t_a and t_b in $\mathbb{S}^1$.

We obtain that $d \cdot t_a = t_a + l(t_a, t_b) \bmod 1$, or that one choice for t_a is $t_a = l(t_a, t_b)/(d-1)$. (Any other choices correspond to an isomorphism of the mating.) Applying this to $p_a = p_2$, we get that the parameter of t corresponding to p_2 is $1/12$. We can then successively add appropriate l_i to this parameter to determine the successive parameters of marked elements of post_f along γ^0: e.g. $1/12 + l_1$ gives the parameter for p_3', adding l_2 to this gives the parameter for p_0', and so on. We obtain that the parameters for our postcritical markings (starting at p_2 and traveling in a positive direction with respect to the white tile) are $\frac{1}{12}, \frac{1}{6}, \frac{1}{3}, \frac{5}{12}, \frac{2}{3}$, and $\frac{5}{6}$.

(5) We do not picture the following pullbacks here, but the parameterization induced by the lifts of pseudo-isotopies generates an $\mathbb{S}^1$-parameterization of $\mathbb{S}^2$. As such, this map arises as a mating of two polynomials, P_w, and P_b whose critical portraits we find below.

(6) We note that the middle left "pinched spot" on the white collection of 1-tiles maps non homeomorphically to the postcritical point p_1, and as such is one of our critical points. Applying $q_2^{-1} \circ q_d$ to this critical parameter is the same as computing $q_2^{-1}(5/12)$ (since the parameter for p_1 is $5/12$), thus yielding a critical portrait $P_w = \{\frac{5}{24}, \frac{17}{24}\}$.

The similar construction for the black polynomial yields $P_b = \{\frac{1}{24}, \frac{13}{24}\}$.

5. Future study

While this approach aligns the unmating approach more with the cases suggested in [**6**], it does leave something to be desired that the initialization of the algorithm is not as constructive. In particular, we begin with Kameyama's assumption of a curve which is fully f-invariant up to homotopy—but what are sufficient criteria for such a curve to be self-intersecting in a nontrivial manner? The author has interest in whether further study of more general subdivision rules (rather than those of just 2-tilings) could yield a more constructive approach, or if subdivisions could be used to single out which elements of post_f are the "problematic" ones requiring intersections and multiple markings. Alternatively, in the spirit of noting that Thurston's pullback algorithm is frequently known to converge despite obstructions (e.g., [**4**]), there may be stipulations for when it is possible to force a 2-tile decomposition procedure without resorting to iterates of f.

Another potential avenue of investigation for the techniques presented here regard what have sometimes been called *anti-pseudo-equators*. In some cases, the orientation of the pullback curve $f^{-1}(\mathcal{C})$ is reverse that of $\mathcal{C}$, which is one reason for passing through to an iterate of f before attempting an unmating. For instance, in [**9**, Example 6.6] the pullback curve traverses post_f in reverse order from that of $\mathcal{C}$. It has been conjectured by Meyer and Jung [**5**] that these may be a result of functions called anti-matings, and that it may be possible to extend the unmating algorithm in these cases. Notably, similar maps have been studied by Timorin [**14**] and Dastjerdi [**3**] in terms of laminations, and unlike matings where the two "sides" of the lamination are fixed; these are represented by laminations where the two sides swap on each iteration.

Acknowledgments

The author would like to thank Wolf Jung for his assistance in obtaining a difficult to find reference, as well as for a helpful review comment on a previous paper which prompted some of this investigation. The author would also like to acknowledge the editor for his patience in handling this manuscript.

References

[1] Mario Bonk and Daniel Meyer, *Expanding Thurston maps*, Mathematical Surveys and Monographs, vol. 225, American Mathematical Society, Providence, RI, 2017, DOI 10.1090/surv/225. MR3727134

[2] J. W. Cannon, W. J. Floyd, and W. R. Parry, *Constructing subdivision rules from rational maps*, Conform. Geom. Dyn. **11** (2007), 128–136, DOI 10.1090/S1088-4173-07-00167-1. MR2329140

[3] Davood Ahmadi Dastjerdi, *Dynamics of certain rational maps of degree two.*, Ph.D. thesis, University of Liverpool, 1991.

[4] Suzanne Hruska Boyd and Christian Henriksen, *The Medusa algorithm for polynomial matings*, Conform. Geom. Dyn. **16** (2012), 161–183, DOI 10.1090/S1088-4173-2012-00245-7. MR2943594

[5] Wolf Jung, *Quadratic matings and anti-matings*, preprint (2020).

[6] Atsushi Kameyama, *On Julia sets of postcritically finite branched coverings. II. S^1-parametrization of Julia sets*, J. Math. Soc. Japan **55** (2003), no. 2, 455–468, DOI 10.2969/jmsj/1191419126. MR1961296

[7] Daniel Meyer, *Expanding Thurston maps as quotients*, Preprint `arXiv:0910.2003`, (2012).

[8] Daniel Meyer, *Invariant Peano curves of expanding Thurston maps*, Acta Math. **210** (2013), no. 1, 95–171, DOI 10.1007/s11511-013-0091-0. MR3037612

[9] Daniel Meyer, *Unmating of rational maps: sufficient criteria and examples*, Frontiers in complex dynamics, Princeton Math. Ser., vol. 51, Princeton Univ. Press, Princeton, NJ, 2014, pp. 197–233. MR3289912

[10] Alfredo Poirier, *Critical portraits for postcritically finite polynomials*, Fund. Math. **203** (2009), no. 2, 107–163, DOI 10.4064/fm203-2-2. MR2496235

[11] Mary Rees, *A partial description of parameter space of rational maps of degree two. I*, Acta Math. **168** (1992), no. 1-2, 11–87, DOI 10.1007/BF02392976. MR1149864

[12] Lei Tan, *Matings of quadratic polynomials*, Ergodic Theory Dynam. Systems **12** (1992), no. 3, 589–620, DOI 10.1017/S0143385700006957. MR1182664

[13] William P. Thurston, *On the geometry and dynamics of iterated rational maps*, Complex dynamics, A K Peters, Wellesley, MA, 2009, pp. 3–137, DOI 10.1201/b10617-3. Edited by Dierk Schleicher and Nikita Selinger and with an appendix by Schleicher. MR2508255

[14] Vladlen Timorin, *The external boundary of* M_2, Holomorphic dynamics and renormalization, Fields Inst. Commun., vol. 53, Amer. Math. Soc., Providence, RI, 2008, pp. 225–266. MR2477426

[15] Mary Wilkerson, *Subdivision rule constructions on critically preperiodic quadratic matings*, New York J. Math. **22** (2016), 1055–1084. MR3576282

[16] Mary Wilkerson, *Thurston's algorithm and rational maps from quadratic polynomial matings*, Discrete Contin. Dyn. Syst. Ser. S **12** (2019), no. 8, 2403–2433, DOI 10.3934/dcdss.2019151. MR4025990

DEPARTMENT OF MATHEMATICS AND STATISTICS, COASTAL CAROLINA UNIVERSITY, CONWAY, SOUTH CAROLINA 29528-6054

Email address: mwilkerso@coastal.edu

SELECTED PUBLISHED TITLES IN THIS SERIES

797 **Sangita Jha, Mrinal Kanti Roychowdhury, and Saurabh Verma, Editors,** Recent Developments in Fractal Geometry and Dynamical Systems, 2024

796 **John Cremona, John Jones, Jennifer Paulhus, Andrew V. Sutherland, and John Voight, Editors,** LuCaNT: LMFDB, Computation, and Number Theory, 2024

795 **Michael A. Jones, David McCune, and Jennifer M. Wilson, Editors,** Mathematical Analyses of Decisions, Voting and Games, 2024

794 **Jean-Pierre Magnot, Editor,** Recent Advances in Diffeologies and Their Applications, 2024

793 **Abba Gumel, Editor,** Mathematical and Computational Modeling of Phenomena Arising in Population Biology and Nonlinear Oscillations, 2024

792 **Shaoming Guo, Zane Kun Li, and Brian Street, Editors,** Recent Developments in Harmonic Analysis and its Applications, 2024

791 **Mee Seong Im, Bach Nguyen, and Arik Wilbert, Editors,** Algebraic and Topological Aspects of Representation Theory, 2024

790 **Peter Crooks and Alexandru I. Suciu, Editors,** Compactifications, Configurations, and Cohomology, 2023

789 **I. S. Krasil′shchik, A. B. Sossinsky, and A. M. Verbovetsky, Editors,** The Diverse World of PDEs, 2023

788 **I. S. Krasil′shchik, A. B. Sossinsky, and A. M. Verbovetsky, Editors,** The Diverse World of PDEs, 2023

787 **Hemen Dutta, Editor,** Mathematical Modelling, 2023

786 **Hemen Dutta, Editor,** Mathematical Modelling, 2023

785 **A. Leroy and S. K. Jain, Editors,** Algebra and Coding Theory, 2023

784 **Dinh-Liem Nguyen, Loc Hoang Nguyen, and Thi-Phong Nguyen, Editors,** Advances in Inverse Problems for Partial Differential Equations, 2023

783 **Alla Detinko, Michael Kapovich, Alex Kontorovich, Peter Sarnak, and Richard Schwartz, Editors,** Computational Aspects of Discrete Subgroups of Lie Groups, 2023

782 **Galina Filipuk, Alberto Lastra, and Sławomir Michalik, Editors,** Recent Trends in Formal and Analytic Solutions of Diff. Equations, 2023

781 **Emanuel Indrei, Diego Marcon, and Levon Nurbekyan, Editors,** Geometric and Functional Inequalities and Recent Topics in Nonlinear PDEs, 2023

780 **Erik Koelink, Stefan Kolb, Nicolai Reshetikhin, and Bart Vlaar, Editors,** Hypergeometry, Integrability and Lie Theory, 2022

779 **Samuele Anni, Valentijn Karemaker, and Elisa Lorenzo García, Editors,** Arithmetic, Geometry, Cryptography, and Coding Theory 2021, 2022

778 **Carlos Galindo, Alejandro Melle Hernández, Julio José Moyano-Fernández, and Wilson A. Zúñiga-Galindo, Editors,** p-Adic Analysis, Arithmetic and Singularities, 2022

777 **Bang-Yen Chen, Nicholas D. Brubaker, Takashi Sakai, Bogdan D. Suceavă, Makiko Sumi Tanaka, Hiroshi Tamaru, and Mihaela B. Vajiac, Editors,** Differential Geometry and Global Analysis, 2022

776 **Aaron Wootton, S. Allen Broughton, and Jennifer Paulhus, Editors,** Automorphisms of Riemann Surfaces, Subgroups of Mapping Class Groups and Related Topics, 2022

775 **Fernando Galaz-García, Cecilia González-Tokman, and Juan Carlos Pardo Millán, Editors,** Mexican Mathematicians in the World, 2021

774 **Randall J. Swift, Alan Krinik, Jennifer M. Switkes, and Jason H. Park, Editors,** Stochastic Processes and Functional Analysis, 2021

For a complete list of titles in this series, visit the
AMS Bookstore at **www.ams.org/bookstore/conmseries/**.